Universitext

Birger Iversen

Cohomology of Sheaves

Springer-Verlag
Berlin Heidelberg New York Tokyo

Birger Iversen
Mathematisk Institut, Aarhus Universitet
Ny Munkegade, DK-8000 Aarhus C, Denmark

AMS-MOS (1980) Classification numbers:
14C17, 18E30, 18F20, 18G35, 32A27, 55N30, 55U30, 57R20

ISBN-13:978-3-540-16389-3   e-ISBN-13:978-3-642-82783-9
DOI.10:1007/978-3-642-82783-9

Library of Congress Cataloging in Publication Data. Main entry under title:
Iversen, Birger. Cohomology of sheaves. (Universitext). Bibligraphy: p. Includes index.
1. Sheaves, Theory of.   2. Homology theory. I. Title.   QA612.36.I93   1986   514'.224
86–3789

2141/3140–543210

Introduction

This text exposes the basic features of cohomology of sheaves and its applications. The general theory of sheaves is very limited and no essential result is obtainable without turning to particular classes of topological spaces. The most satisfactory general class is that of locally compact spaces and it is the study of such spaces which occupies the central part of this text.

The fundamental concepts in the study of locally compact spaces is cohomology with compact support and a particular class of sheaves, the so-called soft sheaves. This class plays a double role as the basic vehicle for the internal theory and is the key to applications in analysis. The basic example of a soft sheaf is the sheaf of smooth functions on $\mathbb{R}^n$ or more generally on any smooth manifold. A rather large effort has been made to demonstrate the relevance of sheaf theory in even the most elementary analysis. This process has been reversed in order to base the fundamental calculations in sheaf theory on elementary analysis.

The central theme of the text is Poincaré duality or rather its generalizations by Borel and Verdier. In its first form this appears as a duality between cohomology and cohomology with compact support. A more general Poincaré duality theory is developed for a continuous map between locally compact spaces. The important special case of a closed imbedding admits generalization to arbitrary topological spaces and is best understood in the framework of local cohomology. This theory is used for construction of characteristic classes of all sorts: Chern classes, Stiefel-Whitney classes, ...

For further applications to algebraic topology, a homology theory is developed for locally compact spaces and proper maps. This allows one to express Poincaré duality as an isomorphism between homology and cohomology. Applications are given to the classical theory of topological manifolds: fundamental class, diagonal class, Lefschetz fixed point formula ...

This homology theory is particularly suited for the study of algebraic varieties and a detailed introduction to (co)homology classes of algebraic cycles is given, including a topological definition of the local intersection symbol. It is a rather remarkable feature that this homology theory more or less automatically grinds out algebraic cycles.

A word about homological algebra. The first chapter of the text gives an introduction to homological algebra sufficient for most of the text. The last chapter, or appendix if you wish, gives an introduction to derived categories used in the more advanced parts of the text and in the proofs of the basic cup product formulas. It is my hope that this will give some readers motivation for Verdier's rather difficult text (1) on triangulated categories.

It remains for me to thank W. Fulton and R. MacPherson for their encouragement to publish the text, to thank a number of colleagues, who read part of the manuscript, H.H. Andersen, J.P. Hansen, A. Kock, O. Kroll, O.A. Laudal and H.A. Nielsen and to thank Else Yndgaard for excellent typing and cooperation.

Aarhus, Denmark

September 1985   Birger Iversen

Contents

I. Homological Algebra

Consider a category with <u>zero object</u> 0, that is for every object A there is precisely one morphism $A \to 0$ and precisely one $0 \to A$.

A <u>zero morphism</u> $A \to B$ is one which can be factored $A \to 0 \to B$.

A <u>kernel</u>, Ker f   for a morphism $f: A \to B$ is a pair (K,i) where $i: K \to A$ is a monomorphism with $f\,i = 0$ and such that any morphism $g: X \to A$ with $f\,g = 0$ factors through $i: K \to A$.

A <u>cokernel</u>, Cok f   for f is a pair (C,p) where $p: B \to C$ is an epimorphism with $p\,f = 0$ such that any morphism $h: B \to Y$ with $h\,f = 0$ factors through p.

We shall assume that every morphism has kernel and cokernel.

An <u>image</u>, Im f   is a kernel for a cokernel.

A <u>coimage</u>, Coim f   is a cokernel for a kernel.

Every morphism f has a <u>canonical factorization</u>

$$A \to \mathrm{Coim}\, f \xrightarrow{f'} \mathrm{Im}\, f \to B$$

Definition 1.1. An <u>exact category</u> is a category with zero objects, kernels, cokernels and such that $\operatorname{Coim} f \xrightarrow{f'} \operatorname{Im} f$ always is an isomorphism.

In the remaining part of this section we shall work in an exact category.

Definition 1.2. A sequence of morphisms

$$\ldots A^{n-1} \xrightarrow{f^{n-1}} A^n \xrightarrow{f^n} A^{n+1} \xrightarrow{f^{n+1}} \ldots$$

is called <u>exact</u> if $\operatorname{Im}(f^{n-1}) = \operatorname{Ker}(f^n)$, for all n.

Proposition 1.3. Consider the exact, commutative diagram

$$\begin{array}{ccccccc}
A & \longrightarrow & B & \longrightarrow & C & \longrightarrow & D \\
\downarrow a & & \downarrow b & & \downarrow c & & \downarrow d \\
A' & \longrightarrow & B' & \longrightarrow & C' & \longrightarrow & D' \\
\downarrow & & & & & & \\
0 & & & & & &
\end{array}$$

The induced sequence $\operatorname{Ker} b \to \operatorname{Ker} c \to \operatorname{Ker} d$ is exact.

Proof. Break the diagram into two pieces

$$\begin{array}{ccccccc}
0 & \to & E & \to & C & \to & D \\
& & \downarrow e & & \downarrow c & & \downarrow d \\
0 & \to & E' & \to & C' & \to & D'
\end{array}
\qquad
\begin{array}{ccccccc}
A & \to & B & \to & E & \to & 0 \\
\downarrow a & & \downarrow b & & \downarrow e & & \\
A' & \to & B' & \to & E' & \to & 0 \\
& & & & \downarrow & & \\
& & & & 0 & &
\end{array}$$

We have to prove that

$\alpha)$ $0 \to \operatorname{Ker} e \to \operatorname{Ker} c \to \operatorname{Ker} d$ is exact

$\beta)$ $\operatorname{Ker} b \to \operatorname{Ker} e$ is surjective

α) Check that Ker e → Ker c is a kernel for Ker c → D.

β.1) Check that cok b → Cok e is an isomorphism (use the dual statement to α , if necessary).

β.2) The exact commutative diagram

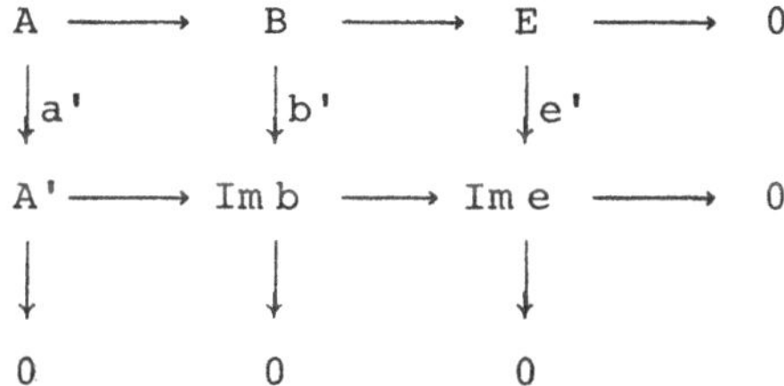

shows that A' → Im b → Im e is exact: replace A' by a kernel for B' → E' and use α).

β.3) This gives an exact commutative diagram

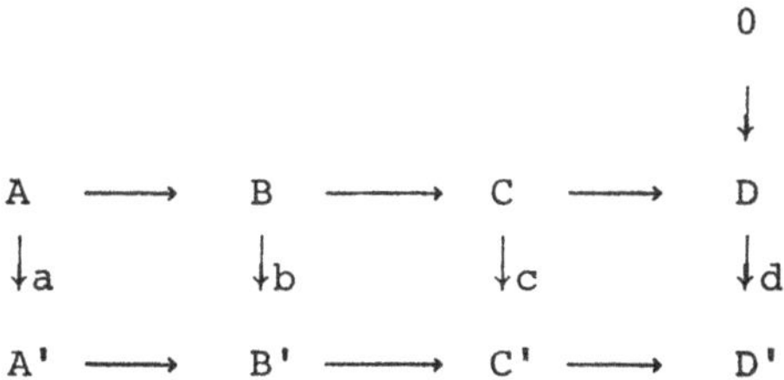

Check that e' is a cokernel for Ker b' → E.

The dual statement is   Q.E.D.

Proposition 1.4. Consider the exact, commutative diagram

$$
\begin{array}{ccccccc}
 & & & & & & 0 \\
 & & & & & & \downarrow \\
A & \longrightarrow & B & \longrightarrow & C & \longrightarrow & D \\
\downarrow a & & \downarrow b & & \downarrow c & & \downarrow d \\
A' & \longrightarrow & B' & \longrightarrow & C' & \longrightarrow & D'
\end{array}
$$

The induced sequence Cok a → Cok b → Cok c is exact.

Corollary 1.5. Consider morphisms $f: X \to Y$ and $g: Y \to Z$.
The following sequence is exact

$$0 \to \text{Ker } f \to \text{Ker } gf \to \text{Ker } g \to \text{Cok } f \to \text{Cok } gf \to \text{Cok } g \to 0.$$

Proof. Apply 1.3 and 1.4 to the two diagrams

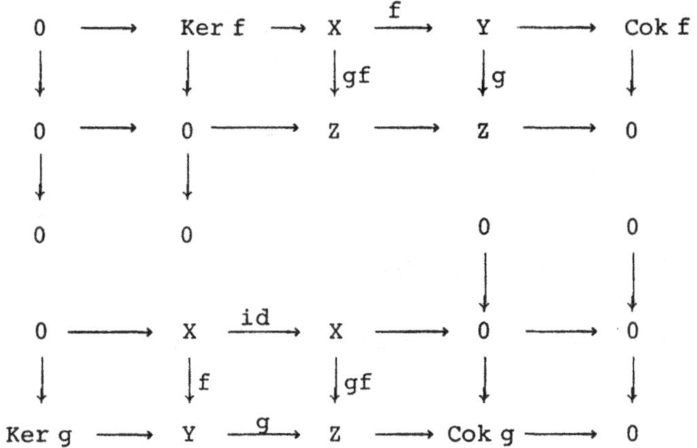

Q.E.D.

Snake Lemma 1.6. Consider the exact commutative diagram

There is an exact sequence

$$\text{Ker } b \to \text{Ker } c \to \text{Ker } d \xrightarrow{\partial} \text{Cok } b \to \text{Cok } c \to \text{Cok } d .$$

More precisely

1) Put $K = \mathrm{Ker}(C \to D')$. $K \to \mathrm{Ker}\, d$ is an epimorphism.

2) Put $K' = \mathrm{Cok}(B \to C')$. $\mathrm{Cok}\, b \to K'$ is a monomorphism.

3) There exists a unique map

$$\partial :\ \mathrm{Ker}\, d \to \mathrm{Cok}\, b$$

such that the two composites

$$K \to C \overset{c}{\to} C' \to K'$$
$$K \to \mathrm{Ker}\, d \overset{\partial}{\to} \mathrm{Cok}\, b \to K'$$

are the same.

4) The six-term sequence above is exact.

$\underline{\text{Proof}}$. Let f denote the morphism $C' \to \mathrm{Ker}(D' \to E')$.
Consider the exact commutative diagram

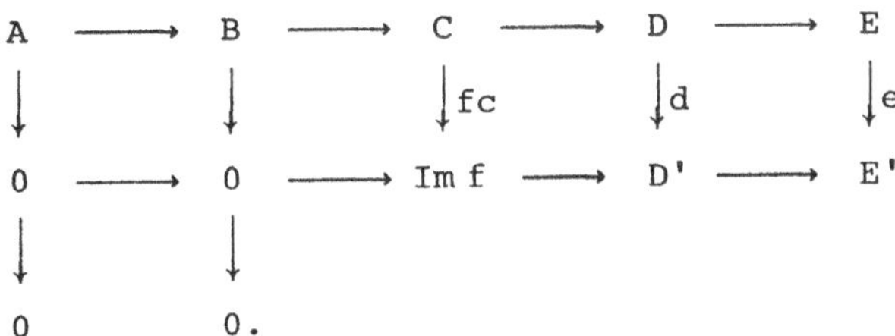

It follows from 1.3 applied twice that

$$B \longrightarrow K \longrightarrow \mathrm{Ker}\, d \longrightarrow 0$$

is exact, and similar, that

$$0 \longrightarrow \mathrm{Cok}\, b \longrightarrow K' \longrightarrow D'$$

is exact. This proves 1), 2), 3). By 1.3, 1.4 and duality it suffices to prove that

$$\mathrm{Ker}\, c \longrightarrow \mathrm{Ker}\, d \longrightarrow \mathrm{Cok}\, b$$

is exact. It suffices to prove exactness of

$$\mathrm{Ker}\, c \longrightarrow \mathrm{Ker}\, d \longrightarrow K'.$$

Consider the diagram

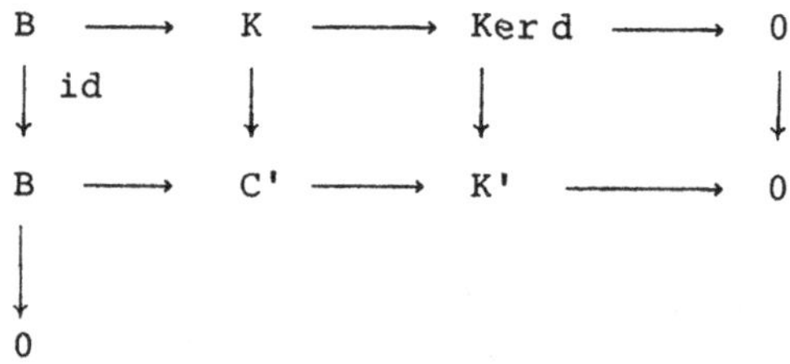

Conclusion by 1.3.

Q.E.D.

Let us record a much used special case

<u>Five lemma 1.7</u>. Given an exact commutative diagram

$$
\begin{array}{ccccccccc}
A & \longrightarrow & B & \longrightarrow & C & \longrightarrow & D & \longrightarrow & E \\
\downarrow a & & \downarrow b & & \downarrow c & & \downarrow d & & \downarrow e \\
A' & \longrightarrow & B' & \longrightarrow & C' & \longrightarrow & D' & \longrightarrow & E'
\end{array}
$$

If a, b, d, e are isomorphism, then c is an isomorphism.

I.2 Homology of complexes

We shall discuss the concept, _homology_ in the framework of an exact category.

By a _complex_ we understand a sequence $C^{\cdot} = (C^n, \partial^n)_{n \in \mathbb{Z}}$ of objects and morphisms

$$\cdots \longrightarrow C^{n-1} \xrightarrow{\ \partial^{n-1}\ } C^n \xrightarrow{\ \partial^n\ } C^{n+1} \xrightarrow{\ \partial^{n+1}\ } C^{n+2} \longrightarrow \cdots$$

with $\partial^{n+1}\partial^n = 0$ for all $n \in \mathbb{Z}$. The ∂'s are called _differentials_ or _boundary operators_.

A _morphism of complexes_ $f\colon C^{\cdot} \longrightarrow D^{\cdot}$ is a sequence $f = (f^n)_{n \in \mathbb{Z}}$ of morphisms $f^n\colon C^n \longrightarrow D^n$ with

$$f^{n+1}\partial^n = \partial^n f^n \quad \text{for all} \quad n \in \mathbb{Z}.$$

For a complex $C^{\cdot}$ we define for $n \in \mathbb{Z}$ the n'th _homology_ object

$$2.1 \qquad\qquad H^n(C^{\cdot}) = \operatorname{Ker} \partial^n / \operatorname{Im} \partial^{n-1}$$

A morphism $f\colon C^{\cdot} \longrightarrow D^{\cdot}$ of complexes will induce a morphism on homology

$$2.2 \qquad\qquad H^n(f) = H^n(C^{\cdot}) \longrightarrow H^n(D^{\cdot})$$

Consider a sequence of complexes

$$0 \longrightarrow P^{\cdot} \xrightarrow{\ f\ } Q^{\cdot} \xrightarrow{\ g\ } R^{\cdot} \longrightarrow 0$$

which is a _chainwise exact_, i.e. with

$$0 \longrightarrow P^n \longrightarrow Q^n \longrightarrow R^n \longrightarrow 0$$

exact for all $n \in \mathbb{Z}$. We shall construct the so called _connecting morphism_

8

2.3
$$c^n: H^n(R^\cdot) \to H^{n+1}(P^\cdot)$$

and derive a long exact sequence

2.4
$$H^n(P^\cdot) \xrightarrow{H^n(f)} H^n(Q^\cdot) \xrightarrow{H^n(g)} H^n(R^\cdot) \xrightarrow{c^n} H^{n+1}(P^\cdot) \xrightarrow{H^{n+1}(f)} H^{n+1}(Q^\cdot)$$

<u>Construction</u>. For a complex $C^\cdot$ we put

2.5
$$Z^{n+1}(C^\cdot) = \mathrm{Ker}\, \partial^{n+1}, \qquad {}'Z^n(C^\cdot) = \mathrm{Cok}\, \partial^{n-1}$$

The boundary $\partial^n: C^n \to C^{n+1}$ induces

$$d^n: {}'Z^n(C^\cdot) \to Z^{n+1}(C^\cdot)$$

As is easily seen we have an exact sequence

2.6
$$0 \to H^n(C^\cdot) \to {}'Z^n(C^\cdot) \xrightarrow{d^n} Z^{n+1}(C^\cdot) \to H^{n+1}(C^\cdot) \to 0$$

We can now derive a commutative diagram

$$
\begin{array}{ccccccc}
{}'Z^n(P^\cdot) & \longrightarrow & {}'Z^n(Q^\cdot) & \longrightarrow & {}'Z^n(R^\cdot) & \longrightarrow & 0 \\
\downarrow & & \downarrow & & \downarrow & & \\
0 \longrightarrow Z^{n+1}(P^\cdot) & \longrightarrow & Z^{n+1}(Q^\cdot) & \longrightarrow & Z^{n+1}(R^\cdot) & &
\end{array}
$$

whose rows are exact as one easily derives from 1.3 and 1.4. We
can now conclude the construction by appealing to the snake lemma
1.6.

The connecting morphism 2.3 has the following functorial
property. Given

$$
\begin{array}{ccccccccc}
0 & \longrightarrow & P^\cdot & \longrightarrow & Q^\cdot & \longrightarrow & R^\cdot & \longrightarrow & 0 \\
& & \downarrow u & & \downarrow v & & \downarrow w & & \\
0 & \longrightarrow & U^\cdot & \longrightarrow & V^\cdot & \longrightarrow & W^\cdot & \longrightarrow & 0
\end{array}
$$

a commutative diagram of complexes whose rows are chainwise exact.

Then the following diagram is <u>commutative</u>

2.7

$$
\begin{array}{ccc}
H^n(R^\cdot) & \xrightarrow{\ c^n\ } & H^{n+1}(P^\cdot) \\
\downarrow{\scriptstyle H^n(w)} & & \downarrow{\scriptstyle H^{n+1}(u)} \\
H^n(W^\cdot) & \xrightarrow{\ c^n\ } & H^{n+1}(U^\cdot)
\end{array}
$$

The commutativity results from the origin of the connecting morphism. We shall at once take a step towards a realization of the connecting morphism.

$\underline{\text{Lemma 2.8.}}$ Given an exact sequence

$$
0 \longrightarrow P^\cdot \xrightarrow{\ f\ } Q^\cdot \xrightarrow{\ g\ } R^\cdot \longrightarrow 0
$$

of complexes and a morphism of complexes $h: R^\cdot \to P^\cdot[1]$. — For $n \in \mathbb{Z}$ let $i^n: L^n \to Q^n$ be a kernel for $\partial^n g^n: Q^n \to R^{n+1}$ and let $p^{n+1}: Q^{n+1} \to {}'L^{n+1}$ be a cokernel for $\partial^n f^n: P^n \to Q^{n+1}$. Then

$$
H^n(h): H^n(R^\cdot) \to H^{n+1}(P^\cdot)
$$

is equal to the connecting morphism if and only if

$$
p^{n+1} \, f^{n+1} \, h^n \, g^n \, i^n \;=\; p^{n+1} \, \partial^n \, i^n
$$

where ∂^n is the differential from $Q^\cdot$.

$\underline{\text{Proof.}}$ Use I.1.5 to see that the sequences

$$
L^n \to {}'Z^n(Q^\cdot) \to Z^{n+1}(R^\cdot)
$$

$$
{}'Z^n(P^\cdot) \to Z^{n+1}(Q^\cdot) \to {}'L^{n+1}
$$

are exact. It follows from 1.6 and the diagram below 2.6, that

$$\mathrm{Im}(L^n \to {}'Z^n(R^{\bullet})) = H^n(R^{\bullet})$$

$$\mathrm{Coim}(Z^{n+1}(P^{\bullet}) \to {}'L^{n+1}) = H^{n+1}(P^{\bullet})$$

We shall first show that the composite

$$L^n \longrightarrow H^n(R^{\bullet}) \xrightarrow{H^n(h)} H^{n+1}(P^{\bullet}) \longrightarrow {}'L^{n+1}$$

equals $p^{n+1}f^{n+1}h^n g^n i^n$. To see this bring three commutative
diagrams together

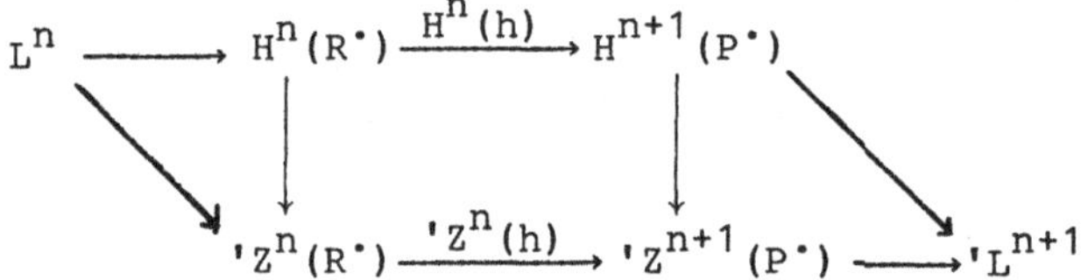

Next bring three more commutative diagrams together

This proves the promised formula. - The result now follows from
1.5.

Q.E.D.

I.3 Additive categories

By an additive category we understand a category in which
the Hom-sets come equipped with abelian group structures making
all compositions bilinear. Besides we require the presence of a
zero object and <u>direct sums</u>, i.e. for objects P and Q,
there exists an object P ⊕ Q and morphisms i: P → P ⊕ Q and
j: Q → P ⊕ Q, such that whenever given

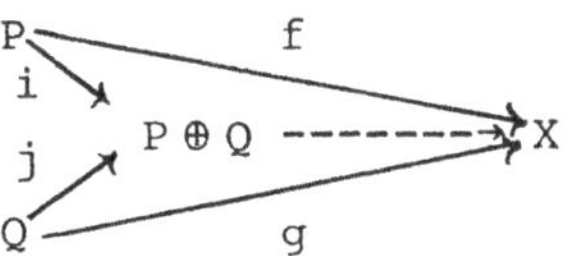

morphisms f and g as above we can fill in the dotted arrow making
the diagram commutative, and this in a unique way. - In particular,
we can find p: P ⊕ Q → P with pi = 1 and pj = 0 and in the
same way q: P ⊕ Q → Q with qi = 0 and qj = 1. Let us prove
that P ⊕ Q, p,q makes up a <u>direct product</u>, i.e. given a diagram

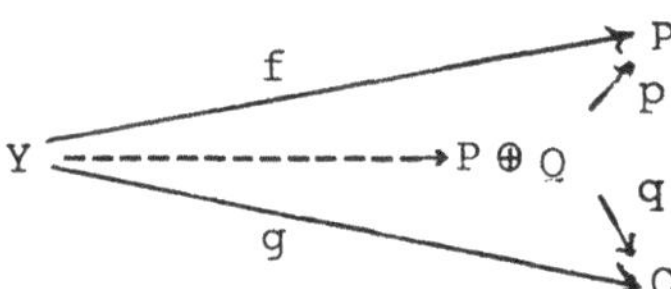

we can fill in the dotted arrow in a unique way making the diagram
commute.- Notice the relation

3.1 $$1 = ip + jq$$

which we can check by composing both sides with i and j using
the relations

3.2 $$pi = 1, \quad pj = 0, \quad qi = 0, \quad qj = 1$$

The result will follow from the following

12

Lemma 3.3. Consider a diagram

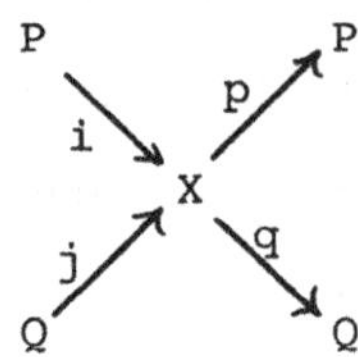

which satisfies relation 3.1 and 3.2. Then the arrows i,j form
a direct sum and the arrows p,q form a direct product.

Proof. Let us prove the second statement. Given f: Y → P
and g: Y → Q. Then

$$F = if + jg : Y → X$$

satisfies pF = f and qF = g as it follows from 3.2. - Conversely, given F with pF = f and qF = g. Then using 3.1 we get

$$F = ipF + jqF = if + jg$$

from which uniqueness follows.

                                                        Q.E.D.

Definition 3.4. A sequence of morphisms

$$C \xrightarrow{\ f\ } D \xrightarrow{\ g\ } E$$

is called a split exact sequence, if for every object X, the
following sequence of abelian groups is exact

3.5 $0 → \mathrm{Hom}(X,C) → \mathrm{Hom}(X,D) → \mathrm{Hom}(X,E) → 0$

Let us notice that, gf = 0 as it follows from 3.5 with X = C.

Example 3.6. With the notation of 3.3

$$P \xrightarrow{\ i\ } X \xrightarrow{\ p\ } Q$$

is a split exact sequence. - It will follow from the proof of 3.7, that all split exact sequences have this form

Proposition 3.7. Given a split exact sequence

$$C \xrightarrow{\ f\ } D \xrightarrow{\ g\ } E$$

Then for every object Y, the induced sequence of abelian groups

$$0 \to \mathrm{Hom}(E,Y) \to \mathrm{Hom}(D,Y) \to \mathrm{Hom}(C,Y) \to 0$$

is exact.

Proof. Let us first notice that the exact sequence 3.5 with $E = X$ shows that we can find a section to g i.e. a morphism $s: E \to D$ such that $gs = 1$. Let s be any such section. Notice that

$$g(1-sg) = 0$$

By the exactness of 3.5 with $X = D$ we can find $r: D \to C$ such that $1-sg = fr$ or

$$1 = sg + fr$$

Let us prove that

$$gf = 0, \quad gs = 1, \quad rf = 1, \quad rs = 0 \ .$$

14

We have already accounted for the two first. To prove the third relation, remark

$$f(rf) = (1-sq)f = f-sqf = f = f1$$

whence $rf = 1$ since f is a monomorphism. To prove the fourth relation notice that

$$r = r(sg + fr) = rsg + rfr = rsg + r$$

which shows that $rsg = 0$ and whence

$$rs = rs(gs) = (rsg)s = 0$$

It is now easy to conclude by Lemma 3.3.

Q.E.D.

Definition 3.8. A morphism $f: C \to D$ is called a split monomorphism if there exists a morphism $g: D \to E$ such that f,g form a split exact sequence. The dual notion is split epimorphism.

Additive Functors 3.9. Let A and B be additive categories. A functor

$$F: A \to B$$

is called additive, if for every pair of objects X and Y in A , F induces linear maps

$$\text{Hom}(X,Y) \to \text{Hom}(F(X),F(Y))$$

Let us notice that the additive functor F preserves split exact sequences as it follows from 3.3.

<u>Matrix notation</u>

A family of objects $(X_t)_{t \in T}$ indexed by a finite set T gives rise to an object X and a family of morphisms

$$X_t \xrightarrow{\ i_t\ } X \xrightarrow{\ p_t\ } X_t \qquad ;\quad t \in T$$

such that for all u and v in T

$$3.10 \qquad \sum_{t \in T} i_t p_t = 1 \qquad p_u i_v = \begin{cases} 1 & \text{for } u = v \\ 0 & \text{for } u \neq v \end{cases}$$

These data can be constructed by iterating the previous process. The object X is often written $\oplus_{t \in T} X_t$ and is called the <u>direct sum</u> of the X_t's. - Given a second finite family of objects $(Y_s)_{s \in S}$ with $Y = \oplus_{s \in S} Y_s$. To a morphism $f \colon X \to Y$ we associate the <u>matrix</u>

$$\{f_{st}\}_{(s,t) \in S \times T} \qquad ;\quad f_{st} = p_s f i_t$$

Conversely to such a matrix with

$$f_{st} \in \mathrm{Hom}(X_t, Y_s) \qquad ;\quad (s,t) \in S \times T$$

corresponds a uniquely determined morphism $f \colon X \to Y$. - Let there be given still another finite family $(Z_r)_{r \in R}$ with $Z = \oplus_{r \in R} Z_r$. For morphisms $f \colon X \to Y$ and $g \colon Y \to Z$ we have

$$3.11 \qquad (gf)_{rt} = \sum_s g_{rs} f_{st}$$

as it follows from the following calculation

$$(gf)_{rt} = p_r g f i_t = p_r g (\sum i_s p_s) f i_t =$$

$$\sum_s p_r g i_s p_s f i_t = \sum_s g_{rs} f_{st}$$

I.4 Homotopy theory of complexes

Throughout this section we shall work in a fixed additive category. A __complex__ $C^{\cdot} = (C^n, \partial^n)$ is a sequence of objects and morphisms

$$\ldots \to C^{n-1} \xrightarrow{\partial^{n-1}} C^n \xrightarrow{\partial^n} C^{n+1} \to \ldots$$

with $\partial^n \partial^{n-1} = 0$ for all $n \in \mathbb{Z}$.

A __morphism of complexes__ $f: C^{\cdot} \to D^{\cdot}$ $f = (f^n)_{n \in \mathbb{N}}$, is a sequence of morphisms $f^n: C^n \to D^n$ with

$$f^{n+1} \partial^n = \partial^n f^n \quad ; \quad n \in \mathbb{Z}$$

Morphisms of complexes are __composed__ in the obvious way, i.e. componentwise.

__Definition 4.1.__ For a complex $C^{\cdot}$ and $p \in \mathbb{Z}$ we find a new complex $C^{\cdot}[p]$ given by

$$(C^{\cdot}[p])^n = C^{n+p}$$

with boundary operator $(-1)^p \partial^p$. - Given a morphism of complexes $f: C^{\cdot} \to D^{\cdot}$. Define $f[n]: C[n] \to D[n]$ by the formula $f[n]^p = f^{n+p}$.

__Definition 4.2.__ Let $f^{\cdot}, g^{\cdot}: C^{\cdot} \to D^{\cdot}$ be morphisms of complexes. A __homotopy__ $s^{\cdot}$ from $f^{\cdot}$ to $g^{\cdot}$ is a sequence $s^{\cdot} = (s^n)_{n \in \mathbb{Z}}$, $s^n: C^n \to D^{n-1}$ of morphisms, such that

$$f^n - g^n = \partial^{n-1} s^n + s^{n+1} \partial^n \quad\quad ; n \in \mathbb{Z}$$

Morphisms f and g are said to be <u>homotopic</u> if there exists a
homotopy from f to g, we write f~g. Homotopy is an equivalence relation
compatible with composition. The additive group of <u>homotopy classes</u>
<u>of morphisms from $C^{\cdot}$ to $D^{\cdot}$ is denoted $[C^{\cdot},D^{\cdot}]$</u>.

We can now form a new additive category by considering the
complexes $P^{\cdot},Q^{\cdot},..$ in our category as objects and $[P^{\cdot},Q^{\cdot}]$ as
the group of morphisms from $P^{\cdot}$ to $Q^{\cdot}$. The new category is called
the <u>homotopy category</u>.

A morphism of complexes $f:P^{\cdot} \to Q^{\cdot}$ is called a <u>homotopy</u>
<u>equivalence</u> if it represents an isomorphism in the homotopy category,
i.e. if there exists a morphism of complexes $g:Q^{\cdot} \to P^{\cdot}$ such that

$$fg \sim 1 \quad \text{and} \quad gf \sim 1$$

<u>Definition 4.3</u>. For complexes $X^{\cdot}$ and $Y^{\cdot}$ let $\mathrm{Hom}^{\cdot}(X^{\cdot},Y^{\cdot})$
denote the complex of <u>abelian groups</u> given by

$$\mathrm{Hom}^n(X^{\cdot},Y^{\cdot}) = \prod_{p \in \mathbb{Z}} \mathrm{Hom}(X^p,Y^{p+n})$$

and boundary operator

$$[\partial^n f]^p = \partial^{n+p} f^p + (-1)^{n+1} f^{p+1} \partial^p$$

Remark the key formula, $n \in \mathbb{Z}$

4.4 $$[X^{\cdot},Y^{\cdot}[n]] = H^n \mathrm{Hom}^{\cdot}(X^{\cdot},Y^{\cdot})$$

The complex $\mathrm{Hom}^{\cdot}(X^{\cdot},Y^{\cdot})$ is obviously functorial, covariant in
$Y^{\cdot}$ and contravariant in $X^{\cdot}$.

18

We shall now introduce the main theme of a homotopy theory of complexes: Consider a <u>chainwise split exact sequence</u> of complexes

$$P^\cdot \xrightarrow{f} Q^\cdot \xrightarrow{g} R^\cdot$$

i.e. two morphisms of complexes such that for each $n \in \mathbb{Z}$, the sequence

$$P^n \xrightarrow{f^n} Q^n \xrightarrow{g^n} R^n$$

is a split exact sequence. Choose for each $n \in \mathbb{Z}$ a section $s^n : R^n \to Q^n$ to g^n, i.e. $g^n s^n = 1$ or just $gs = 1$, where $s: R \to Q$. We use the notation that R, which is $R^\cdot$ with the dot removed, denotes the graded object underlying the complex $R^\cdot$.

$$g(\partial s - s\partial) = \partial(gs) - g(s\partial) = 0$$

Thus we can find $h: R \to P[1]$ such that

$$4.5 \qquad\qquad fh = \partial s - s\partial$$

Using this relation twice, we find

$$f(h\partial + \partial h) = (fh)\partial + \partial(fh) =$$
$$(\partial s - s\partial)\partial + \partial(\partial s - s\partial) = \partial s\partial - \partial s\partial = 0$$

Using that f is a monomorphism we find

$$h\partial + \partial h = 0$$

which shows that $h: R \to P[1]$ is in fact a morphism of complexes

$$4.6 \qquad\qquad h: R^\cdot \to P^\cdot[1]$$

The <u>morphism h is unique up to homotopy</u>:

Let $\sigma:R \to Q$ be an alternative section of g and let $k:R^{\bullet} \to P^{\bullet}[1]$ be given by

$$fk = \partial\sigma - \sigma\partial$$

Since $g(s-\sigma) = 0$ we can find $t:Q \to P$ such that $s-\sigma = ft$. We find that

$$f(h-k) = \partial(s-\sigma) - (s-\sigma)\partial = \partial ft - ft\partial = f(\partial t - t\partial)$$

Using that f is a monomorphism we find

$$h - k = \partial t - t\partial = \partial t + t(-\partial)$$

which shows that $h \sim k$.

<u>Definition 4.7</u>. With the notation above, the class of h in $[R^{\bullet}, P^{\bullet}[1]]$ is called the <u>homotopy invariant</u> of the chainwise split exact sequence $P^{\bullet} \to Q^{\bullet} \to R^{\bullet}$.

For a given complex $X^{\bullet}$ our chainwise split exact sequence yields a short exact sequence of complexes of abelian groups

$$0 \to \operatorname{Hom}^{\bullet}(X^{\bullet}, P^{\bullet}) \to \operatorname{Hom}^{\bullet}(X^{\bullet}, Q^{\bullet}) \to \operatorname{Hom}^{\bullet}(X^{\bullet}, R^{\bullet}) \to 0$$

which in turn gives rise to a long exact homology sequence with <u>connecting morphism</u>

4.8 $$[1, h[n]]:[X^{\bullet}, R^{\bullet}[n]] \to [X^{\bullet}, P^{\bullet}[n+1]]$$

Proof. With the notation of 4.5 given morphism of complexes $x^n : X^{\cdot} \to R^{\cdot}[n]$. Lift this to $sx^n : X \to Q[n]$. The result of applying the boundary operator from the complex $\mathrm{Hom}^{\cdot}(X^{\cdot}, Q^{\cdot}[n])$ is

$$\partial sx^n + (-1)^{n+1} sx^n \partial = (\partial s - s\partial)x^n = fhx^n$$

where we have used the formula 4.5.

Q.E.D.

For a complex $Y^{\cdot}$ our split exact sequence yields a short exact sequence of complexes of abelian groups

$$0 \to \mathrm{Hom}^{\cdot}(R^{\cdot}, Y^{\cdot}) \to \mathrm{Hom}^{\cdot}(Q^{\cdot}, Y^{\cdot}) \to \mathrm{Hom}^{\cdot}(P^{\cdot}, Y^{\cdot}) \to 0$$

which gives rise to a long exact homology sequence with <u>connecting</u> <u>morphism</u>

4.9 $\qquad (-1)^{n+1}[h,1] : [P^{\cdot}, Y^{\cdot}[n]] \to [R^{\cdot}, Y^{\cdot}[n+1]]$

Proof. With the notation of 4.5 let $r : Q \to P$ denote the retraction satisfying

$$rf = 1, \quad rs = 0, \quad 1 = sg + fr$$

Let us first prove formula

4.10 $\qquad\qquad hg = r\partial - \partial r$

It suffices to prove the formula after we have applied the monomorphism f on both sides

$$fgh = (\partial s - s\partial)g = \partial sg - sg\partial$$

$$= \partial(1-fr) - (1-fr)\partial = f(r\partial - \partial r)$$

Let us now consider an element of $[P^\cdot, Y^\cdot[n]]$ represented by a morphism of complexes $y^n : P^\cdot \to Y^\cdot[n]$. Lift this to $y^n r : Q \to Y[n]$ and apply the boundary operator from $\operatorname{Hom}^\cdot(Q^\cdot, Y^\cdot)$

$$\partial y^n r + (-1)^{n+1} y^n r\partial = (-1)^n y^n \partial r + (-1)^{n+1} y^n r\partial =$$

$$(-1)^{n+1} y^n (r\partial - \partial r) = (-1)^{n+1} y^n hg$$

which proves the formula 4.9

Q.E.D.

Let us consider a <u>commutative</u> diagram

$$
\begin{array}{ccccc}
P^\cdot & \xrightarrow{\;f\;} & Q^\cdot & \xrightarrow{\;g^\cdot\;} & R^\cdot \\
\downarrow{\scriptstyle u} & & \downarrow{\scriptstyle v} & & \downarrow{\scriptstyle w} \\
U^\cdot & \xrightarrow{\;a\;} & V^\cdot & \xrightarrow{\;b\;} & W^\cdot
\end{array}
$$

of complexes, involving chainwise split exact sequences f,g and a,b. The resulting homotopy invariants k and c makes the following diagram <u>homotopy commutative</u>

4.11
$$
\begin{array}{ccc}
R^\cdot & \xrightarrow{\;k\;} & P^\cdot[1] \\
\downarrow{\scriptstyle w} & & \downarrow{\scriptstyle u[1]} \\
W^\cdot & \xrightarrow{\;c\;} & U^\cdot[1]
\end{array}
$$

<u>Proof</u>. Let $X^\cdot$ be an arbitrary complex. The diagram above give rise to a commutative diagram of complexes of abelian groups

$$
\begin{array}{ccccccccc}
0 & \to & \operatorname{Hom}^\cdot(X^\cdot, P^\cdot) & \to & \operatorname{Hom}^\cdot(X^\cdot, Q^\cdot) & \to & \operatorname{Hom}^\cdot(X^\cdot, R^\cdot) & \to & 0 \\
& & \downarrow & & \downarrow & & \downarrow & & \\
0 & \to & \operatorname{Hom}^\cdot(X^\cdot, U^\cdot) & \to & \operatorname{Hom}^\cdot(X^\cdot, V^\cdot) & \to & \operatorname{Hom}^\cdot(X^\cdot, W^\cdot) & \to & 0
\end{array}
$$

It follows from 2.7 and 4.8 that the following diagram is commutative

$$\begin{array}{ccc}
[X^\cdot,R^\cdot] & \xrightarrow{\ [1,k]\ } & [X^\cdot,P^\cdot[1]] \\
\downarrow{\scriptstyle[1,w]} & & \downarrow{\scriptstyle[1,u[1]]} \\
[X^\cdot,W^\cdot] & \xrightarrow[\ [1,c]\ ]{} & [X^\cdot,U^\cdot[1]]
\end{array}$$

Now specify to $X^\cdot = R^\cdot$ and evaluate the maps on the identity of $R^\cdot$.

Q.E.D.

We shall now introduce a very fundamental terminology, that of

Triangles

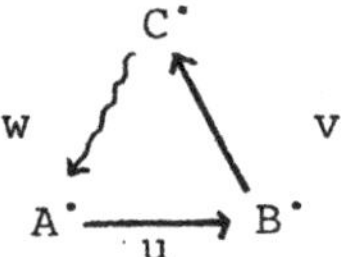

By a triangle is understood three complexes $A^\cdot, B^\cdot, C^\cdot$ and three morphisms of complexes

$$A^\cdot \xrightarrow{\ u\ } B^\cdot \xrightarrow{\ v\ } C^\cdot \xrightarrow{\ w\ } A^\cdot[1]$$

such that there exists a chainwise split exact sequence of complexes

$$P^\cdot \xrightarrow{\ f^\cdot\ } Q^\cdot \xrightarrow{\ g^\cdot\ } R^\cdot$$

with homotopy invariant h and homotopy equivalences a,b,c making the following diagram homotopy commutative

$$\begin{array}{ccccccc}
A^\cdot & \xrightarrow{u} & B^\cdot & \xrightarrow{v} & C^\cdot & \xrightarrow{w} & A^\cdot[1] \\
\downarrow{\scriptstyle a} & & \downarrow{\scriptstyle b} & & \downarrow{\scriptstyle c} & & \downarrow{\scriptstyle a[1]} \\
P^\cdot & \xrightarrow{f} & Q^\cdot & \xrightarrow{g} & R^\cdot & \xrightarrow{h} & P^\cdot[1]
\end{array}$$

For complexes X˙ and Y˙ the triangle above gives rise to long
exact sequences

$$\to [X^\cdot,A^\cdot] \xrightarrow{[1,u]} [X^\cdot,B^\cdot] \xrightarrow{[1,v]} [X^\cdot,C^\cdot] \xrightarrow{[1,w]} [X^\cdot,A^\cdot[1]] \xrightarrow{[1,u[1]]} [X^\cdot,B^\cdot[1]]-$$

4.12

$$\to [B^\cdot[1],Y^\cdot] \xrightarrow{[u[1],1]} [A^\cdot[1],Y^\cdot] \xrightarrow{[w,1]} [C^\cdot,Y^\cdot] \xrightarrow{[v,1]} [B^\cdot,Y^\cdot] \xrightarrow{[u,1]} [A^\cdot,Y^\cdot]$$

<u>The mapping cone</u>. Let $u:P^\cdot \to Q^\cdot$ denote a morphism of complexes
we define the mapping cone $Con^\cdot(u)$ by

$$Con(u) = P[1]\oplus Q, \quad \partial = \begin{pmatrix} -\partial & 0 \\ -u & \partial \end{pmatrix}$$

Notice that the sequence

$$Q^\cdot \xrightarrow{\binom{0}{1}} P^\cdot[1]\oplus Q^\cdot \xrightarrow{(1,0)} P^\cdot[1]$$

is chainwise split exact. Using the splitting $s = \binom{1}{0}$ we get

$$\partial s - s\partial = \begin{pmatrix} -\partial & 0 \\ -u & \partial \end{pmatrix} \binom{1}{0} - \binom{1}{0}(-\partial) = \binom{0}{-u} = \binom{0}{1}(-u)$$

which shows that the homotopy invariant is

$$-u[1]: P^\cdot[1] \longrightarrow Q^\cdot[1]$$

as a consequence we have a <u>triangle</u>

4.13

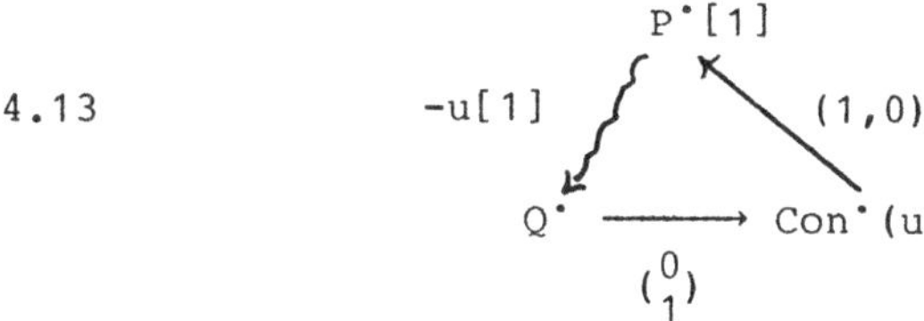

The mapping cylinder

Given a morphism $u: P^{\bullet} \to Q^{\bullet}$ of complexes. From this we derive the morphism

$$\begin{pmatrix} 1 \\ -u \end{pmatrix} : P^{\bullet} \to P^{\bullet} \oplus Q^{\bullet}$$

The cone over this morphism is the **mapping cylinder of** u, $\mathrm{Cyl}^{\bullet}(u)$. Explicitly

$$\mathrm{Cyl}^{\bullet}(u) = P[1] \oplus P \oplus Q \quad ; \quad \begin{pmatrix} -\partial & 0 & 0 \\ -1 & \partial & 0 \\ u & 0 & \partial \end{pmatrix}$$

The mapping cylinder $\mathrm{Cyl}^{\bullet}(u)$ is homotopy equivalent to $Q^{\bullet}$ in a canonical way: the two morphisms of complexes

$$Q^{\bullet} \xrightarrow{\hspace{2cm}} \mathrm{Cyl}^{\bullet}(u) \xrightarrow{\hspace{2cm}} Q^{\bullet}$$

$$\begin{pmatrix} 0 \\ 0 \\ 1 \end{pmatrix} \qquad\qquad (0,u,1)$$

composes to the identity of $Q^{\bullet}$. The composite of these two morphisms in the opposite order is homotopic to the identity on $\mathrm{Cyl}^{\bullet}(u)$ as it follows from the formula

$$\begin{pmatrix} 0 \\ 0 \\ 1 \end{pmatrix}(0,u,1) - \begin{pmatrix} 1 & 0 & 0 \\ 0 & 1 & 0 \\ 0 & 0 & 1 \end{pmatrix} = \begin{pmatrix} -\partial & 0 & 0 \\ -1 & \partial & 0 \\ u & 0 & \partial \end{pmatrix}\begin{pmatrix} 0 & 1 & 0 \\ 0 & 0 & 0 \\ 0 & 0 & 0 \end{pmatrix} + \begin{pmatrix} 0 & 1 & 0 \\ 0 & 0 & 0 \\ 0 & 0 & 0 \end{pmatrix}\begin{pmatrix} -\partial & 0 & 0 \\ -1 & \partial & 0 \\ u & 0 & \partial \end{pmatrix}$$

Consider the chainwise split exact sequence

$$P^{\bullet} \xrightarrow{\hspace{2cm}} \mathrm{Cyl}^{\bullet}(u) \xrightarrow{\hspace{2cm}} \mathrm{Con}^{\bullet}(u)$$

explicitly given by

$$\partial \qquad \begin{pmatrix} 0 \\ 1 \\ 0 \end{pmatrix} \qquad \begin{pmatrix} -\partial & 0 & 0 \\ -1 & \partial & 0 \\ u & 0 & \partial \end{pmatrix} \qquad \begin{pmatrix} -1 & 0 & 0 \\ 0 & 0 & 1 \end{pmatrix} \qquad \begin{pmatrix} -\partial & 0 \\ -u & \partial \end{pmatrix}$$

$$P \longrightarrow P[1] \oplus P \oplus Q \longrightarrow P[1] \oplus Q$$

The second of the two morphisms has the section $\begin{pmatrix} -1 & 0 \\ 0 & 0 \\ 0 & 1 \end{pmatrix}$, which we will use to calculate the homotopy invariant of the sequence. We have

$$\begin{pmatrix} -\partial & 0 & 0 \\ -1 & \partial & 0 \\ u & 0 & \partial \end{pmatrix}\begin{pmatrix} -1 & 0 \\ 0 & 0 \\ 0 & 1 \end{pmatrix} - \begin{pmatrix} -1 & 0 \\ 0 & 0 \\ 0 & 1 \end{pmatrix}\begin{pmatrix} -1 & 0 \\ 0 & 0 \\ 0 & 1 \end{pmatrix}\begin{pmatrix} -\partial & 0 \\ -u & \partial \end{pmatrix} = \begin{pmatrix} 0 & 0 \\ 1 & 0 \\ 0 & 0 \end{pmatrix} = \begin{pmatrix} 0 \\ 1 \\ 0 \end{pmatrix}(1,0)$$

Thus we have constructed a <u>triangle</u>

$$P^{\cdot} \longrightarrow \mathrm{Cyl}^{\cdot}(u) \longrightarrow \mathrm{Con}^{\cdot}(u) \xrightarrow{\ (1,0)\ } P[1]$$

Consider the following diagram

$$
\begin{array}{ccccccc}
P^{\cdot} & \xrightarrow{\ u\ } & Q^{\cdot} & \xrightarrow{\binom{0}{1}} & \mathrm{Con}^{\cdot}(u) & \xrightarrow{(1,0)} & P^{\cdot}[1] \\[4pt]
\downarrow{\scriptstyle 1} & & \downarrow{\binom{0}{0}{1}} & & \downarrow{\scriptstyle 1} & & \downarrow{\scriptstyle 1} \\[4pt]
P^{\cdot} & \xrightarrow{\binom{0}{1}{0}} & \mathrm{Cyl}^{\cdot}(u) & \xrightarrow{\left(\begin{smallmatrix}-1&0&0\\0&0&1\end{smallmatrix}\right)} & \mathrm{Con}^{\cdot}(u) & \xrightarrow{(1,0)} & P^{\cdot}[1]
\end{array}
$$

The first square is homotopy commutative, since $\begin{pmatrix} 0 \\ 0 \\ 1 \end{pmatrix}$ has the homotopy inverse $(0,u,1)$ and

$$(0,u,1)\begin{pmatrix} 0 \\ 1 \\ 0 \end{pmatrix} = u$$

From this we conclude that a morphism $u: P^{\cdot} \to Q^{\cdot}$ of complexes give rise to a <u>triangle</u>

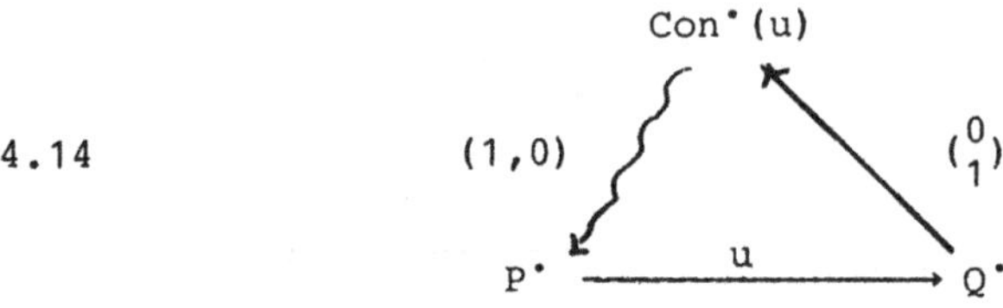

4.14

In fact, any triangle has this form.

Proposition 4.15. Given a chainwise split exact sequence of complexes

$$P^{\cdot} \xrightarrow{\ u\ } Q^{\cdot} \xrightarrow{\ v\ } R^{\cdot}$$

with homotopy invariant $w: R^{\cdot} \to P^{\cdot}[1]$. Then the following diagram[*)]

$$
\begin{array}{ccccccc}
P^{\cdot} & \xrightarrow{\ u\ } & Q^{\cdot} & \xrightarrow{\binom{0}{1}} & \text{Con}^{\cdot}(u) & \xrightarrow{(1,0)} & P^{\cdot}[1] \\
\downarrow{\scriptstyle 1} & & \downarrow{\scriptstyle 1} & & \downarrow{\scriptstyle (0,v)} & & \downarrow{\scriptstyle 1} \\
P^{\cdot} & \xrightarrow{\ u\ } & Q^{\cdot} & \xrightarrow{\ v\ } & R^{\cdot} & \xrightarrow{\ w\ } & P[1]
\end{array}
$$

is homotopy commutative and the columns are homotopy equivalences.

Proof. According to the results preceding 4.14 it suffices to prove a similar result for the diagram

$$
\begin{array}{ccccccc}
P^{\cdot} & \xrightarrow{\binom{0}{1}{0}} & \text{Cyl}^{\cdot}(u) & \xrightarrow{\left(\begin{smallmatrix} -1 & 0 & 0 \\ 0 & 0 & 1 \end{smallmatrix}\right)} & \text{Con}^{\cdot}(u) & \xrightarrow{(1,0)} & P^{\cdot}[1] \\
\downarrow{\scriptstyle 1} & & \downarrow{\scriptstyle (0,u,1)} & & \downarrow{\scriptstyle (0,v)} & & \downarrow{\scriptstyle 1} \\
P^{\cdot} & \xrightarrow{\ u\ } & Q^{\cdot} & \xrightarrow{\ v\ } & R^{\cdot} & \xrightarrow{\ w\ } & P^{\cdot}[1]
\end{array}
$$

---

*) It is the relation $vu = 0$ which ensures that $(0,v): \text{Con}^{\cdot}(u) \to R^{\cdot}$ is a morphism of complexes.

The first square is obviously commutative. The second square
is commutative since vu = 0. From the functoriality of the
homotopy invariant, 4.11 follows that the third square is homotopy commutative.

It remains to prove that (0,v) is a homotopy equivalence.
It suffices to prove that (0,v) induces an isomorphism after
transformation by any of the functors of the form $[X^{\cdot},-]$,
where $X^{\cdot}$ is a complex. We can apply the five lemma 1.7 to the
diagram

$$[X^{\cdot},P^{\cdot}] \longrightarrow [X^{\cdot},Q^{\cdot}] \longrightarrow [X^{\cdot},\mathrm{Con}(u)] \longrightarrow [X^{\cdot},P^{\cdot}[1]] \longrightarrow [X^{\cdot},Q^{\cdot}[1]]$$

$$\downarrow 1 \qquad\qquad \downarrow 1 \qquad\qquad \downarrow [1,(0,v)] \qquad\qquad \downarrow 1 \qquad\qquad \downarrow 1$$

$$[X^{\cdot},P^{\cdot}] \longrightarrow [X^{\cdot},Q^{\cdot}] \longrightarrow [X^{\cdot},R^{\cdot}] \longrightarrow [X^{\cdot},P^{\cdot}] \longrightarrow [X^{\cdot},Q^{\cdot}[1]]$$

to conclude the proof.

$$\text{Q.E.D.}$$

<u>Turning triangles 4.16</u>. Given a triangle

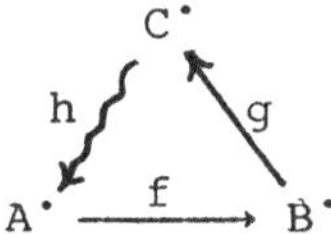

then the following diagram is a triangle

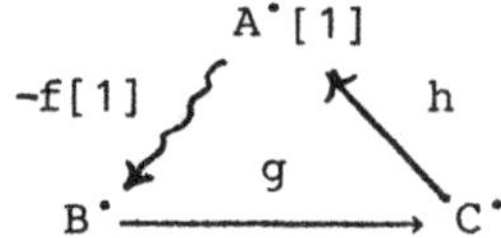

Proof. According to Lemma 4.15 and 4.14 we can replace the
triangle above with a mapping cone triangle. The result of
turning the mapping cone triangle 4.14 is

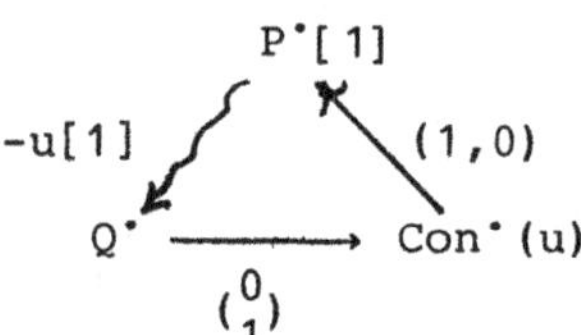

which is a triangle, see 4.13.

Q.E.D.

Corollary 4.17. For any morphism $u: P^{\cdot} \to Q^{\cdot}$ of complexes
there exists a homotopy commutative diagram

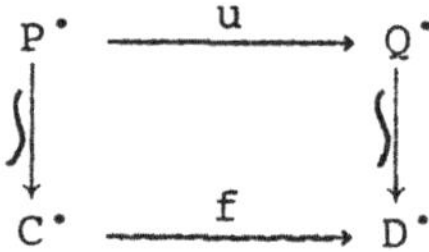

where f is a chainwise split epimorphism and the vertical
arrows are homotopy equivalences.

Proof. Turn the mapping cone 4.14 twice.

Q.E.D.

<u>Twisting triangles 4.18</u>. Given a triangle

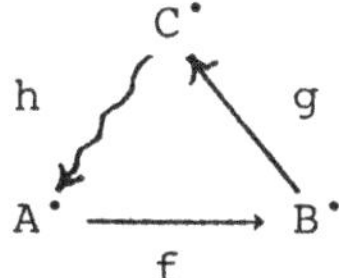

Then for any $n \in \mathbb{Z}$, the diagram

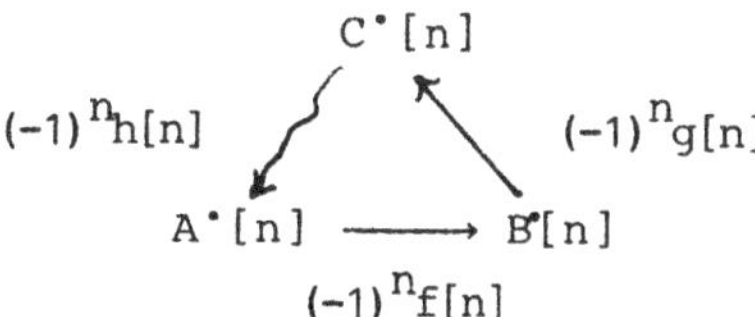

is a triangle.

<u>Proof</u>. Note first, that the second triangle above is isomorphic to

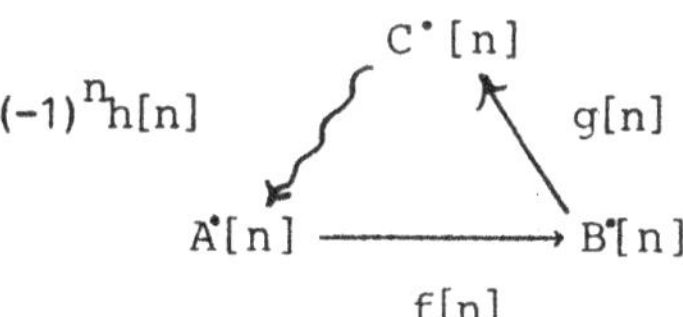

via $(-1)^n, 1, (-1)^n$. The result now follows rather immediately from formula 4.5.

Q.E.D.

30

<u>Filling in the third arrow 4.19</u>. Consider the diagram

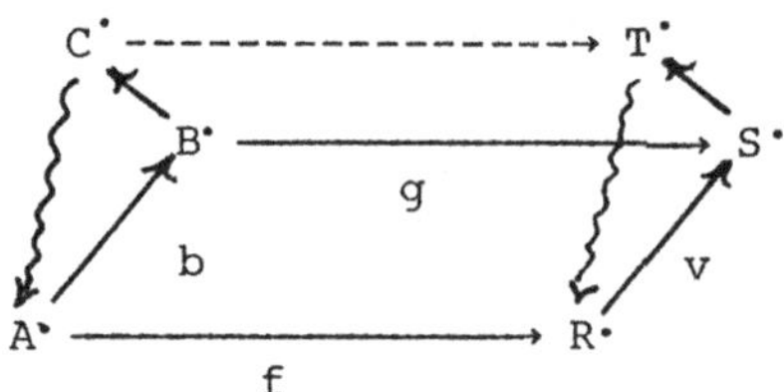

consisting of triangles ABC and RST and arrows f and g
making the bottom square homotopy commutative. Then the dotted
arrow may be filled in making the remaining two squares homotopy
commutative.

   <u>Proof</u>. We may assume that the triangles are mapping cones, 4.14.
Consider a homotopy, realizing that the bottom square is homotopy
commutative

$$gu - vf = k\partial + \partial k$$

Define $h: \mathrm{Con}^{\bullet}(u) \to \mathrm{Con}^{\bullet}(v)$ to be

$$\begin{pmatrix} f & 0 \\ -k & g \end{pmatrix} \;:\; A[1] \oplus B \to R[1] \oplus S$$

Notice that

$$\begin{pmatrix} -\partial & 0 \\ -v & \partial \end{pmatrix}\begin{pmatrix} f & 0 \\ -k & g \end{pmatrix} = \begin{pmatrix} f & 0 \\ -k & g \end{pmatrix}\begin{pmatrix} -\partial & 0 \\ -u & \partial \end{pmatrix}$$

$$\begin{pmatrix} f & 0 \\ -k & g \end{pmatrix}\begin{pmatrix} 0 \\ 1 \end{pmatrix} = \begin{pmatrix} 0 \\ 1 \end{pmatrix} g \;\; ; \;\; f(1,0) = (1,0)\begin{pmatrix} f & 0 \\ -k & g \end{pmatrix}$$

Q.E.D.

<u>Signs</u>

Let there be given a triangle

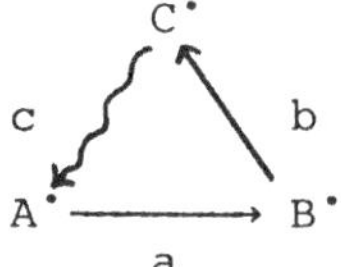

If we replace a by $-a$ and b by $-b$ we still have a
triangle as it follows by using an isomorphism of the form
$(-1,1,-1)$. More generally we can change sign on any two of the
arrows and we will still have a triangle. However, if we replace
c by $-c$ we will in general not have a triangle. An example is
given on the basis of the following lemma.

<u>Lemma 4.20.</u> Given a triangle

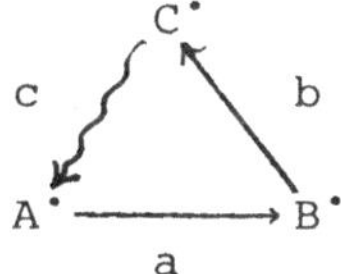

If $[A^{\cdot}[1],C^{\cdot}] = 0,$ then c is the only morphism from $C^{\cdot}$ to
$A^{\cdot}[1]$ in the homotopy category that makes a,b,c into a triangle.

<u>Proof</u>. Let $x \in [C^{\cdot},A^{\cdot}[1]]$ be such that a,b,x form a triangle.
Then according to 4.19 we can find $\theta: C^{\cdot} \to C^{\cdot}$ such that the
following diagram is commutative

$$
\begin{array}{ccccccc}
A^{\cdot} & \xrightarrow{\ a\ } & B^{\cdot} & \xrightarrow{\ b\ } & C^{\cdot} & \xrightarrow{\ c\ } & A^{\cdot}[1] \\
\downarrow{\scriptstyle 1} & & \downarrow{\scriptstyle 1} & & \downarrow{\scriptstyle \theta} & & \downarrow{\scriptstyle 1} \\
A^{\cdot} & \xrightarrow{\ a\ } & B^{\cdot} & \xrightarrow{\ b\ } & C^{\cdot} & \xrightarrow{\ x\ } & A^{\cdot}[1]
\end{array}
$$

Notice that $(1-\theta)\,b = 0$ whence we can find $h \in [A^{\cdot}[1],C^{\cdot}]$, such that $hc = 1-\theta$. Since $[A^{\cdot}[1],C^{\cdot}] = 0$ we must have $\theta = 1$ and from the equation $x\theta = 1c$, we conclude that $x = c$.

Q.E.D.

Example 4.21. Consider the following triangle of complexes of abelian groups

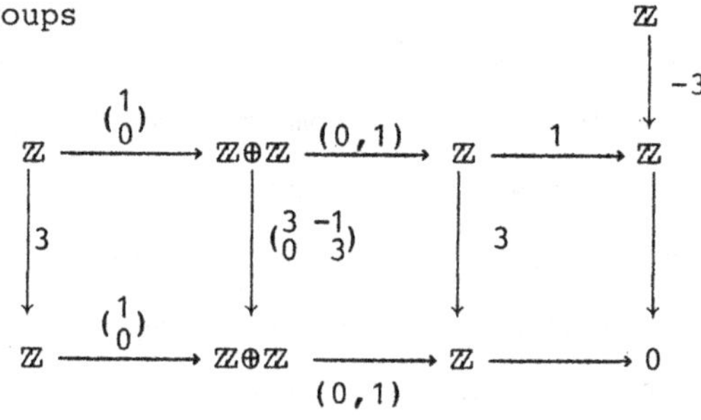

$$A^{\cdot} \xrightarrow{\;a\;} B^{\cdot} \xrightarrow{\;b\;} C^{\cdot} \xrightarrow{\;c\;} A^{\cdot}[1]$$

Notice that $[A^{\cdot}[1],B^{\cdot}] = 0$ and $[B^{\cdot},A^{\cdot}[1]] = \mathbb{Z}/(3)$ generated by c. In particular c is not homotopic to $-c$ and we conclude from Lemma 4.20 that $a,b,-c$ is not a triangle.

The triangle of two composable morphisms

Starting from two composable morphisms of complexes

$$u: X^{\cdot} \to Y^{\cdot}, \qquad v: Y^{\cdot} \to Z^{\cdot}$$

we can build a new triangle

4.22

<u>Verification</u>[*]

    We shall derive this triangle from the mapping cone triangle

$$\begin{pmatrix} 1 & 0 & 0 & 0 \\ 0 & 1 & 0 & 0 \end{pmatrix} \qquad \mathrm{Con}^{\bullet}(V) \qquad \begin{pmatrix} 0 & 0 \\ 0 & 0 \\ 1 & 0 \\ 0 & 1 \end{pmatrix}$$

$$\begin{pmatrix} -\partial & 0 \\ -u & \partial \end{pmatrix}; \ \mathrm{Con}^{\bullet}(u) \xrightarrow{\hspace{3cm}} \mathrm{Con}^{\bullet}(vu); \ \begin{pmatrix} -\partial & 0 \\ -vu & \partial \end{pmatrix}$$

$$V = \begin{pmatrix} 1 & 0 \\ 0 & v \end{pmatrix}$$

and the following identfication $\mathrm{Con}^{\bullet}(v)$ and $\mathrm{Con}^{\bullet}(V)$

$$\begin{pmatrix} 0 & 0 \\ 1 & 0 \\ 0 & 0 \\ 0 & 1 \end{pmatrix}$$

$$\begin{pmatrix} -\partial & 0 \\ -v & \partial \end{pmatrix}; \quad \mathrm{Con}^{\bullet}(v) \rightleftarrows \mathrm{Con}^{\bullet}(V) \quad ; \quad \begin{pmatrix} \partial & 0 & 0 & 0 \\ u & -\partial & 0 & 0 \\ -1 & 0 & -\partial & 0 \\ 0 & -v & -vu & \partial \end{pmatrix}$$

$$\begin{pmatrix} 0 & 1 & u & 0 \\ 0 & 0 & 0 & 1 \end{pmatrix}$$

using the explicit homotopy

$$\begin{pmatrix} 0 & 0 \\ 1 & 0 \\ 0 & 0 \\ 0 & 1 \end{pmatrix} \begin{pmatrix} 0 & 1 & 0 & 0 \\ 0 & 0 & 0 & 1 \end{pmatrix} - \begin{pmatrix} 1 & 0 & 0 & 0 \\ 0 & 1 & 0 & 0 \\ 0 & 0 & 1 & 0 \\ 0 & 0 & 0 & 1 \end{pmatrix} = \begin{pmatrix} -1 & 0 & 0 & 0 \\ 0 & 0 & u & 0 \\ 0 & 0 & -1 & 0 \\ 0 & 0 & 0 & 0 \end{pmatrix} =$$

$$\begin{pmatrix} \partial & 0 & 0 & 0 \\ u & -\partial & 0 & 0 \\ -1 & 0 & -\partial & 0 \\ 0 & -v & -vu & \partial \end{pmatrix} \begin{pmatrix} 0 & 0 & 1 & 0 \\ 0 & 0 & 0 & 0 \\ 0 & 0 & 0 & 0 \\ 0 & 0 & 0 & 0 \end{pmatrix} + \begin{pmatrix} 0 & 0 & 1 & 0 \\ 0 & 0 & 0 & 0 \\ 0 & 0 & 0 & 0 \\ 0 & 0 & 0 & 0 \end{pmatrix} \begin{pmatrix} \partial & 0 & 0 & 0 \\ u & -\partial & 0 & 0 \\ -1 & 0 & -\partial & 0 \\ 0 & -v & -vu & \partial \end{pmatrix}$$

---

[*] To simplify the notation we write u instead of u[1]. The
same convention applies whenever u[1] causes notational in-
conveniences.

I.5 Abelian categories

A category which is <u>additive</u> and <u>exact</u> is called an abelian category. Throughout this section we shall work in a fixed abelian category. A close analysis of the cocartesian diagrams will be of great use when it comes to construction of injective resolutions.

<u>Definition 5.1</u>. A commutative diagram

$$
\begin{array}{ccc}
A & \xrightarrow{\;f\;} & B \\
\downarrow{\scriptstyle a} & & \downarrow{\scriptstyle b} \\
C & \xrightarrow{\;g\;} & D
\end{array}
$$

is called <u>cocartesian</u> if whenever given a pair of morphisms $p: B \to X$ and $q: C \to X$ with $pf = qa$ there exists a unique morphism $r: D \to X$ with $rb = p$ and $rg = q$.

Given morphisms a and f as above we can always find b and g making the diagram cocartesian: For D take a co-kernel for the morphism

$$(a,-f): A \longrightarrow C \oplus B$$

The dual notion is <u>cartesian diagram</u>.

<u>Proposition 5.2</u>. Consider a <u>cocartesian</u> diagram

$$
\begin{array}{ccc}
A & \xrightarrow{\;f\;} & B \\
\downarrow{\scriptstyle a} & & \downarrow{\scriptstyle b} \\
C & \xrightarrow{\;g\;} & D
\end{array}
$$

1) $\operatorname{Cok} a \cong \operatorname{Cok} b$, $\operatorname{Cok} f \cong \operatorname{Cok} g$.

2) The morphisms Ker a → Ker b, Ker f → Ker g are
epimorphisms. – These two epimorphisms have isomorphic
kernels.

3) Our cocartesian diagram is in addition <u>cartesian</u> if
and only if

$$\text{Ker } a \to \text{Ker } b \quad \text{and} \quad \text{Ker } f \to \text{Ker } g$$

are isomorphisms.

<u>Proof</u>. 1) Follows by direct verification. 2) We can express
the assumption that the diagram is cocartesian, by saying that the
sequence

$$A \xrightarrow{\binom{a}{-f}} C \oplus B \xrightarrow{(g,b)} D \longrightarrow 0$$

is exact. Let $\widetilde{A}$ denote the coimage for $\begin{pmatrix} a \\ -f \end{pmatrix}$. There results
an exact sequence

$$0 \longrightarrow \widetilde{A} \xrightarrow{\left(\begin{smallmatrix} \widetilde{a} \\ -\widetilde{f} \end{smallmatrix}\right)} C \oplus B \xrightarrow{(g,b)} D \longrightarrow 0$$

which proves that the diagram

$$\begin{array}{ccc}
\widetilde{A} & \xrightarrow{\widetilde{f}} & B \\
\downarrow{\widetilde{a}} & & \downarrow{b} \\
C & \xrightarrow{g} & D
\end{array}$$

is cartesian. By the statement dual to 1) we find isomorphisms

$$\text{Ker } \widetilde{a} \xrightarrow{\sim} \text{Ker } b, \quad \text{Ker } \widetilde{f} \xrightarrow{\sim} \text{Ker } g.$$

It is easy to verify that the morphism

$$\text{Ker } a \to \text{Ker } \widetilde{a}, \quad \text{Ker } f \to \text{Ker } \widetilde{f}$$

are epimorphisms, 1.5. The last statement of 2) follows from 1.3. - 3) If the diagram is cartesian, it follows by a statement dual to 1), that the kernels are isomorphic. - To prove the converse, one checks that the morphism $\begin{pmatrix} a \\ -f \end{pmatrix} : A \to C \oplus B$ is a monomorphism.

Q.E.D.

Homology and homotopy 5.3. Let $P^{\cdot}$ and $Q^{\cdot}$ be chain complexes. Homotopic chain map $f, g : P^{\cdot} \to Q^{\cdot}$ will induce the same map on homology, i.e.

$$H^n(f) = H^n(g) \qquad ; \qquad n \in \mathbb{N}.$$

Proof. Let $s = (s^n)_{n \in \mathbb{Z}}$ be a homotopy from f to g, i.e.

$$g^n - f^n = s^{n+1} \partial^n + \partial^{n-1} s^n ; \qquad n \in \mathbb{Z}$$

Notice that the following diagram is commutative

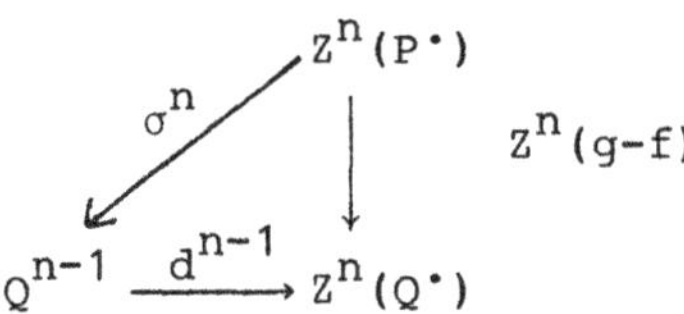

where σ^n is the restriction of s^n to $Z^n(P^{\cdot})$ and d^{n-1} is the result of factoring ∂^{n-1} through the canonical monomorphism $Z^n(Q^{\cdot}) \to Q^n$. Notice that

$$\mathrm{Cok}(d^{n-1}) = H^n(Q^{\cdot})$$

from which the result follows.

Q.E.D.

<u>The connecting morphism 5.4</u>. Let there be given a chainwise split exact sequence of complexes

$$0 \longrightarrow P^{\cdot} \xrightarrow{\ f\ } Q^{\cdot} \xrightarrow{\ g\ } R^{\cdot} \longrightarrow 0$$

The homotopy invariant $h: R^{\cdot} \to P^{\cdot}[1]$ from 4.7 will induce a morphism in homology

$$H^n(h): H^n(R^{\cdot}) \to H^{n+1}(P^{\cdot})$$

which is identical with the connecting morphism introduced in 2.3.

<u>Proof</u>. Let us introduce the objects

$$K^n = \mathrm{Ker}(\partial^n g^n) \qquad\qquad C^{n+1} = \mathrm{Cok}(\partial^n f^n)$$

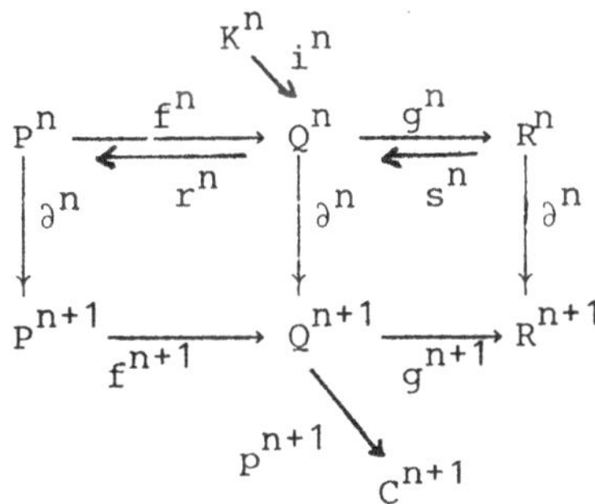

According to Lemma 2.8 it must be proved that

$$p\,f\,h\,g\,i = p\,\partial\,i$$

Choose a splitting $s: R \to Q$ of g. We can write $fh = \partial s - s\partial$ and whence

$$p\,f\,h\,g\,i = p(\partial s - s\partial)gi = p\,\partial\,s\,g\,i$$

let now $r: Q \to P$ be a retraction of f such that $1 = fr+sg$. Then

$$p f h g i = p\partial(1-fr)i = p\,\partial\,i - p\,\partial\,f\,r\,i$$

Notice that $p\,\partial\,f = 0$ by the definition of p.

Q.E.D.

<u>Corollary 5.5</u>. A triangle of complexes

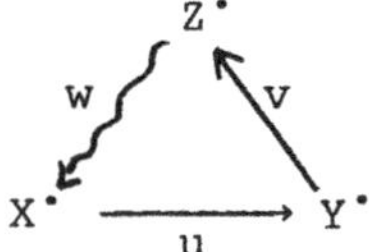

will induce a long <u>exact</u> homology sequence

$$\longrightarrow H^n(X^{\cdot}) \xrightarrow{H^n(u)} H^n(Y^{\cdot}) \xrightarrow{H^n(v)} H^n(Z^{\cdot}) \longrightarrow$$
$$\xrightarrow{H^n(w)}$$
$$\longrightarrow H^{n+1}(X^{\cdot}) \xrightarrow{H^{n+1}(u)} H^{n+1}(Y^{\cdot}) \xrightarrow{H^{n+1}(v)} H^{n+1}(Z^{\cdot}) \longrightarrow$$

<u>Definition 5.6</u>. A morphism of complexes $f: X^{\cdot} \to Y^{\cdot}$ is called a <u>quasi-isomorphism</u> if $H^n(f): H^n(X^{\cdot}) \to H(Y^{\cdot}_j)$ is an isomorphism for all $n \in \mathbb{Z}$.

Notice that $f: X^{\cdot} \to Y^{\cdot}$ is a quasi-isomorphism if and only if $H^{\cdot}(Con^{\cdot}(f)) = 0$ as it follows by considering the mapping cone triangle 4.14

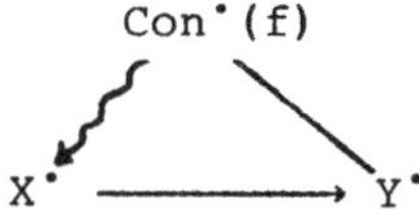

and the long homology sequence 5.5.

$\underline{\text{Truncation}}$. Let $K^{\cdot}$ denote a complex and $n \in \mathbb{Z}$. We form a new complex

$$\tau_{\leq n} K^{\cdot} \quad \ldots \to K^{n-2} \to K^{n-1} \to \operatorname{Ker}\partial^{n} \to 0 \ldots$$

Notice the canonical inclusion

5.7
$$\tau_{\leq n} K^{\cdot} \longrightarrow K^{\cdot}$$

induces an isomorphism on homology in levels $\leq n$. Notice also that homotopic morphisms $f,g: K^{\cdot} \to L^{\cdot}$ will induce $\underline{\text{homotopic}}$ morphisms

$$f,g: \tau_{\leq n} K^{\cdot} \longrightarrow \tau_{\leq n} L^{\cdot}$$

Similarly we define the complex

$$\tau_{\geq n} K^{\cdot}: \qquad 0 \to \operatorname{Cok}\partial^{n-1} \to K^{n+1} \to K^{n+2} \to \ldots$$

which comes equipped with a canonical morphism

5.8
$$K^{\cdot} \longrightarrow \tau_{\geq n} K^{\cdot}$$

which induces an isomorphism on homology in levels $\geq n$. Notice again that formation of $\tau_{\geq n}$ carries over to the homotopy category. - We can establish a relationship between these by noticing that the composite

$$\tau_{\leq n} K^{\cdot} \longrightarrow K^{\cdot} \longrightarrow \tau_{\geq n+1} K^{\cdot}$$

is zero and induces a $\underline{\text{quasi-isomorphism}}$

5.9
$$K^{\cdot}/\tau_{\leq n} K^{\cdot} \longrightarrow \tau_{\geq n+1} K^{\cdot}$$

I.6 Injective resolutions

In this section we shall work in a fixed abelian category A .
We say that an object I of A is <u>injective</u> if for any monomorphism $f: X \to Y$ and any morphism $x: X \to I$, there exists a
morphism $y: Y \to I$ with $x = yf$.

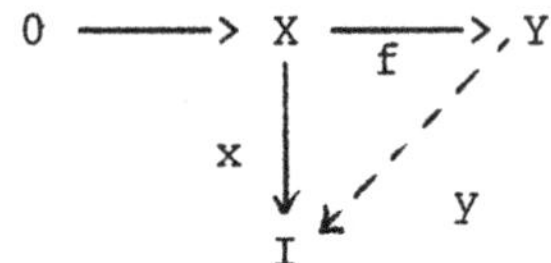

Let I be an injective object of A . Any short exact sequence in A

$$0 \to X \to Y \to Z \to 0$$

give rise to an exact sequence

$$0 \to \mathrm{Hom}(Z,I) \to \mathrm{Hom}(Y,I) \to \mathrm{Hom}(X,I) \to 0$$

as it follows from the definition.

In the rest of this section we shall assume that A has
<u>enough injectives</u>, i.e. that any object of A admits a mono-
morphism into an injective object. - We shall mostly be concerned
with <u>bounded below complexes</u>, i.e. complexes $K^{\cdot}$ with $K^n = 0$
for $n \ll 0$.

<u>Theorem 6.1</u>. Any bounded below complex $X^{\cdot}$ admits a quasi-iso-
morphism $f: X^{\cdot} \to I^{\cdot}$ into a bounded below complex of injective
objects (<u>an injective resolution of $X^{\cdot}$</u>)

<u>Proof</u>. We shall proceed by increasing induction on $n \in \mathbb{Z}$.
Suppose f has already been constructed up to level $n \in \mathbb{Z}$. That
is we have the solid diagram

$$\begin{array}{ccccccccc}
\longrightarrow & X^{n-1} & \longrightarrow & X^n & \longrightarrow & 'Z^n(X^{\boldsymbol{\cdot}}) & \overset{\bar{\partial}^n}{\longrightarrow} & Z^{n+1}(X^{\boldsymbol{\cdot}}) & \longrightarrow & X^{n+1} \\
& \downarrow{\scriptstyle f^{n-1}} & & \downarrow{\scriptstyle f^n} & & \downarrow & \bigcirc\ \downarrow & & \downarrow \\
\longrightarrow & I^{n-1} & \overset{\partial^{n-1}}{\longrightarrow} & I^n & \longrightarrow & \mathrm{Cok}(\partial^{n-1}) & \dashrightarrow & Y^{n+1} & \longrightarrow & I^{n+1}
\end{array}$$

We shall assume this done in such a way that

$$0 \to H^n(X^{\boldsymbol{\cdot}}) \to \mathrm{Cok}(\partial^{n-1})$$

is exact, here we identify $H^n(X^{\boldsymbol{\cdot}})$ with the kernel of $\bar{\partial}^n$, 2.6. Now fill in the two dotted arrows to make the circled square <u>cocartesian</u>. It follows from 5.2 that the circled square gives rise to an exact sequence

$$0 \to H^n(X^{\boldsymbol{\cdot}}) \to \mathrm{Cok}(\partial^{n-1}) \to Y^{n+1} \to H^{n+1}(X^{\boldsymbol{\cdot}}) \to 0$$

Next, choose a monomorphism $Y^{n+1} \to I^{n+1}$ with I^{n+1} injective and fill in the last dotted arrow at random. Notice that

$$0 \to \mathrm{Cok}(I^n \to Y^{n+1}) \to \mathrm{Cok}(I^n \to I^{n+1})$$

is exact, which concludes the inductive step.

Q.E.D.

<u>Theorem 6.2.</u> Let $I^{\boldsymbol{\cdot}}$ be a bounded below complex of injective objects. Any quasi-isomorphism $c: X^{\boldsymbol{\cdot}} \to Y^{\boldsymbol{\cdot}}$ induces an isomorphism

$$[c,1]: [Y^{\boldsymbol{\cdot}}, I^{\boldsymbol{\cdot}}] \overset{\sim}{\to} [X^{\boldsymbol{\cdot}}, I^{\boldsymbol{\cdot}}]$$

Proof. Consider the mapping cone triangle 4.14

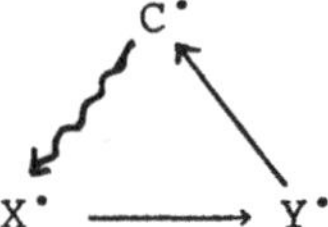

It follows from the long homology sequence I.5.5

$$\to H^n(X^{\cdot}) \to H^n(Y^{\cdot}) \to H^n(C^{\cdot}) \to H^{n+1}(X^{\cdot}) \to H^{n+1}(Y^{\cdot})$$

that $H^n(C^{\cdot}) = 0$ for all $n \in \mathbb{Z}$. Next consider the homotopy sequence induced by $[-,I^{\cdot}]$

$$\to [C^{\cdot},I^{\cdot}] \to [Y^{\cdot},I^{\cdot}] \to [X^{\cdot},I^{\cdot}] \to [C^{\cdot}[-1],I^{\cdot}] \to$$

to see that it suffices to prove that

6.3 $\qquad [C^{\cdot},I^{\cdot}] = 0$ _whenever_ $C^{\cdot}$ _is a complex with_

$\qquad\qquad\qquad\quad H^{\cdot}(C^{\cdot}) = 0,$ _and_ $I^{\cdot}$ _is a bounded_

$\qquad\qquad\qquad\quad$ _below complex of injectives._

We must prove that any morphism of complexes $f: C^{\cdot} \to I^{\cdot}$ is homotopic to zero. We are going to construct a homotopy by induction on $n \in \mathbb{Z}$. Suppose we have already constructed $s^p: C^p \to I^{p-1}$ for all $p \leq n$ with

$$s^p \partial^{p-1} + \partial^{p-2} s^{p-1} = f^{p-1} \qquad ; \quad p \leq n$$

We shall proceed to construct $s^{n+1}: C^{n+1} \to I^n$ such that the relation above holds with $p = n+1$.

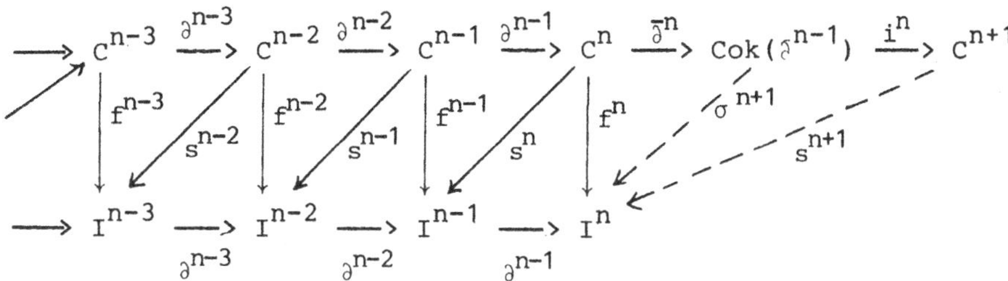

The following calculation shows that $(f^n - \partial^{n-1} s^n) \partial^{n-1} = 0$

$$(f^n - \partial^{n-1} s^n) \partial^{n-1} = \partial^{n-1} f^{n-1} - \partial^{n-1} s^n \partial^{n-1} =$$

$$= \partial^{n-1} s^n \partial^{n-1} + \partial^{n-1} \partial^{n-2} s^{n-1} - \partial^{n-1} s^n \partial^{n-1} = 0$$

Let $\partial^n : C^n \to C^{n-1}$ be factorized

$$C^n \xrightarrow{\bar{\partial}^n} \mathrm{Cok}(\partial^{n-1}) \xrightarrow{i^n} C^{n+1}$$

We can find $\sigma^{n+1} : \mathrm{Cok}(\partial^{n-1}) \longrightarrow I^n$ such that

$$f^n - \partial^{n-1} s^n = \sigma^{n+1} \bar{\partial}^n$$

Note, that $i^n : \mathrm{Cok}(\partial^{n-1}) \to C^{n+1}$ is a monomorphism. Any extension s^{n+1} of σ^{n+1} to C^{n+1} will do.

Q.E.D.

We shall now derive some consequences of 6.1 and 6.2 in terms of the following two categories

6.4

$K^+(A)$ the homotopy category of bounded below complexes in A,

$D^+(A)$ the homotopy category of bounded below complexes of injectives in A

44

Let there for each $X^{\bullet}$ in $K^+(A)$ be <u>chosen</u> a quasi-isomorphism

6.5 $$c: X^{\bullet} \longrightarrow \rho X^{\bullet}$$

where $\rho X^{\bullet}$ is an object of $D^+(A)$. For a morphism $f: X^{\bullet} \to Y^{\bullet}$
in $K^+(A)$, there exists one and only one morphism $\rho f: \rho X^{\bullet} \to \rho Y^{\bullet}$
making the following diagram commutative

$$
\begin{array}{ccc}
X^{\bullet} & \xrightarrow{\quad c \quad} & \rho X^{\bullet} \\
\downarrow{\scriptstyle f} & & \downarrow{\scriptstyle \rho f} \\
Y^{\bullet} & \xrightarrow{\quad c \quad} & \rho Y^{\bullet}
\end{array}
$$

It follows easily from 6.2, that this construction yields a
canonical isomorphism

6.6 $$[X^{\bullet}, I^{\bullet}] = [\rho X^{\bullet}, I^{\bullet}]$$

where X varies through $K^+(A)$ and $I^{\bullet}$ varies through $D^+(A)$.
Otherwise expressed the functor

6.7 $$\rho: K^+(A) \to D^+(A)$$

is a <u>left adjoint</u> to the inclusion $i: D^+(A) \to K^+(A)$. In this
interpretation the resolution morphism c from 6.5 plays the
role of <u>adjunction</u> morphism.

<u>Theorem 6.8</u>. The additive functor

$$\rho: K^+(A) \to D^+(A)$$

transforms triangles into triangles.

<u>Proof</u>. Let us start with a triangle in $K^+(A)$

$$X^{\bullet} \xrightarrow{f} Y^{\bullet} \xrightarrow{g} Z^{\bullet} \xrightarrow{h} X^{\bullet}[1]$$

We can form the mapping cone triangle in $D^+(A)$ based on
$\rho f: \rho X^\bullet \to \rho Y^\bullet$

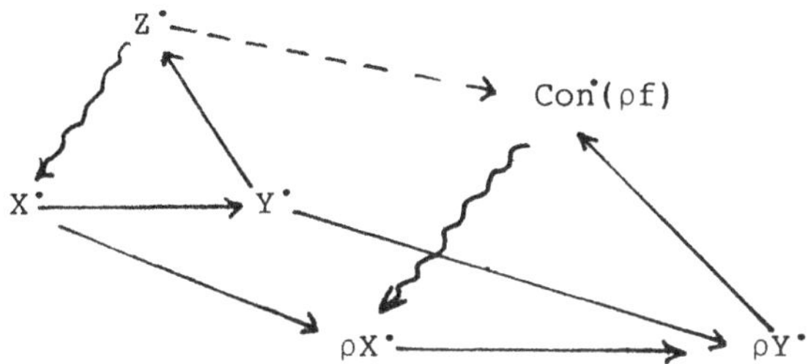

and fill in the dotted arrow in $K^+(A)$, making the diagram
commutative. Writing out the long homology ladder 2.7 we conclude from the five Lemma 1.7, that the dotted arrow is a
quasi-isomorphism. It is now easy to conclude the proof by means
of 6.2.

Q.E.D.

<u>Construction of triangles from short exact sequences of complexes.</u>

Let there be given a short exact sequence of bounded below
complexes in A

$$0 \longrightarrow P^\bullet \xrightarrow{f} Q^\bullet \xrightarrow{g} R^\bullet \longrightarrow 0$$

By an <u>injective resolution</u> of this sequence we understand a
<u>commutative</u>, not merely homotopy commutative diagram

6.9
$$\begin{array}{ccccccccc}
0 & \longrightarrow & P^\bullet & \xrightarrow{f} & Q^\bullet & \xrightarrow{g} & R^\bullet & \longrightarrow & 0 \\
& & \downarrow{a} & & \downarrow{b} & & \downarrow{c} & & \\
0 & \longrightarrow & I^\bullet & \xrightarrow{u} & J^\bullet & \xrightarrow{v} & K^\bullet & \longrightarrow & 0
\end{array}$$

where the bottom sequence is exact and consists of bounded below
complexes of injectives, and the vertical arrows are quasi-isomorphisms.

<u>Proposition 6.10</u>. Any short exact sequence of bounded below complexes has an injective resolution.

<u>Proof</u>. Let us first choose quasi-isomorphisms $b: Q^{\bullet} \to J^{\bullet}$ and $c: R^{\bullet} \to K^{\bullet}$ where $J^{\bullet}$ and $K^{\bullet}$ are bounded below complexes of injectives, 6.1. Next, choose $v: J^{\bullet} \to K^{\bullet}$ such that the following diagram is <u>homotopy</u> commutative, 6.2

$$
\begin{array}{ccc}
Q^{\bullet} & \xrightarrow{\ g\ } & R^{\bullet} \\
\downarrow{\scriptstyle b} & & \downarrow{\scriptstyle c} \\
J^{\bullet} & \xrightarrow{\ v\ } & K^{\bullet}
\end{array}
$$

By 4.17 we may assume that v is a chainwise split epimorphism. The diagram is homotopy commutative, so choose a homotopy $s: Q \to K[-1]$ from cg to vb

$$cg-vb = s\partial + \partial s$$

Next, choose $t: Q \to J[-1]$ such that $s = vt$. The formula above may be rewritten

$$cg = vb + vt\partial + \partial vt = v(b + t\partial + \partial t)$$

which shows that we can modify b within its homotopy class making the diagram commutative. We can now quite simply put $I^{\bullet} = \mathrm{Ker}(v)$ to obtain a commutative diagram

$$
\begin{array}{ccccccccc}
0 & \longrightarrow & P^{\bullet} & \xrightarrow{\ f\ } & Q^{\bullet} & \xrightarrow{\ g\ } & R^{\bullet} & \longrightarrow & 0 \\
& & \downarrow{\scriptstyle a} & & \downarrow{\scriptstyle b} & & \downarrow{\scriptstyle c} & & \\
0 & \longrightarrow & I^{\bullet} & \xrightarrow{\ u\ } & J^{\bullet} & \xrightarrow{\ v\ } & K^{\bullet} & \longrightarrow & 0
\end{array}
$$

It remains to prove that a is a quasi-isomorphism. To see this write out the long homology sequences and apply the five lemma.

$$\text{Q.E.D.}$$

A resolution as the one in 6.7 gives rise to a triangle in $D^{+}(A)$

6.11

$$
\begin{array}{ccc}
 & K^{\cdot} & \\
w \nearrow & & \nwarrow v \\
I^{\cdot} \xrightarrow{\quad u \quad} & & J^{\cdot}
\end{array}
$$

<u>this triangle is unique up to homotopy</u>.

<u>Proof</u>. Let $X^{\cdot}$ be an arbitrary bounded below complex of injective objects. A diagram like 6.9 will give rise to two exact sequences of complexes of abelian groups and a commutative diagram

$$
\begin{array}{ccccccc}
0 \to \text{Hom}^{\cdot}(K^{\cdot},X^{\cdot}) & \to & \text{Hom}^{\cdot}(J^{\cdot},X^{\cdot}) & \to & \text{Hom}^{\cdot}(I^{\cdot},X^{\cdot}) & \to & 0 \\
\downarrow & & \downarrow & & \downarrow & & \\
0 \to \text{Hom}^{\cdot}(R^{\cdot},X^{\cdot}) & \to & \text{Hom}^{\cdot}(Q^{\cdot},X^{\cdot}) & \to & \text{Hom}^{\cdot}(P^{\cdot},X^{\cdot}) & \to & 0
\end{array}
$$

By taking homology we obtain two sets of connecting morphism, making the following diagram commute

$$
\begin{array}{ccc}
[I^{\cdot},X^{\cdot}] & \longrightarrow & [K^{\cdot},X^{\cdot}[1]] \\
\Big\updownarrow [c,1] & & \Big\updownarrow [a,1] \\
[P^{\cdot},X^{\cdot}] & \longrightarrow & [R^{\cdot},X^{\cdot}[1]]
\end{array}
$$

According to formula 4.9 the bottom homomorphism is $-[w,1]$,

where w is the third side in the triangle. This description fixes w and thereby shows that the triangle is unique.

$$\text{Q.E.D.}$$

In terms of the resolution functor ρ we have assigned to the short exact sequence of complexes a definite triangle

6.12

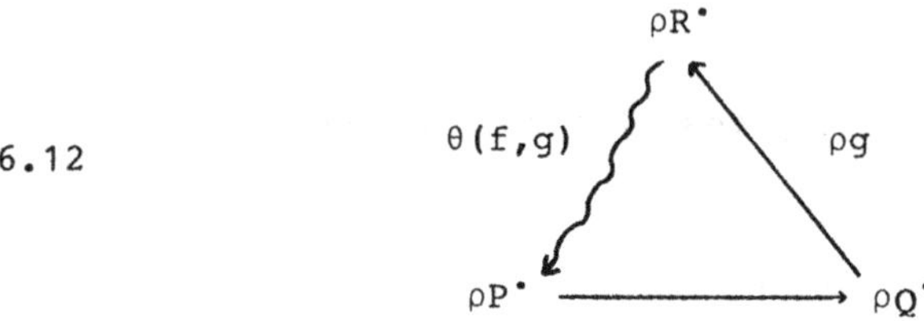

Given a commutative diagram of complexes in A

$$0 \to P^{\cdot} \xrightarrow{f} Q^{\cdot} \xrightarrow{g} R^{\cdot} \to 0$$
$$\downarrow u \qquad \downarrow v \qquad \downarrow w$$
$$0 \to U^{\cdot} \xrightarrow{a} V^{\cdot} \xrightarrow{b} W^{\cdot} \to 0$$

with exact rows. Then the following diagram is commutative

6.13

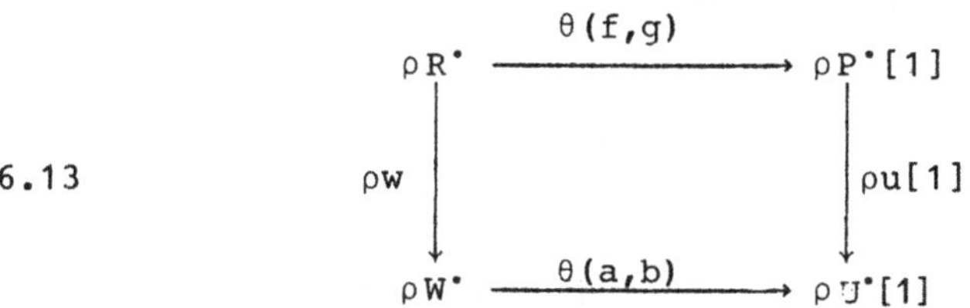

as it follows by a consideration similar to the one made in the proof of 6.11.

Truncation in the derived category

For an integer n consider the two endofunctors $\tau_{\leq n}$ and $\tau_{\geq n+1}$ of $K^{+}(A)$. These give rise to two endofunctors $\rho\tau_{\leq n}$ and $\rho\tau_{\geq n+1}$ of $D^{+}(A)$ which we shall still denote by $\tau_{\leq n}$ and $\tau_{\geq n+1}$ when no confusion is possible. For $X^{\cdot}$ in $D^{+}(A)$ we shall construct a triangle in $D^+(A)$

6.14

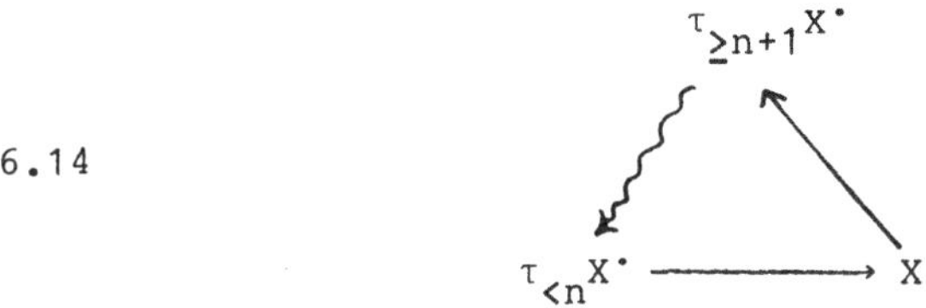

which depends functorially on $X^\cdot$. To do so we start with the

short exact sequence of complexes in A

$$0 \to \tau_{\leq n}X^\cdot \to X^\cdot \to X^\cdot/\tau_{\leq n}X^\cdot \to 0$$

this we can combine with the quasi-isomorphism 5.9

$$X^\cdot/\tau_{\leq n}X^\cdot \longrightarrow \tau_{\geq n+1}X^\cdot$$

and apply the previous procedure.

Let us analyse this a bit closer: Let $\mathcal{D}^{\leq n}$ and $\mathcal{D}^{\geq n+1}$ denote the full subcategories of $\mathcal{D} = D^+(A)$ whose objects are complexes $X^\cdot$ with

6.15
$$\mathcal{D}^{\leq n}: \qquad H^i(X^\cdot) = 0 ; \quad i > n$$
$$\mathcal{D}^{\geq n+1}: \qquad H^i(X^\cdot) = 0 ; \quad i < n+1$$

These two categories satisfies

6.16 $\qquad [I^\cdot, J^\cdot] = 0 \qquad$ for $\quad I^\cdot \in \mathcal{D}^{\leq n} \quad$ and $\quad J^\cdot \in \mathcal{D}^{\geq n+1}$

Proof. The complex $J^\cdot$ is a bounded below complex of injectives with $H^i(J^\cdot) = 0$ for $i \leq n$. It follows easily, that $\text{Cok}(\partial^i)$ is an injective object for $i \leq n$, whence the complex $\tau_{\geq n+1}J^\cdot$ is a complex of injectives, homotopy equivalent to $J^\cdot$. The assumption on $I^\cdot$ implies that $\tau_{\leq n}I^\cdot \to I^\cdot$

50

is a quasi-isomorphism. Whence by 6.2

$$[I^\cdot, J^\cdot] = [\tau_{\leq n} I^\cdot, \tau_{\geq n+1} J^\cdot]$$

the last homotopy group is zero for trivial reasons.

Q.E.D.

With $J^\cdot \in \mathcal{D}^{\geq n+1}$ we deduce from 6.14 an exact sequence

$$[\tau_{\leq n} X^\cdot[1], J^\cdot] \to [\tau_{\geq n+1} X^\cdot, J^\cdot] \to [X^\cdot, J^\cdot] \to [\tau_{\leq n} X^\cdot, J^\cdot] \to$$

Let us notice that quite generally, $n \in \mathbb{Z}$

6.17
$$\mathcal{D}^{\leq n}[+1] = \mathcal{D}^{\leq n-1} \subseteq \mathcal{D}^{\leq n}$$

$$\mathcal{D}^{\geq n+1}[-1] = \mathcal{D}^{\geq n+2} \subseteq \mathcal{D}^{\geq n+1}$$

Combining this with the sequence above and 6.16 we get

6.18
$$[\tau_{\leq n+1} X^\cdot, J^\cdot] = [X^\cdot, J^\cdot] \quad ; \quad J \in \mathcal{D}^{\geq n+1}$$

which shows that $\tau_{\leq n+1}$ is <u>left adjoint</u> to the inclusion of $\mathcal{D}^{\geq n+1}$ in $\mathcal{D}$. In a similar way we deduce from 6.14, 6.16, 6.17 that

6.19
$$[I^\cdot, \tau_{\leq n} X^\cdot] = [I^\cdot, X^\cdot] \quad ; \quad I^\cdot \in \mathcal{D}^{\leq n}$$

which shows that $\tau_{\leq n}$ is a <u>right adjoint</u> to the inclusion of $\mathcal{D}^{\leq n}$ into $\mathcal{D},$. Let us finally remark that the <u>curly arrow</u> in 6.14 <u>is unique</u>, see 4.20.

<u>Remark 6.20</u>. Given $K^\cdot$ in $\mathcal{D}$, such that

$$[K^\cdot, J^\cdot] = 0 \quad \text{for all } J^\cdot \text{ in } \mathcal{D}^{\geq n+1}$$

then $K^\cdot$ is in $\mathcal{D}^{\leq n}$ as it follows from 6.18 and 6.16.

I.7   Right derived functors

We shall first develop the classical part of the theory of derived functors.

Let $T: A \to B$ be a covariant additive functor between abelian categories A and B of which A is assumed to have enough injectives.

For an object X of A choose an <u>injective resolution of</u> X, i.e. a long exact sequence with $I^0, I^1, \ldots$ injective

$$0 \to X \xrightarrow{\varepsilon} I^0 \xrightarrow{\partial^0} I^1 \xrightarrow{\partial^1} I^2 \ldots\ldots$$

Consider X as a complex concentrated in degree zero, and let $I^\cdot$ denote the complex

$$0 \to I^0 \xrightarrow{\partial^0} I^1 \xrightarrow{\partial^1} I^2 \ldots\ldots$$

We can then interprete the resolution above as a <u>quasi-isomorphism</u> of complexes $\varepsilon: X \to I^\cdot$. Given a second resolution $X \to J^\cdot$, according to 6.2 there exists one and up to homotopy only one $\theta: I^\cdot \to J^\cdot$ making the following diagram homotopy commutative [*]

$$\begin{array}{ccc} & & I^\cdot \\ & \nearrow & \\ X & \downarrow\theta & \uparrow\varphi \\ & \searrow & \\ & & J^\cdot \end{array}$$

In the same way we construct $\varphi: J^\cdot \to I^\cdot$. It follows from 6.2, that $\theta\varphi \sim 1$ and $\varphi\theta \sim 1$. In view of this uniqueness it makes sense to define for $n \in \mathbb{Z}$

$$7.1 \qquad\qquad R^n T(X) = H^n T(I^\cdot)$$

---

[*] Since the complexes $I^\cdot$ and $J^\cdot$ are positive and the complex X is concentrated in degree zero, the diagram is necessarily commutative.

where $X \to I^\bullet$ is an injective resolution. - Given a morphism $f:X \to Y$ in A we shall construct a morphism

$$R^n T(f): R^n T(X) \to R^n T(Y)$$

Let $X \to I^\bullet$ and $Y \to J^\bullet$ be injective resolutions. According to 6.2 we can choose $\psi:I^\bullet \to J^\bullet$, unique up to homotopy, making the following diagram homotopy[*)] commutative:

$$
\begin{array}{ccc}
X^\bullet & \longrightarrow & I^\bullet \\
\Big\downarrow f & & \Big\downarrow \psi \\
Y^\bullet & \longrightarrow & J^\bullet
\end{array}
$$

In this way we have for each $n \in \mathbb{Z}$ constructed an additive covariant functor

$$R^n T: A \to B$$

Notice, that $R^n T = 0$ for $n < 0$

Connecting morphisms. Let there be given a short exact sequence in $\underline{A}$

$$0 \to X \to Y \to Z \to 0$$

Then according to 6.10 we can find a commutative diagram with exact rows

$$
\begin{array}{ccccccccc}
0 & \to & X & \to & Y & \to & Z & \to & 0 \\
 & & \downarrow & & \downarrow & & \downarrow & & \\
0 & \to & I^\bullet & \to & J^\bullet & \to & K^\bullet & \to & 0
\end{array}
$$

[*)] Since the complexes $I^\bullet$ and $J^\bullet$ are positive and the complex X is concentrated in degree zero, the diagram is necessarily commutative.

where the vertical arrows are injective resolutions. Remark that the bottom sequence is chainwise split, and derive a short exact sequence of complexes in B

$$0 \to T(I^{\cdot}) \to T(J^{\cdot}) \to T(K^{\cdot}) \to 0$$

By taking homology we obtain a long exact sequence

7.2 $$\to R^n T(X) \to R^n T(Y) \to R^n T(Z) \xrightarrow{\partial^n} R^{n+1} T(X) \to R^{n+1} T(Y)$$

Given a commutative diagram in A with exact rows

$$
\begin{array}{ccccccccc}
0 & \to & X & \to & Y & \to & Z & \to & 0 \\
  &   & \downarrow & & \downarrow & & \downarrow & & \\
0 & \to & U & \to & V & \to & W & \to & 0
\end{array}
$$

then there results a commutative ladder

7.3 $$
\begin{array}{ccccccccccc}
\to & R^n T(X) & \to & R^n T(Y) & \to & R^n T(Z) & \xrightarrow{\partial^n} & R^{n+1} T(X) & \to & R^{n+1}(Y) & \to \\
    & \downarrow & & \downarrow & & \downarrow & & \downarrow & & \downarrow & \\
\to & R^n T(U) & \to & R^n T(V) & \to & R^n T(W) & \xrightarrow{\partial^n} & R^{n+1} T(U) & \to & R^{n+1} T(W) &
\end{array}
$$

this follows from 6.13.

<u>Left exact functors</u>. Let us now assume that our additive functor $T: A \to B$ is <u>left exact</u>, i.e. preserve kernels. This means that we may identify T and $R^0 T$.

<u>Definition 7.4</u>. Let T: $A \to B$ be a left exact functor. An object X of A is called <u>T-acyclic</u>, if

$$R^i T(X) = 0 \quad \text{for all} \quad i \geq 1$$

54

Acyclicity theorem 7.5. Let $T: A \to B$ be a left exact functor. A quasi-isomorphism $f: X^{\cdot} \to Y^{\cdot}$ between bounded below complexes of T-acyclic objects is transformed into a quasi-isomorphism $T(f): T(X^{\cdot}) \to T(Y^{\cdot})$.

Proof. Consider the mapping cone 4.14 to see that it suffices to treat the case where $Y^{\cdot} = 0$. Let us prove that $\mathrm{Ker}(\partial^m)$, the kernel of the m'th differential in $X^{\cdot}$, is T-acyclic. By a translation argument it suffices to prove that $\mathrm{Ker}(\partial^0)$ is T-acyclic. Let us prove that

$$R^i T(\mathrm{Ker}\, \partial^0) = R^{i+p}(\mathrm{Ker}(\partial^{-p}) \quad ; \quad i \geq 1, \quad p \in \mathbb{N}$$

This is done by induction on $p \in \mathbb{N}$ using the exact sequence

$$0 \longrightarrow \mathrm{Ker}(\partial^{-p-1}) \to X^{-p-1} \to \mathrm{Ker}(\partial^{-p}) \longrightarrow 0$$

The formula with p large shows that $\mathrm{Ker}(\partial^0)$ is T-acyclic. Let us now prove the following statement for all $n \in \mathbb{Z}$ "any complex $X^{\cdot}$ of T-acyclic objects with $H^{\cdot}(X^{\cdot}) = 0$ and $X^i = 0$ for all $i < n$ has $H^0(X^{\cdot}) = 0$ ". This is done by decreasing induction on $n \in \mathbb{Z}$. To see that "$n \Rightarrow n-1$" consider the short exact sequence

$$0 \to \tau_{\leq n} X^{\cdot} \to X^{\cdot} \to \tau_{\geq} X^{\cdot} \to 0$$

Transform this by T to get an exact sequence

$$0 \to T(\tau_{\leq n} X^{\cdot}) \to T(X^{\cdot}) \to T(\tau_{\geq n} X^{\cdot}) \to 0$$

The inductive step can now be deduced from the sequence

$$H^0(T(\tau_{\leq n} X^\cdot)) \to H^0(T(X^\cdot)) \to H^0(T(\tau_{\geq n} X^\cdot))$$

We can now conclude the proof by a translation argument.

$$Q.E.D.$$

Theorem on unbounded complexes 7.6. Let $T: A \to B$ be a left exact functor with $R^n T = 0$ for some $n \in \mathbb{N}$. Then any complex $E^\cdot$ in A admits quasi-isomorphism $E^\cdot \to X^\cdot$ into a complex of T-acyclic objects.- Moreover, any quasi-isomorphism $f: X^\cdot \to Y^\cdot$ between complexes of T-acyclic objects is transformed into a quasi-isomorphism $T(f): T(X^\cdot) \to T(Y^\cdot)$.

Proof. Choose a short exact sequence of complexes

$$0 \longrightarrow E^\cdot \xrightarrow{v} F^\cdot \xrightarrow{u} G^\cdot \longrightarrow 0$$

where $F^\cdot$ is a complex of T-acyclic objects. This can for example be done by choosing for each $n \in \mathbb{N}$ a monomorphism $i^n: E^n \to I^n$ into an injective object. Put $F^n = I^n \oplus I^{n+1}$, $\partial^n = \begin{pmatrix} 0 & 1 \\ 0 & 0 \end{pmatrix}$ and $v^n = \begin{pmatrix} i^n \\ i^{n+1}\partial^n \end{pmatrix}$ and let $(u, G^\cdot)$ be a cokernel for v. The morphism

$$\begin{pmatrix} v \\ 0 \end{pmatrix}: \quad E^\cdot \longrightarrow \mathrm{Con}^\cdot(u)[-1]$$

is a quasi-isomorphism as it follows from the proof of 4.15, where the dual statement is treated, see also XI.3.5. Put $X^{(1)} = \mathrm{Con}^\cdot(u)[-1]$ and notice that this is a complex of objects with vanishing $R^{n-1}T$. Iterate this construction to obtain a sequence of quasi-isomorphisms

$$E^{\cdot} \to X^{(1)} \to X^{(2)} \to \ldots \to X^{(n)}$$

The composite of these is a resolution of $E^{\cdot}$ by T-acyclic objects.

To prove the second part of the theorem we may assume that $Y^{\cdot} = 0$. It follows from the proof of 7.5, that any truncation $\tau_{\geq p}X^{\cdot}$ consists of T-acyclic objects. Notice that

$$H^p(T(X^{\cdot})) = H^p(T(\tau_{\geq p-2}X^{\cdot}))$$

and deduce the result from 7.5.

Q.E.D.

Corollary 7.7. Suppose every object of the abelian category A admits an injective resolution of length n, where n is a fixed integer. Then any complex $E^{\cdot}$ admits a quasi-isomorphism $E^{\cdot} \to I^{\cdot}$ into a complex of injective objects.

Proof. We shall use the notation of the proof above and the theory of Ext from I.8. - For any objects R of A of the functor $\mathrm{Ext}^{n+1}(R,-)$ is zero by assumption. It follows that all functors of the form $\mathrm{Ext}^n(R,-)$ vanish on the chain objects of $X^{(1)}$. More generally, all functors of the form $\mathrm{Ext}^{n+1-p}(R,-)$ vanish on the chain objects of $X^{(p)}$. Take $p = n$ to see that $X^{(n)}$ is a complex of injectives.

Q.E.D.

<u>Corollary 7.8</u>. Let $T = A \to B$ and $U: B \to C$ be left exact addive functors between abelian categories with enough injectives. If T is exact and transforms injectives into U-acyclics

$$R^n(U \circ T) = (R^n U) \circ T$$

<u>Proof</u>. Given an object X of A with injective resolution $X \to I^{\cdot}$. Choose an injective resolution $T(X) \to J^{\cdot}$ in B. According to 6.2, we can find $f: T(I^{\cdot}) \to J^{\cdot}$ making the following deagram commutative

$$T(X) \nearrow \begin{matrix} T(I^{\cdot}) \\ \downarrow f \\ J^{\cdot} \end{matrix}$$

According to 7.5, $U(f)$ is a quasi-isomorphism.

Q.E.D.

We shall now introduce the derived functor in the sense of Verdier. - We can first extend our functor T in the obvious way to a functor

$$T^+: K^+(A) \to K^+(B)$$

We define the <u>derived functor</u> $RT: D^+(A) \to D^+(B)$

by $RT = \rho\, T^+ i$ as displayed in the diagram, i denote the inclusion.

7.9
$$\begin{matrix} D^+(A) & \xrightarrow{i} & K^+(A) \\ \downarrow{\scriptstyle RT} & & \downarrow{\scriptstyle T^+} \\ D^+(B) & \xleftarrow{\rho} & K^+(B) \end{matrix}$$

<u>Theorem 7.10</u>. The derived functor

$$RT : D^+(A) \to D^+(B)$$

transforms triangles into triangles.

<u>Proof</u>. By the formula $RT = \rho T^+ i$ we need to check that each of these three functors transforms triangles into triangles. This is trivial for i, the reference for T^+ is 3.9 and the reference for ρ is 6.8.

Q.E.D.

Given additive functors $U,V: A \to B$ and a natural transformation $\gamma: U \to V$ of additive functors. We can in an obvious way extend γ to a transformation $\gamma^+: U^+ \to V^+$ and finally to a natural transformation, compare 7.9

7.11
$$R\gamma: RU \longrightarrow RV$$

Consider additive functors $T: A \to B$ and $U: B \to C$. Let us establish a natural transformation

7.12
$$R(U \circ T) \to RU \circ RT$$

To do so consider the commutative diagram

$$
\begin{array}{ccccc}
D^+(A) & \xrightarrow{RT} & D^+(B) & \xrightarrow{RU} & D^+(C) \\
\rho \uparrow \downarrow i & & \rho \uparrow \downarrow i & & \rho \uparrow \downarrow i \\
K^+(A) & \xrightarrow{T^+} & K^+(B) & \xrightarrow{U^+} & K^+(C)
\end{array}
$$

Notice that $(U \circ T)^+ = U^+ \circ T^+$ and consequently

$$R(U \circ T) = \rho U^+ T^+ i \, , \qquad RU \circ RT = \rho U^+ i \rho T^+ i$$

The adjunction morphism $1 \to i\rho$ relative to B will induce the natural transformation 7.12.

The adjunction $1 \to i\rho$ is always a quasi-isomorphism, from which we conclude

7.13 $\qquad\qquad\qquad R(U \circ T) \overset{\sim}{\to} RU \circ RT \qquad$ if U is exact

Another simple case is the following

7.14 $\qquad\qquad\qquad R(U \circ T) \overset{\sim}{\to} RU \circ RT \qquad$ if T preserves injectives

The all important case is the folllowing

Theorem 7.15. Let $T: A \to B$ and $U: B \to C$ be additive functors between abelian categories with enough injectives. If T transforms injectives into U-acyclics and U is left exact, then

$$R(U \circ T) \overset{\sim}{\to} RU \circ RT$$

Proof. Consider an object $I^{\boldsymbol{\cdot}}$ of $D^+(A)$ and an injective resolution $c: T(I^{\boldsymbol{\cdot}}) \to J^{\boldsymbol{\cdot}}:$ We must prove that

$$U(c): UT(I^{\boldsymbol{\cdot}}) \to U(J^{\boldsymbol{\cdot}})$$

is a quasi-isomorphism. This is a consequence of 7.5.

Q.E.D.

I.8 Composition products

Let A be an abelian category with enough injective objects.
For objects M and N in A, choose an injective resolution J$^\cdot$
of N and define for each $q \in \mathbb{Z}$

$$8.1 \qquad \mathrm{Ext}^q(M,N) = H^q(\mathrm{Hom}(M,J^\cdot))$$

This amounts to consider $\mathrm{Ext}^q(M,-)$ as the q'th derived functor
of the Hom-functor

$$\mathrm{Hom}(M,-): \quad A \to Ab$$

If I$^\cdot$ is an injective resolution of M we can use Theorem 6.2
to write Ext as a homotopy group in $D^+(A)$

$$8.2 \qquad \mathrm{Ext}^q(M,N) = [I^\cdot, J^\cdot[q]]$$

This representation reveals that composition in the derived category
gives rise to a pairing

$$\mathrm{Ext}^p(N,P) \times \mathrm{Ext}^q(M,N) \to \mathrm{Ext}^{p+q}(M,P)$$

Let namely K$^\cdot$ be an injective resolution of P and let o denote
composition in the derived category.

$$[J^\cdot,K^\cdot[p]] \times [I^\cdot,J^\cdot[q]] \to [I^\cdot,K^\cdot[p+q]]$$

$$8.3 \qquad (\alpha,\beta) \to \alpha[q] \circ \beta$$

The result $\alpha \cup \beta$ of pairing α with β is called the __composition
product__. From the categorical origin of the product follows
immediately that it is associative and bilinear, and that the
pairing for $p = q = 0$ is composition in the category A.

We shall now relate the composition product to the connecting

morphism arizing from a short exact sequence in A

$$0 \to M_1 \to M_2 \to M_3 \to 0$$

Consider a resolution of this sequence

by a triangle 6.11 in $D^+(A)$. The element

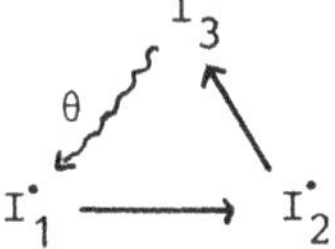

8.4 $\qquad \theta \in [I_3^{\cdot}, I_1^{\cdot}[1]] = \mathrm{Ext}^1(M_3, M_1)$

is called the <u>characteristic class</u> of the short exact sequence

above. - Let $J^{\cdot}$ be an injective resolution of N. The short

exact sequence of complexes

$$0 \to \mathrm{Hom}(M_3, J^{\cdot}) \to \mathrm{Hom}(M_2, J^{\cdot}) \to \mathrm{Hom}(M_1, J^{\cdot}) \to 0$$

give rise to a connecting morphism, $n \in \mathbb{N}$

$$\partial^n \colon \mathrm{Ext}^n(M_1, N) \to \mathrm{Ext}^{n+1}(M_3, N)$$

which is related to θ by the formula

8.5 $\qquad \partial^n(\alpha) = (-1)^{n+1} \alpha \cup \theta \qquad ; \quad \alpha \in \mathrm{Ext}^n(M_1, N)$

<u>Proof</u>. Consider the commutative, exact diagram

$$0 \to \mathrm{Hom}^{\cdot}(I_3^{\cdot}, J^{\cdot}) \to \mathrm{Hom}^{\cdot}(I_2^{\cdot}, J^{\cdot}) \to \mathrm{Hom}^{\cdot}(I_1^{\cdot}, J^{\cdot}) \to 0$$
$$\downarrow \qquad\qquad\qquad \downarrow \qquad\qquad\qquad \downarrow$$
$$0 \to \mathrm{Hom}(M_3, J^{\cdot}) \to \mathrm{Hom}^{\cdot}(M_2, J^{\cdot}) \to \mathrm{Hom}^{\cdot}(M_1, J^{\cdot}) \to 0$$

recall that the vertical arrows are quasi-isomorphisms. The result

follows from 2.7 and 4.9.

$$\text{Q.E.D.}$$

Let us now consider a <u>covariant</u> additive functor $T: A \to B$ from the abelian category A to the abelian category B. Given objects M and N in A and $p, q \in \mathbb{Z}$, let us introduce a pairing

$$\text{Ext}^p(M,N) \times R^q T(M) \to R^{p+q}(N)$$

With the notation from 8.2 we can represent $\alpha \in \text{Ext}^p(M,N)$ as a morphism of complexes $\alpha : I^{\cdot} \to J^{\cdot}[p]$. This will induce a morphism of complexes

$$T(\alpha) : T(I^{\cdot}) \to T(J^{\cdot})[p]$$

and consequently a morphism on homology

$$H^q T(\alpha) : H^q T(I^{\cdot}) \to H^{p+q} T(R^{\cdot})$$

which we can evaluate on $\beta \in H^q T(I^{\cdot})$:

8.6 $$\alpha \cup \beta = H^q T(\alpha)\beta$$

From the categorical origin follows that the pairing is associative

8.7 $$\alpha \cup (\beta \cup \gamma) = (\alpha \cup \beta) \cup \gamma$$
$$\gamma \in R^p T(M), \beta \in \text{Ext}^q(M,N), \alpha \in \text{Ext}^p(N,P)$$

In particular this makes $R^{\cdot} T(M)$ a graded left $\text{Ext}^{\cdot}(M,M)$-module.

We shall now proceede to investigate the relationship between the product and the connecting morphism arizing from an exact sequence in A
$$0 \to D \to E \to F \to 0$$

The connecting homomorphism, 7.3 $\qquad \partial^n : R^n T(F) \to R^{n+1} T(D)$

is related to the characteristic class $\qquad \theta \in \text{Ext}^1(F,D)$

of the sequence through the formula

8.8
$$\partial^n(\alpha) = \theta U \alpha \quad ; \quad \alpha \in R^n T(F)$$

<u>Proof</u>. Consider a resolution of the sequence above
by a triangle in $D^+(A)$.
It follows from 4.8 that $\partial^n = H^n T(\theta)$
which proves the result.

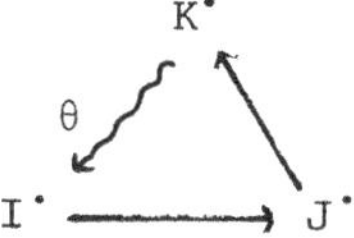

Let us assume that the category A has enough <u>injectives</u>
and enough <u>projectives</u>.*) With the notation of 8.2 let $K˙ \to M$ and
$L˙ \to N$ be projective resolutions. We can represent $Ext^q(M,N)$ by

8.9
$$Ext^q(M,N) = [K˙, L˙[q]]$$

    <u>Proof</u>. According to 8.1 we can write

$$Ext^q(M,N) \xrightarrow{\sim} [M, J˙[q]]$$

By Theorem 6.2, the resolution $K˙ \to M$ will induce an isomorphism

$$[M, J˙[q]] \xrightarrow{\sim} [K˙, J˙[q]]$$

By the dual of 6.2 the resolutions $L˙ \to N$ and $N \to J˙$ will induce
isomorphisms

$$[K˙, L˙[q]] \xrightarrow[\sim]{} [K˙, N[q]] \xrightarrow[\sim]{} [K˙, J˙[q]]$$

Compose these isomorphisms to get the result.

                  Q.E.D.

---

*) Compare the following section I.9.

I.9   Resumé of the projective case

Throughout this chapter we shall work in an abelian category A. We say that an object P is <u>projective</u>, if for any epimorphism $f: X \to Y$ and any morphism $y: P \to Y$, there exists a morphism $x: P \to X$, such that $y = fx$

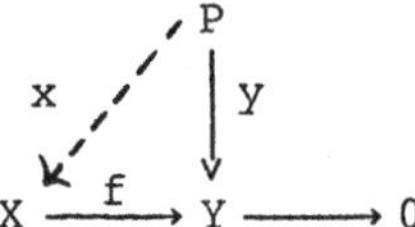

We shall assume that the category A has enough <u>projectives</u>, i.e. that there for every object X of A exists an epimorphism $P \to X$ where P is projective.

The notion of projective is <u>dual</u> to that of injective in the sense of category theory. We shall list a number of propositions obtained by applying duality to previous results. The duality is maintained in the notation. Thus for example Theorem 6.2° is dual to Theorem 6.2.

Let us now turn to complexes. In this context it is standard to change the notation such that a complex $C^{\cdot} = (C^n, \partial^n)_{n \in \mathbb{Z}}$ is rewritten $C. = (C_n, \partial_n)_{n \in \mathbb{Z}}$ according to the convention

$$C_n = C^{-n}, \qquad \partial_n = \partial^{-n}$$

Thus a complex is a sequence of objects and morphisms

$$C_{n+1} \xrightarrow{\partial_{n+1}} C_n \xrightarrow{\partial_n} C_{n-1}$$

subjected to $\partial_{n+1}\partial_n = 0$ for all $n \in \mathbb{Z}$. The same convention applies to morphisms of complexes: $f_n = f^{-n}$.

We shall mostly be concerned with <u>bounded above</u> complexes, i.e. complexes C. with $C_n = 0$ for $n \ll 0$.

<u>Theorem 6.1^o</u>. For any bounded above complex X., there exists a bounded above complex P. of projectives and a quasi-isomorphism P. → X.

<u>Theorem 6.2^o</u>. Let P. be a bounded above complex of projectives. Any quasi-isomorphism c: X. → Y. induces an isomorphism

$$[1,c]: [P.,X.] \xrightarrow{\sim} [P.,Y.]$$

<u>The derived category</u>. On the basis of 6.1^o and 6.2^o we shall introduce the categories

6.4^o

$K^-(A)$	the homotopy category of bounded above complexes in A
$D^-(A)$	the homotopy category of bounded above complexes of projectives in A

The inclusion of the subcategory $D^-(A)$ in $K^-(A)$ is denoted

$$i: D^-(A) \longrightarrow K^-(A)$$

For each X. in K(A) choose a quasi-isomorphism

6.5^o
$$p: \lambda X. \longrightarrow X.$$

where $\lambda X.$ is an object of $D^-(A)$. For a morphism f: X. → Y. in $K^-(A)$ there exists one and only one morphism $\lambda f: \lambda X. \to \lambda Y.$ making the following diagram commutative

$$\lambda X. \xrightarrow{\;\;p\;\;} X.$$

$$\downarrow \lambda f \qquad\qquad \downarrow f$$

$$\lambda Y. \xrightarrow{\;\;p\;\;} Y.$$

This construction yields a canonical isomorphism

$$6.6^{\circ} \qquad\qquad [P.,X.] = [P., \lambda X.]$$

where $X.$ varies through $K^-(A)$ and $P.$ varies through $D^-(A)$. Otherwise expressed the functor

$$6.7^{\circ} \qquad\qquad \lambda: K^-(A) \to D^-(A)$$

is a <u>right adjoint</u> to the inclusion $i: D^-(A) \to K^-(A)$. In this interpretation the resolution morphism p from 6.5° plays the role of adjunction morphism.

<u>Theorem 6.8°</u>. The resolution functor

$$\lambda: K^-(A) \to D^-(A)$$

transforms triangles into triangles.

We shall next refine theorem 6.8° by performing a construction which assigns a triangle in $D^-(A)$ to a short exact sequence of bounded above complexes in A.

$$0 \to C. \to D. \to E. \to 0$$

The key result is

<u>Proposition 6.10°</u>. There exists a <u>commutative</u>, not mere-
ly homotopy commutative diagram of complexes in A

$$
\begin{array}{ccccccccc}
0 & \longrightarrow & P. & \longrightarrow & Q. & \longrightarrow & R. & \longrightarrow & 0 \\
 & & \downarrow & & \downarrow & & \downarrow & & \\
0 & \longrightarrow & C. & \longrightarrow & D. & \longrightarrow & E. & \longrightarrow & 0
\end{array}
$$

where the top row is a short exact sequence of bounded above
complexes of projectives and the vertical arrows are quasi-
isomorphism.

The short exact sequence of projectives constructed in
6.10° gives rise to a triangle

6.11°

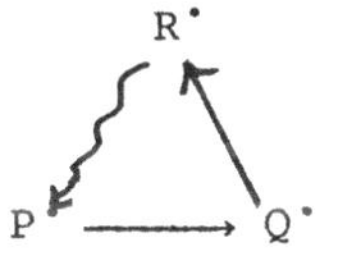

$$
\begin{array}{ccc}
 & R^{\cdot} & \\
 & \nearrow & \\
P^{\cdot} & \longrightarrow & Q^{\cdot}
\end{array}
$$

<u>which is unique up to homotopy</u>.

I.10 Complexes of free abelian groups

We shall be concerned with complexes $L.$ of free abelian groups.

10.1. Structure theorem. Any complex of free abelian groups is isomorphic to a complex of the form

$$\underset{i \in \mathbb{Z}}{\oplus} \; L.^{(i)}$$

where for each $i \in \mathbb{Z}$, the complex $L.^{(i)}$ can be written

$$L.^{(i)}: \; 0 \to L_{i+1}^{(i)} \to L_i^{(i)} \to 0 \qquad \text{with} \;\; H_{i+1}(L.^{(i)}) = 0.$$

Proof. Let $L.$ be a complex of free abelian groups. For i fixed let us notice that $\text{Ker } \partial_i$ is a direct summand of L_i as it follows from the fact that $\text{Im } \partial_i$ is a free abelian group, in fact <u>any subgroup of a free abelian group is free</u>. Choose for each $i \in \mathbb{Z}$ a direct summand N_i to $\text{Ker } \partial_i$ in L_i and let $L.^{(i)}$ denote the complex $d_{i+1}: N_{i+1} \to \text{Ker } \partial_i$ where d_{i+1} is the restriction of ∂_{i+1} to N_{i+1}. Quite obviously $L. \overset{\sim}{\to} \oplus \, L.^{(i)}$.

Q.E.D.

Corollary 10.2. For any complex $L.$ of free abelian groups there is a morphism of complexes $L. \to H.(L.)$ inducing the indentity in homology.

Let us remark that neither this morphism nor its homotopy class is canonical as one sees from example 10.6 below.

Corollary 10.3. For a complex L. of free abelian groups
and a complex X. of abelian groups X. with H.(X.) = 0
we have

$$[L.,X.] = 0$$

Proof. According to 10.1 it suffices to treat the case
where L. has only two non vanishing terms. This case follows
from 6.2^{O}.

Q.E.D.

Corollary 10.4. Two complexes L. and N. of free abelian
groups are homotopy equivalent if and only if H.(N.) $\tilde{\rightarrow}$ H.(L.).

Proof. Choose morphisms as in 10.2

$$b: N. \rightarrow H.(N.) \qquad a: L. \rightarrow H.(L.)$$

and let φ: H.(L.) $\rightarrow$ H.(N.) be an isomorphism.

Consider the triangle

and the resulting long exact sequence

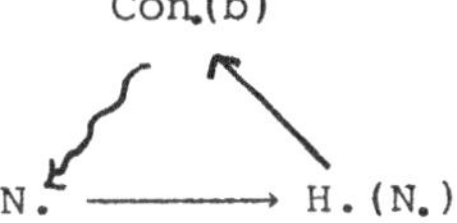

$$[L.,N.] \rightarrow [L.,H.(N.)] \rightarrow [L.,Con.(b)]$$

Since H.(Con.(b)) = 0 it follows from 10.3 that [L.,Con.(b)] = 0.
Consequently we can find f: L. $\rightarrow$ N. such that bf $\sim \varphi$a.
Notice that f is a quasi-isomorphism and consequently has a
contractible mapping cone by 10.3 It follows that f is a homotopy equivalence.

Q.E.D.

<u>Corollary 10.5</u>. Let $f: L_{\cdot} \to N_{\cdot}$ be a morphism of complexes. If all homology groups of $L_{\cdot}$ are free, then

$$H_{\cdot}(\mathrm{Con.}(f)) \xrightarrow{\sim} H_{\cdot}(\mathrm{Con.}(H(f)))$$

<u>Proof</u>. Choose morphisms of complexes

$$a: L_{\cdot} \to H_{\cdot}(L_{\cdot}) \qquad b: N_{\cdot} \to H_{\cdot}(N_{\cdot})$$

It follows from 10.3 that a is a homotopy equivalence. Thus we can find a homotopy commutative diagram

$$
\begin{array}{ccc}
L_{\cdot} & \xrightarrow{\ f\ } & N_{\cdot} \\
{\scriptstyle a}\downarrow & & \downarrow{\scriptstyle b} \\
H_{\cdot}(L_{\cdot}) & \xrightarrow{\ g\ } & H_{\cdot}(N_{\cdot})
\end{array}
$$

Since $H(a) = 1$ and $H(b) = 1$ we find $g = H(f)$. According to 4.19 we can choose a morphism $c: \mathrm{Con.}(f) \to \mathrm{Con.}(H(f))$ such that a,b,c is a morphism of triangles. It follows from the long exact homology ladder and the five lemma 1.7, that c is a quasi-isomorphism.

Q.E.D.

<u>Example 10.6</u>. Consider the two complexes

$$L_{\cdot}: \mathbb{Z} \xrightarrow{\ 2\ } \mathbb{Z} \qquad N_{\cdot}: \mathbb{Z} \to 0$$

and the morphism $f: L_{\cdot} \to N_{\cdot}$ induced by multiplication by 3. Notice that $H_{\cdot}(\mathrm{Con.}(f)) = \mathbb{Z} \qquad H_{\cdot}(\mathrm{Con.}(H(f))) = \mathbb{Z} \oplus \mathbb{Z}/2$ which shows that the assumption "H.(L.) free" is important in 10.5.

I.11 Sign rules

Let us consider a fixed additive category C. By a _double complex_ we understand data $C^{\cdot\cdot} = (C, \partial_1, \partial_2)$ where

$$C = (C^{p,q})_{(p,q)\in\mathbb{Z}\times\mathbb{Z}} \qquad \partial_1 = (\partial_1^{p,q})_{(p,q)\in\mathbb{Z}\times\mathbb{Z}} \qquad \partial_2 = (\partial_2^{p,q})_{(p,q)\in\mathbb{Z}\times\mathbb{Z}}$$

is a family of objects and two families of morphisms

$$\partial_1^{p,q}: C^{p,q} \to C^{p+1,q} \qquad \partial_2^{p,q}: C^{p,q} \to C^{p,q+1}$$

subjected to the following conditions

11.1 $$\partial_1\partial_1 = 0 \qquad \partial_2\partial_2 = 0 \qquad \partial_1\partial_2 = \partial_2\partial_1$$

Let us assume that for all $n \in \mathbb{Z}$ the direct sum

11.2 $$\mathrm{Tot}^n C^{\cdot\cdot} = \bigoplus_{p+q=n} C^{p,q}$$

is finite. Then we define $\partial^n: \mathrm{Tot}^n C^{\cdot\cdot} \to \mathrm{Tot}^{n+1} C^{\cdot\cdot}$ by the formula

11.3 $$\partial^n = \bigoplus_{p+q=n} \partial_1^{p,q} + (-1)^p \partial_2^{p,q}$$

This defines the _total complex_ associated to the double complex

$$\mathrm{Tot}^{\cdot} C^{\cdot\cdot} = ((\mathrm{Tot}^n C^{\cdot\cdot})_{n\in\mathbb{Z}}, (\partial^n)_{n\in\mathbb{Z}})$$

By a morphism $f: C^{\cdot\cdot} \to D^{\cdot\cdot}$ of couble complexes we understand an indexed family of morphisms $f^{p,q}: C^{p,q} \to D^{p,q}$ which commutes with ∂_1 and ∂_2. Such a morphism will in the obvious manner induce a morphism $\mathrm{Tot}\, f: \mathrm{Tot}^{\cdot} C^{\cdot\cdot} \to \mathrm{Tot}^{\cdot} D^{\cdot\cdot}$

Given a second morphism of double complexes $g: C^{\cdot\cdot} \to D^{\cdot\cdot}$. By a homotopy from f to g we understand a pair (s_1, s_2) where s_1 is a family of morphisms $s_1^{p,q}: C^{p,q} \to C^{p-1,q}$ and s_2 a family of morphism $s_2^{p,q}: C^{p,q} \to C^{p,q-1}$ such that

$$11.4 \qquad\qquad \partial_2 s_1 = s_1 \partial_2 \qquad \partial_1 s_2 = s_2 \partial_1$$

$$11.5 \qquad\qquad f-g = s_1 \partial_1 + \partial_1 s_1 + \partial_2 s_2 + s_2 \partial_2$$

To these data we can associate $S^n: \mathrm{Tot}^n C^{\cdot\cdot} \to \mathrm{Tot}^{n-1} D^{\cdot\cdot}$ by

$$11.6 \qquad\qquad s^n = \bigoplus_{p+q=n} s_1^{p,q} + (-1)^p s_2^{p,q}$$

We leave it to the reader to show that

$$11.7 \qquad\qquad \mathrm{Tot}\, f - \mathrm{Tot}\, g = \partial s + s \partial$$

<u>Bivariant functor 11.8.</u> Let $T: A \times B \to C$ be a biadditive functor contravariant in the first variable and covariant in the second variable. To a complex $A^{\cdot}$ in A and a complex $B^{\cdot}$ in B we associate a double complex $T(A^{\cdot}, B^{\cdot})$ in C given by

$$T(A^{\cdot}, B^{\cdot}) = (T(A^{-q}, B^p), T(1, \partial^p), (-1)^{q+1} T(\partial^{-q-1}, 1))_{(p,q) \in \mathbb{Z} \times \mathbb{Z}}$$

$$
\begin{array}{ccc}
T(A^{-q-1}, B^p) & \xrightarrow{\quad T(1, \partial^p) \quad} & T(A^{-q-1}, B^{p+1}) \\[2mm]
{\scriptstyle (-1)^{q+1}} \Big\uparrow {\scriptstyle T(\partial^{-q-1}, 1)} & & {\scriptstyle (-1)^{q+1}} \Big\uparrow {\scriptstyle T(\partial^{-q-1}, 1)} \\[2mm]
T(A^{-q}, B^p) & \xrightarrow{\quad T(1, \partial^p) \quad} & T(A^{-q}, B^{p+1})
\end{array}
$$

The total complex associated to this double complex is denoted $T^{\cdot}(B^{\cdot},A^{\cdot})$. If we apply these conventions to the Hom-functor we recover the complex 4.3.

<u>Contravariant functor 11.9.</u> Let $T\colon A \to C$ be a contravariant additive functor. To a complex $C^{\cdot}$ we associate the complex $T(C^{\cdot})$ given by

$$T(C^{\cdot}) = (T(C^{-q}), (-1)^{q+1}T(\partial^{-q-1}))$$

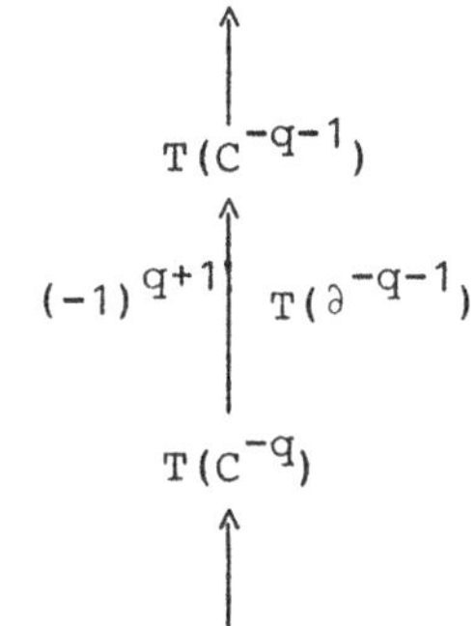

Translation with $m \in \mathbb{Z}$ gives rise to an isomorphism whose effect on $T(C^q)$ is $(-1)^{qm}$

11.10
$$\nu_m\colon T(C^{\cdot}[-m]) \xrightarrow{\sim} T(C^{\cdot})[m]$$

11.11 A triangle in the category A is transformed by the contravariant functor $T\colon A \to C$ into a triangle in C

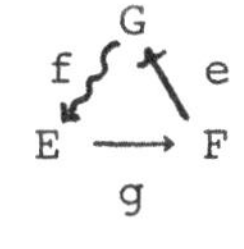

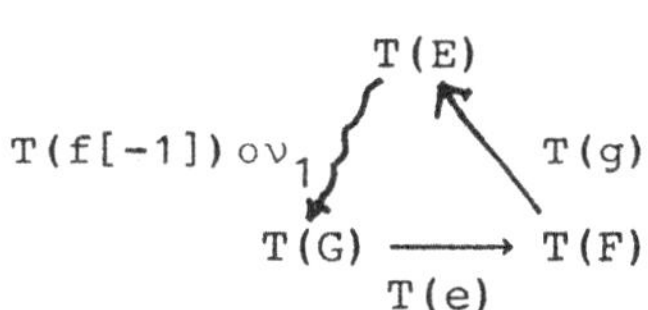

If $U\colon C \to D$ is a second covariant functor then the automorphism of $U(T(C^p))$ given by $(-1)^p$ gives a canonical isomorphism

11.12
$$\rho\colon U(T(C^{\cdot})) \xrightarrow{\sim} U \circ T(C^{\cdot})$$

II. Sheaf Theory

<u>II.0.</u>   <u>Direct limits of abelian groups</u>

By a <u>directed set</u> we understand a non-empty set I equipped
with a preorder, i.e. a relation satisfying

$$a \leq a \quad \text{for all} \quad a \in I$$

$$a \leq b \quad \text{and} \quad b \leq c \quad \text{implies} \quad a \leq c$$

subject to the condition that for any $a \in I$ and $b \in I$
there exists $c \in I$ with $a \leq c$ and $b \leq c$.

By a <u>direct system</u> of abelian groups over I we understand
data (F_a, f_{ba}) consisting of an abelian group F_a for each
$a \in I$ and a linear map $f_{ba} : F_a \to F_b$ for each pair a,b with $a \leq b$
subject to the conditions

$$f_{cb} f_{ba} = f_{ca} \qquad ; \quad a \leq b, \quad b \leq c$$

$$f_{aa} = 1, \quad \text{the identity on} \quad F_a \quad ; \quad a \in I$$

We shall now construct the <u>direct limit</u> $\varinjlim F_a$ of our <u>direct
system</u>: We form the direct sum of the groups F_a,

$$\underset{a \in I}{\oplus} F_a$$

For each $a \in I$ we let i_a denote the canonical map from F_a
to the direct sum. We put

$$0.1 \qquad \varinjlim F_a = \underset{a \in I}{\oplus} F_a \Big/ \genfrac{}{}{0pt}{}{i_b f_{ba} x_a - i_a x_a}{; \ a \leq b, \ x_a \in F_a}$$

i.e. the quotient group of the direct sum with respect to the subgroup generated by the elements written under the bar. - The composite of i_a and the projection of $\oplus F_a$ onto $\varinjlim F_a$ is denoted

$$f_a : F_a \to \varinjlim F_a$$

It follows at once from the construction that

$$f_b f_{ba} = f_a \qquad ; \qquad a \leq b$$

and that we have the following universal

0.2 <u>Mapping property of</u> $\varinjlim F_a$. Let there be given an abelian group G and for each $a \in I$ a linear map $g_a : F_a \to G$ subject to the conditions

$$g_b \, f_{ba} = g_a \qquad ; \qquad a \leq b$$

then there exists a unique $g : \varinjlim F_a \to G$ with

$$g f_a = g_a \qquad ; \qquad a \in I .$$

<u>Proposition 0.3.</u> 1) Any element of $\varinjlim F_a$ has the form $f_u(x_u)$ for some $u \in I$ and some $x_u \in F_u$.

2) For $u \in I$ and $x_u \in F_u$ we have $f_u(x_u) = 0$ if and only if there exists $v \geq u$ with $f_{vu}(x_u) = 0$.

$\underline{\text{Proof.}}$ 1) According to the construction of $\varinjlim F_a$, its typical element is a finite sum

$$\sum f_a(x_a) \quad ; \quad x_a \in F_a$$

Choose $u \in I$ larger than all indices occurring in this sum. Then the following relation will justify 1)

$$\sum_a f_a(x_a) = \sum_a f_u f_{ua}(x_a) = f_u(\sum_a f_{ua} x_a)$$

2) Let us assume that $f_u(x_u) = 0$. Then we can write $i_u(x_u)$ as a finite sum of the form

$$i_u x_u = \sum i_b f_{ba} x_a - i_a x_a$$

Choose $v \in I$ larger than any of the indices occurring in the sum above. We can write

$$i_v f_{vu} x_u = i_v f_{vu} x_u - i_u x_u + \sum i_b f_{ba} x_a - i_a x_a$$

The typical element under Σ can be rewritten

$$i_b f_{ba} x_a - i_a x_a = i_v f_{va} x_a - i_a x_a + i_v f_{vb}(-f_{ba} x_a) - i_b(-f_{ba} x_a)$$

A moment's reflection shows that we can write

$$i_v f_{vu} x_u = \sum_{a;\, a<v} i_v f_{va} z_a - i_a z_a$$

with $z_a \in F_a$ zero except for finitely many $a \in I$. According to the basic character of $\oplus F_u$ this implies $z_a = 0$ for all $a \in I$ and consequently $f_{vu} x_u = 0$.

Q.E.D.

The following Corollary is the <u>key to practical evaluation</u> of the symbol $\varinjlim F_a$.

<u>Corollary 0.4</u>. With the notation from 0.2 suppose that

1) every element of G has the form $g_u(x_u)$ for some $u \in I$ and some $x_u \in F_u$

2) $g_u(x_u) = 0 \Rightarrow \exists\, v \geq u \quad f_{vu} x_u = 0$.

Then $\varinjlim F_a$ is isomorphic to G, through g.

<u>Excactness</u>. Given direct systems of abelian groups (E_a, e_{ba}) and (F_a, f_{ba}) over I. By a morphism from the first system to the second we understand a family of linear maps (φ_a) where for each $a \in I$ $\varphi_a: E_a \to F_a$, subject to the condition that

$$f_{ba} \varphi_a = \varphi_b\, e_{ba} \qquad ;\quad a \leq b$$

By the mapping property of direct limits 0.2, we can find one and only one $\varphi: \varinjlim E_a \to \varinjlim F_a$ such that

$$\varphi \circ e_a = f_a \circ \varphi_a \qquad ;\quad a \in I\ .$$

This map is the <u>direct limit</u> of the morphism and is denoted

$$\varinjlim \varphi_a : \varinjlim E_a \to \varinjlim F_a \ .$$

<u>Proposition 0.5</u>. Given morphisms of direct systems over I

$$(E_a, E_{ba}) \xrightarrow{\ (\varphi_a)\ } (F_a, F_{ba}) \xrightarrow{\ (\psi_a)\ } (G_a, G_{ba})$$

such that the sequence

$$E_a \xrightarrow{\ \varphi_a\ } F_a \xrightarrow{\ \psi_a\ } G_a$$

is exact for all $a \in I$. Then the following sequence is exact

$$\varinjlim E_a \xrightarrow{\ \varinjlim \varphi_a\ } \varinjlim F_a \xrightarrow{\ \varinjlim \psi_a\ } \varinjlim G_a$$

<u>Proof</u>. Straightforward application of the previous pro-
position.

Q.E.D.

<u>General categories</u> 0.6. Let C be an arbitrary category.
By a direct system in C we understand data (F_a, f_{ba}) consisting of an object F_a of C for each $a \in I$ and morphism
$f_{ba} : F_a \to F_b$ for each pair $a \leq b$ subjected to the conditions

$$f_{cb} f_{ba} = f_{ca} \qquad ; \quad a \leq b \leq c$$

$$f_{aa} = 1, \ \text{the identity on } F_a \qquad ; \quad a \in I.$$

By a <u>direct limit</u> of the system we understand data (f_a, F)
consisting of an object F and morphisms $f_a: F_a \to F$ such
that

$$f_b F_{ba} = f_a \qquad ; \; a \leq b$$

and which has a universal mapping property similar to 0.2.

<u>Proposition 0.7</u> Let C be an additive category in which
arbitrary direct limits exist. Then arbitrary <u>direct sums</u>
exist, i.e. systems with a mapping property generalizing I.3.

<u>Proof</u>. Let $(C_s)_{s \in S}$ be a family of objects in C. For
any finite subset A of S we put $C_A = \oplus_{s \in A} C_s$, and for
finite subsets $A \subseteq B$ we define $f_{BA}: C_A \to C_B$ in a rather
obvious way. In this manner we have created a direct system
in C over the directed set of finite subsets of S. It is
easy to see that the direct limit of this system is a direct
sum.

Q.E.D.

<u>Remark 0.8</u>. To a preorered set I we can associate the
category I whose objects are the points of I. The morphisms
from a to b are given by

$$\mathrm{Hom}(a,b) \;=\; \begin{cases} \{(b,a)\} & \text{for } a \leq b \\ \varnothing & \text{otherwise} \end{cases}$$

With this notation a direct system in C over I may be interpreted as a covariant functor $F: I \to C$, and a morphism $F \to G$
of direct systems as a natural transformation of functors.

II.1   Presheaves and sheaves

Let X denote a topological space. By a <u>presheaf</u> F on X we understand a collection of data as follows

1) For each open subset U of X an abelian group $F(U)$, <u>the (group) of sections of</u> F <u>over</u> U.

2) For each pair $U \supseteq V$ of open subsets of X a linear map $r_{VU}: F(U) \to F(V)$, <u>the restriction from</u> U <u>to</u> V.

These data are subject to the conditions that for open sets $U \supseteq V \supseteq W$ we have

$$1.1 \qquad r_{WV}r_{VU} = r_{WU} \qquad ; r_{UU} = 1 \, .$$

This definition can be rephrased if we organize the ordered set of open subsets of X into a category. A presheaf is contravariant functor F from this category to the category of abelian groups. A morphism $f: F \to G$ of <u>presheaves</u> on X is a natural transformation of functors, i.e. for each open set U we have given a linear map $f(U): F(U) \to G(U)$ such that whenever $U \supseteq V$ are open subsets, the following diagram is commutative

$$
1.2 \qquad
\begin{array}{ccc}
F(U) & \xrightarrow{\;r_{VU}\;} & F(V) \\
\downarrow{\scriptstyle f(U)} & & \downarrow{\scriptstyle f(V)} \\
G(U) & \xrightarrow{\;r_{VU}\;} & G(V)
\end{array}
$$

Two morphisms of presheaves $f: F \to G$ and $g: G \to K$ can be composed to $g \circ f: F \to K$ according to the convention

$$1.3 \qquad (g \circ f)(U) = g(U) \circ f(U)$$

Let us define the <u>sum</u> $f + g$ of two morphisms $f, g: F \to G$
by the convention

$$1.4 \qquad (f+g)(U) = f(U) + g(U)$$

With this convention <u>the category of presheaves on</u> X <u>is an</u>
additive category, as the reader easily verifies by construct-
ing the direct sum $F \oplus G$ of two presheaves by the formula

$$1.5 \qquad (F \oplus G)(U) = F(U) \oplus G(U)$$

In fact the <u>category of presheaves on</u> X <u>is an abelian cate-</u>
<u>gory</u> as one sees by constructing kernel and cokernel for a
morphism of sheaves $f: F \to G$ by the rules

$$\mathrm{Ker}(f)(U) = \mathrm{Ker}(f(U))$$
$$1.6$$
$$\mathrm{cok}(f)(U) = \mathrm{Cok}(f(U))$$

   <u>The section functor</u>. Given a presheaf F on X and an
open set U of X. Put

$$1.7 \qquad \Gamma(U,F) = F(U)$$

the (group of) <u>sections of</u> F over U. – The elements of
$\Gamma(X,F)$ are called the <u>global sections</u> of F.

   <u>Definition 1.8</u>. A presheaf F on X is called a <u>sheaf</u>
if for any family $(U_i)_{i \in I}$ of open subsets of X and any

family of sections $(s_i)_{i \in I}$, $s_i \in F(U_i)$, with

$$r_{U_i \cap U_j, U_i}(s_i) = r_{U_i \cap U_j, U_j}(s_j) \qquad ; \quad i,j \in I$$

there exists one and only one $s \in F(U)$, where $U = \bigcup_{i \in I} U_i$, with

$$r_{U_i, U}(s) = s_i \qquad ; \quad i \in I$$

<u>Example 1.9</u>. For an open set U of X let $C(U)$ denote the set of $\mathbb{R}$-valued continuous functions on X. For open subsets $U \supseteq V$, restriction of functions in the ordinary sense of the word defines a linear map $r_{VU}: C(U) \to C(V)$. In fact we have defined a sheaf C on X.

<u>Example 1.10</u>. An abelian group D gives rise to a sheaf $\underset{\sim}{D}$ on the topological space X: For an open set U we let $\underset{\sim}{D}(U)$ denote the group of locally constant functions on U with values in D. Restriction maps in the presheaf $\underset{\sim}{D}$ is ordinary restriction of functions. $\underset{\sim}{D}$ is a sheaf on X.

<u>Example 1.11</u>. Let us describe the sheaf 0 of analytic functions of the complex plane $\mathbb{C}$: $\Gamma(U, 0)$ is the group of analytic functions $f: U \to \mathbb{C}$. Restriction in the presheaf 0 is that of restriction of functions. This defines the sheaf of analytic functions. Consider the morphisms of sheaves

$$0 \longrightarrow \underset{\sim}{\mathbb{C}} \overset{j}{\longrightarrow} 0 \overset{D}{\longrightarrow} 0 \longrightarrow 0$$

where j is the inclusion of the sheaf of locally constant functions into the sheaf of analytic functions. The morphism D assigns to an open set U and $s \in \Gamma(U, 0)$ the complex derivative of s, $Ds \in \Gamma(U, 0)$.

II.2   Localization

In this section we shall study the process of localizing a problem formulated in terms of sheaves. This technique will be applied to show that the category of sheaves on a topological space X is an <u>abelian</u> category.

Let us fix a point x of X. The set of open neighbourhoods U of x in X ordered by inclusion form a <u>directed</u> set. A presheaf F on X will induce a <u>direct system</u> of abelian groups on this set. We define the <u>stalk</u> of F at x by

$$2.1 \qquad\qquad F_x = \varinjlim F(U)$$

A morphism of $f: F \to G$ of presheaves on X will induce a morphism of direct systems and consequently give rise to a map

$$f_x: F_x \longrightarrow G_x$$

In this way we have for each $x \in X$ constructed a functor from the category of presheaves to the category of abelian groups. The following lemma allows us to pass information from local to global.

<u>Lemma 2.2</u>. Let F and G be sheaves on X.

i) If two morphisms of sheaves $f, g: F \to G$ are such that $f_x = g_x$ for all $x \in X$, then $f = g$.

ii) A morphism of sheaves $f: F \to G$ satisfies the condition

$$f_x: F_x \to G_x \quad \text{is injective for all} \quad x \in X$$

if and only if, the map $f(U): F(U) \to G(U)$ is injective for all open subsets U of X.

iii) A morphism of sheaves $f: F \to G$ such that

$$f_x: F_x \longrightarrow G_x \quad \text{is an isomorphism for all} \quad x \in X$$

is an isomorphism of sheaves on X.

Proof. i) We shall prove that $f(U) = g(U)$ for any open subset of X. Consider the commutative diagram

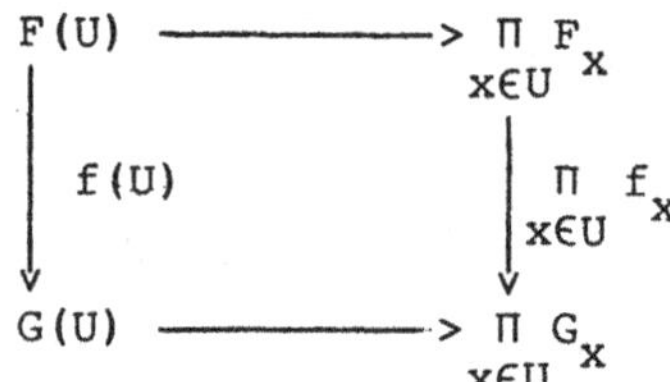

Note that the horizontal arrows are injective by the fact that F and G are sheaves. Since $f_x = g_x$ for all $x \in U$, the diagram remains commutative if we replace $f(U)$ with $g(U)$. Thus $f(U) = g(U)$.

ii) Assume f_x injective for all $x \in X$. From the commutative diagram above we conclude that $f(U)$ is injective. The opposite implication follows from exactness of $\varinjlim$ II.0.5.

iii) Let us first prove that for a given open subset U of X, $f(U): F(U) \to G(U)$ is an isomorphism. It follows from ii that $f(U)$ is injective. So let us prove that any $t \in G(U)$ belongs to the image of $f(U): F(U) \to G(U)$. For any $x \in X$ choose an open subset U^x containing $x \in X$ and a section

$r^x \in F(U^x)$ such that $f(r^x)$ and t has the same stalk of x. Next, choose an open neighbourhood V^x of x in U^x such that if we let s^x denote the restriction of r^x to V^x, then

$$f(V^x)(s^x) = r_{V^x,U}(t)$$

Notice, that for $x,y \in U$, s^x and s^y have the same restriction to $V^x \cap V^y$ as it follows from the formula above and the injectivity of $f(V^x \cap V^y)$. Thus we can find $s \in F(U)$ with restriction s^x to V^x. Clearly, $f(U)(s) = t$.

Let us finally notice that $f: F \to G$ is an isomorphism: The inverse is given by $f(U)^{-1}$, as U varies through the open subsets of X.

$$Q.E.D.$$

<u>Sheafification of a presheaf</u>. Let F be a presheaf on X. We shall associate a sheaf $\tilde{F}$ to the presheaf F. The sections of $\tilde{F}$ over the open set U is the set of those

$$s \in \prod_{z \in U} F_z$$

which for every $x \in U$ satisfies the following condition: "there exists an open neighbourhood W of x in U and a section t of F over W such that $pr_w s = t_{\bar{w}}$ for all $w \in W$".

In order to define restriction maps in the presheaf $\tilde{F}$ notice that for open subsets $U \supseteq V$ of X the natural maps

$$\prod_{x \in U} F_x \longrightarrow \prod_{x \in V} F_x$$

will map $\Gamma(U,\widetilde{F})$ into $\Gamma(V,\widetilde{F})$. It is left to the reader
to show that $\underline{\widetilde{F}\ \text{is a sheaf}}$. Given a section s of F over
an open set U of X, the collection of its stalks at the
points of U will define a section of $\widetilde{F}$ over U. We have
in fact established a morphism of presheaves

$$2.3 \qquad\qquad i_F\colon F \longrightarrow \widetilde{F}$$

$\underline{\text{Proposition 2.4}}$. Let F be a presheaf on the space X.

i) The canonical morphism $i_F\colon F \to \widetilde{F}$ induces an isomorphism on all stalks.

ii) Given a morphism $f\colon F \to G$ from F into a sheaf G.
Then there exists one and only one morphism $\varphi\colon \widetilde{F} \to G$, such
that $f = \varphi \circ i_F$.

$\underline{\text{Proof}}$. i) Given $x \in X$ and an open neighbourhood U of
x. Projection induces a map

$$\Gamma(U,\widetilde{F}) \longrightarrow F_x$$

compatible with restriction maps in $\widetilde{F}$. Passing to the direct
limit we define a map $\widetilde{F}_x \to F_x$ which is easily seen to be a
left inverse to $(i_F)_x$. Thus it remains to prove that $(i_F)_x$
is surjective, which is left to the reader.

ii) Let us first notice that $f\colon F \to G$ will induce a
morphism $\widetilde{f}\colon \widetilde{F} \to \widetilde{G}$ which for each open subset U of X will
yield a commutative diagram

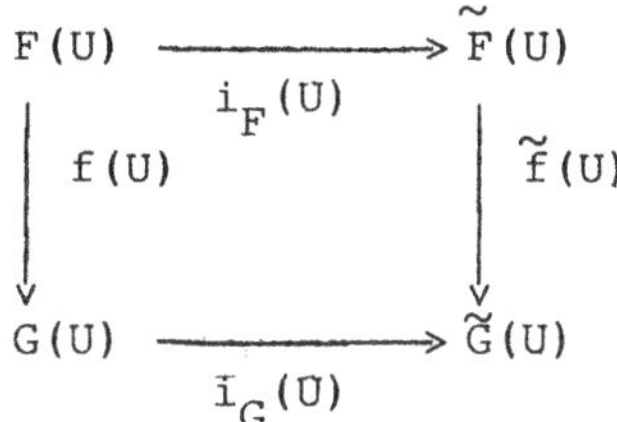

It follows from i and 2.2.iii that $i_G(U)$ is an isomorphism.
Thus we can define $\varphi(U): \widetilde{F}(U) \to G(U)$ by putting

$$\varphi(U) = i_G(U)^{-1} \widetilde{f}(U)$$

This defines a morphism $\varphi: \widetilde{F} \to G$ with $f = \varphi \circ i_F$. Uniqueness
of φ follows from 2.2.

Q.E.D.

The category Sh(X) of sheaves on X is the full subcategory of presheaves whose objects are the sheaves on X.
Notice that the direct sum of two sheaves, calculated in the
category of presheaves, is a sheaf. Thus Sh(X) is an additive
category.

Theorem 2.5. The category Sh(X) of sheaves on the topological space X is an abelian category.

Proof. Given a morphism $f: F \to G$ of sheaves. Let Ker(f)
and cok(f) denote kernel and cokernel presheaf as calculated
by the formula 1.6. It is easily seen that Ker(f) is in fact
a sheaf, and that the inclusion of Ker(f) in F is a kernel
in the sense of I.1 for the morphism f in the category Sh(X).
Put

$$\mathrm{Cok}(f) = \widetilde{\mathrm{cok}(f)}$$

88

The composite of the two canonical morphisms

$$G \to \text{cok}(f) \to \widetilde{\text{cok}}(f) = \text{Cok}(f)$$

is easily seen to be a cokernel in the sense of I.1 for the morphism f in Sh(X). With the terminology of I.1 consider the canonical factorization of $f: F \to G$ in the category Sh(X)

$$F \longrightarrow \text{Coim}(f) \overset{\widetilde{f}}{\to} \text{Im}(f) \to G$$

We will prove by localization that $\widetilde{f}$ is an isomorphism in Sh(X), so let there be given $x \in X$. Notice that the localization functor preserves kernels and cokernels 2.4. From this we conclude that $(\widetilde{f})_x$ is an isomorphism. From 2.2 we conclude that $\widetilde{f}$ is an isomorphism.

Q.E.D.

It is now time to adopt the general notions from the theory of abelian categories I.5.

Theorem 2.6. Given a sequence of morphisms of sheaves on X

$$E \overset{f}{\longrightarrow} F \overset{g}{\longrightarrow} G$$

This sequence is exact in Sh(X) if and only if

$$E_x \overset{f_x}{\longrightarrow} F_x \overset{g_x}{\longrightarrow} G_x \qquad \text{is exact for all} \quad x \in X.$$

Proof. Let us first fix an $x \in X$. Localization at x is an additive functor $\text{Sh}(X) \longrightarrow Ab$

which preserves Kernels and Cokernels as it follows from the
proof of 2.5. Thus the localisation functor is exact.

To prove the opposite inclu
sion let us make the general
remark that the sequence f,g
is exact if and only if

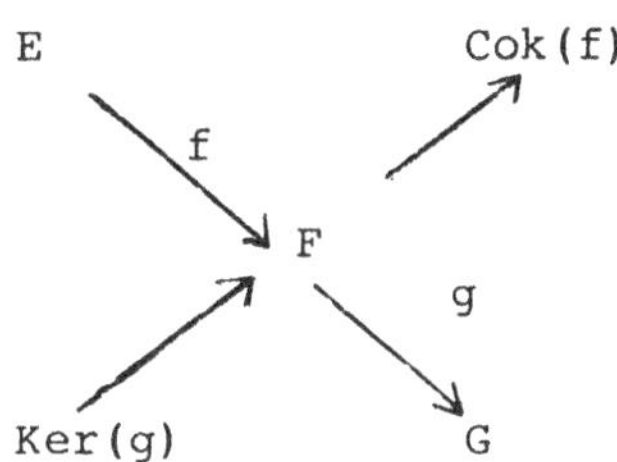

the two compositions along the diagonals in the diagram above
are zero. Now, suppose the local sequences f_x, g_x are exact
for all $x \in X$, then f,g is exact as it follows from the
first part of 2.2 and the previous remark.

$$Q.E.D.$$

Complexes of sheaves

Let $F^{\bullet}$ denote a complex of sheaves on X. For $n \in \mathbb{Z}$
let $H^n F^{\bullet}$ denote the n'th <u>cohomology sheaf</u>, defined according
to general principles for abelian categories.

To describe this let us exhibit a presheaf on X: Given
an open set U of X we can form the complex of abelian
groups $\Gamma(U,F^{\bullet})$. For open sets $U \supseteq V$, restriction from U
to V will induce a chain map $\Gamma(U,F^{\bullet}) \longrightarrow \Gamma(V,F^{\bullet})$
Thus we have established a presheaf on X

$$2.7 \qquad\qquad U \longmapsto H^n \Gamma(U,F^{\bullet})$$

<u>$H^n F^{\bullet}$ is the sheaf associated to the presheaf 2.7</u> as it follows
from the fact that the functor $G \longmapsto \widetilde{G}$ is an exact functor from
the category of presheaves to that of sheaves.

Direct limit of sheaves

Let $(F_\lambda, F_{\mu\lambda})$ denote a direct system of sheaves on X over the directed set I, compare II.0.4. Let us prove that such a system has a direct limit $(f_\lambda, \varinjlim F_\lambda)$.

To do so we shall first construct a presheaf F on X. An open subset U gives rise to a direct system of abelian groups $(F_\lambda(U), f_{\mu\lambda}(U))$ and we put $F(U) = \varinjlim F_\lambda(U)$ For open sets $U \supseteq V$ restriction will define a morphism of directed systems

$$r_{VU} \colon (F_\lambda(U), f_{\mu\lambda}(U)) \longrightarrow (F_\lambda(V), f_{\mu\lambda}(V))$$

whose direct limit is restriction from U to V in the presheaf F. Our construction provides morphisms $f_\lambda \colon F_\lambda \to F$, which makes (f_λ, F) a direct limit of the system $(F_\lambda, f_{\mu\lambda})$ in the category of presheaves on X.

Let $\varinjlim F_\lambda$ denote the sheaf associated the presheaf F above. With the notion of 2.3 the data $(\varinjlim F_\lambda, i_F \circ f_\lambda)$ is a direct limit in $Sh(X)$ of the direct system $(F_\lambda, f_{\mu\lambda})$ as it follows from 2.4.

Let us describe the stalk of $\varinjlim F_\lambda$ at a given point $x \in X$. We get a canonical isomorphism

2.8
$$\boxed{\ \varinjlim F_{\lambda, x} \xrightarrow{\sim} (\varinjlim F_\lambda)_x\ }$$

Proof. According to 2.4 it suffices to prove a similar statement with $\varinjlim F_\lambda$ replaced by the presheaf F, constructed above. The verification is straightforward and left to the reader.

Q.E.D.

II.3   Cohomology of sheaves

In this section we shall introduce cohomology groups for sheaves on a fixed topological space X. This is based on homological algebra in the category $Sh(X)$ of sheaves on X.

__Theorem 3.1.__ The abelian category $Sh(X)$ has enough injectives.

__Proof.__ We shall use that the category Ab of abelian groups has enough injectives: an abelian group D is injective in Ab if and only if it is a divisible group, - any abelian group can be imbedded into a divisible group.

We shall describe an auxiliary construction: To a family $D = (D_x)_{x \in X}$ of abelian groups we can associate a sheaf D_* on X by the formula

$$\Gamma(U, D_*) = \prod_{x \in U} D_x$$

The restriction maps in the sheaf D_* are induced by projections. We leave it to the reader to establish a natural isomorphism

$$\mathrm{Hom}(F, D_*) \xrightarrow{\sim} \prod_{x \in X} \mathrm{Hom}(F_x, D_x)$$

as F varies through $Sh(X)$. This shows that a family $D = (D_x)_{x \in X}$ of divisible groups gives rise to an injective sheaf D_* on X.

Consider a fixed sheaf E on X. For each $x \in X$ choose an embedding $E_x \to D_x$ of the stalk of E at x into a

divisible group D_x and put $D = (D_x)_{x \in X}$. The construction
above provides an imbedding $E \to D_*$ of E into an injective
sheaf D_*.

Q.E.D.

The basic functor for homological algebra on $Sh(X)$ is
the global section functor

$$\Gamma(X,-): Sh(X) \to Ab$$

This is a left exact functor. We shall now apply the general
notions of I.7. - The i'th derived functor of $\Gamma(X,-)$ evaluated
on a sheaf F will be denoted $H^i(X,F)$ and is called the
<u>i'th cohomology group of the space X with coefficients in
the sheaf</u> F. To calculate this we must take an injective
resolution $F \to I^{\boldsymbol{\cdot}}$ of F to get

3.2
$$H^i(X,F) = H^i\Gamma(X,I^{\boldsymbol{\cdot}})$$

<u>Example 3.3</u>. On an open subset X of the complex plane $\mathbb{C}$
we have a fundamental exact sequence of sheaves, compare 1.11

$$0 \longrightarrow \underset{\sim}{\mathbb{C}} \longrightarrow 0 \overset{D}{\longrightarrow} 0 \longrightarrow 0$$

This gives rise to a long exact sequence

$$0 \longrightarrow \Gamma(X,\underset{\sim}{\mathbb{C}}) \longrightarrow \Gamma(X,0) \overset{D}{\longrightarrow} \Gamma(X,0) \longrightarrow$$
$$\longrightarrow H^1(X,\underset{\sim}{\mathbb{C}}) \longrightarrow H^1(X,0) \longrightarrow H^1(X,0) \longrightarrow$$

In particular, we see that if $H^1(X,\underset{\sim}{\mathbb{C}}) = 0$ then for any
$f \in \Gamma(X,0)$ we can find $g \in \Gamma(X,0)$ with $Dg = f$.

<u>Definition 3.4</u>. A sheaf F on X is called <u>flabby</u>, if for any open set U of X , $r_{UX}\colon F(X) \to F(U)$ is surjective.

<u>Theorem 3.5</u>. A flabby sheaf F on X is $\Gamma(X,-)$ -acyclic. Any sheaf I on X which is injective in $Sh(X)$ is flabby.

<u>Proof</u>. Let I be a sheaf on X which is injective in $Sh(X)$. According to the proof of 3.1 we can find a monomorphism $i\colon I \to E$ from I into a flabby sheaf E . Let $f\colon E \to I$ denote a retraction of i , i.e. $f \circ i = 1$. For an open set U of X consider the commutative diagram

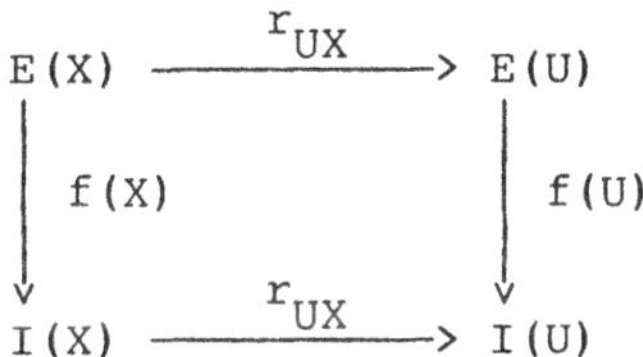

$$\begin{array}{ccc}
E(X) & \xrightarrow{\ r_{UX}\ } & E(U) \\
\downarrow{\scriptstyle f(X)} & & \downarrow{\scriptstyle f(U)} \\
I(X) & \xrightarrow{\ r_{UX}\ } & I(U)
\end{array}$$

Notice, that the vertical arrows and the upper horizontal arrow are epimorphisms. It follows that the lower horizontal arrow is an epimorphism.

Let us now consider a short exact sequence of sheaves on X with E flabby

$$0 \longrightarrow E \xrightarrow{\ e\ } F \xrightarrow{\ f\ } G \longrightarrow 0$$

Let us prove that f induces a surjective map on the global sections. - So let there be given a section s of G over X . Let us introduce the ordered set I of pairs (U,u) where U is an open subset of X and u a section of F over U with $f(u) = r_{UX}(s)$. The symbol $(U,u) \le (V,v)$ means that $U \subseteq V$ and that $u = r_{UV}(v)$.

The set I has the property that any totally ordered sub-set J ($a \leq b$ or $b \leq a$ for any pair of elements of J) has an upper bound. The content of _Zorn's lemma_ is that ordered sets of this type have _maximal elements_.

Let (U,u) be a maximal element of I. Suppose $U \neq X$, and pick $x \in X-U$. Choose an open neighbourhood V of x and a section v of F over V with $f(v) = r_{VX}(s)$. The difference $r_{V \cap U, U}(u) - r_{V \cap U, V}(v)$ represents a section of E over $V \cap U$. Extend this to a section of E over V and add the result to v. After this modification, u and v will have the same restriction to $U \cap V$ and thereby define a section of F over $U \cup V$ contradicting the maximality of (u,U).

Let us return to the exact sequence above. We leave it to the reader to prove "E and F flabby implies G flabby".

We can now prove that $H^1(X,E) = 0$ for a flabby sheaf E:
Choose an exact sequence as above with F injective to get

$$H^0(X,F) \to H^0(X,G) \to H^1(X,E) \to H^1(X,F)$$

from which we can draw the conclusion, since the first map is surjective and the last group is zero.

Let us finally prove by induction on $n > 1$ that "$H^n(X,E) = 0$ for all flabby sheaves E on X". Choose an exact sequence as above with F injective. This makes G flabby by our earlier work. Consider the cohomology sequence

$$\to H^n(X,G) \to H^{n+1}(X,E) \to H^{n+1}(X,F) \to$$

to conclude the inductive step of the proof.

$$\text{Q.E.D.}$$

<u>The Godement resolution 3.6</u>. Let F be a sheaf on the topological space X. We have a natural imbedding $F \to C^0 F$, described in the proof of 3.1:

$$3.7 \qquad \Gamma(U, C^0 F) = \prod_{x \in U} F_x$$

Let F^1 denote the cokernel of $F \to C^0 F$ and put $C^1 = C^0 F^1$. Iterate this process to get a flabby resolution $F \to C^{\cdot} F$, where $C^{\cdot} F$ is given by

$$3.8 \qquad 0 \to C^0 F \to C^1 F \to C^2 F \to \ldots \to C^n F \to \ldots$$

The construction is functorial: a morphism of sheaves $f: F \to G$ will induce a commutative diagram of complexes of sheaves

$$3.9 \qquad \begin{array}{ccc} F & \longrightarrow & C^{\cdot} F \\ \downarrow {\scriptstyle f} & & \downarrow {\scriptstyle C^{\cdot} f} \\ G & \longrightarrow & C^{\cdot} G \end{array}$$

Another virtue of the resolution is that for each $x \in X$ the localization

$$3.10 \qquad F_x \longrightarrow (C^{\cdot} F)_x$$

is a <u>homotopy equivalence</u> in Ab. This is a consequence of the fact that the imbedding $F \to C^0 F$ comes equipped with <u>canonical local retractions</u>, namely the projections of ΠF_x onto its factors.

II.4   Direct and inverse image of sheaves. f_*, f^*.

In this section we consider a fixed continuous map $f: X \to Y$. We shall first describe a procedure __pull back__ or __inverse image__, which to a sheaf G on Y associates a sheaf f^*G on X. The sections of f^*G over an open subset U of X is the set of those

$$s \in \prod_{z \in U} G_{f(z)}$$

which for each $x \in U$ satisfies the following condition: "there exists an open neighbourhood W of x in U, an open set V containing $f(W)$ and a section t of G over V, such that

$$4.1 \qquad \mathrm{pr}_w s = t_{f(w)} \qquad ; \quad w \in W$$

where $\mathrm{pr}_w: \prod_{z \in U} G_{f(z)} \to G_{f(w)}$ is the projection".

For open sets $U \supseteq V$ in X the natural projection

$$\prod_{z \in U} G_{f(z)} \longrightarrow \prod_{z \in V} G_{f(z)}$$

will transform $\Gamma(U, f^*G)$ into $\Gamma(V, f^*G)$. It is left to the reader to verify that in fact we have constructed __a sheaf__ __f^*G on X__.

Given an open subset V of Y and a section t of G over V. The stalks of t will define a section of f^*G over $f^{-1}(V)$, i.e. we have the socalled __adjunction map__

$$4.2 \qquad a(V): \Gamma(V, G) \to \Gamma(f^{-1}(V), f^*G)$$

Given a point $x \in X$. As V varies through the open neighbourhoods of $f(x)$ the adjunction map induces an isomorphism

$$4.3 \qquad G_{f(x)} \xrightarrow{\sim} (f^*G)_x \qquad ; \quad x \in X$$

$\underline{Proof}$. Let $a_x: G_{f(x)} \to (f^*G)_x$ denote the map obtained
by the procedure above. A look at the definition of f^*G
shows that a_x is surjective. To prove that a_x is injective
let us construct a map $p_x: (f^*G)_x \to G_{f(x)}$ with $p_x a_x = 1$:
For an open neighbourhood U of x we can consider the composite

$$\Gamma(U, f^*G) \longrightarrow \prod_{z \in U} G_{f(z)} \xrightarrow{\text{pr}_x} G_{f(x)}$$

and pass to the limit over all such U's to obtain p_x.

$$Q.E.D.$$

The construction of f^*G from G is easily seen to define
an additive functor which is $\underline{exact}$, as it follows from 4.3

$$4.4 \qquad f^*: Sh(Y) \longrightarrow Sh(X)$$

Starting from a sheaf F on X we shall construct a
sheaf f_*F on Y, the $\underline{direct\ image}$ of F by f. The sections
over the open subset V of Y are given by

$$4.5 \qquad \Gamma(V, f_*F) = \Gamma(f^{-1}(V), F)$$

the restriction maps are those induced from F. - This
construction is seen to define a $\underline{left\ exact}$ additive functor

$$4.6 \qquad f_*: Sh(X) \longrightarrow Sh(Y)$$

98

Let us now prove the fundamental formula

4.7 $\mathrm{Hom}(f^*G,F) = \mathrm{Hom}(G,f_*F)$

for all sheaves G on Y and all sheaves F on X, i.e.

   $\underline{\text{Theorem 4.8}}$. The functor f^* is left adjoint to the
functor f_*.

   $\underline{\text{Proof}}$. Let us first record that the adjunction maps
4.2 fits together to form the $\underline{\text{adjunction morphism}}$

4.9 $a:\ G \longrightarrow f_*f^*G$

Starting with $\varphi:\ f^*G \to F$ we can construct a morphism
$\psi:\ G \to f_*F$ given by $\psi = f_*\varphi \circ a$.   Notice that for an open
set V of Y we have the commutative diagram

$$
\begin{array}{ccc}
\Gamma(V,G) & \xrightarrow{\ \psi(V)\ } & \Gamma(V,f_*F) \\[2mm]
\Big\downarrow{\scriptstyle a(V)} & & \Big\downarrow{\scriptstyle =} \\[2mm]
\Gamma(f^{-1}(V),f^*G) & \xrightarrow{\ \varphi(f^{-1}(V))\ } & \Gamma(f^{-1}(V),f)
\end{array}
$$

4.10

Remark that the diagram 4.10 can serve us as definition of
the transform ψ of φ. Let us remark that the map we have
constructed is injective

4.11 $\mathrm{Hom}(f^*G,F) \to \mathrm{Hom}(G,f_*F)$ $;\ \varphi \mapsto f_*\varphi \circ a$

To see this let us fix a point $x \in X$ and take the direct
limit over all open neighbourhoods V of $f(x)$ to obtain
from 4.10 a commutative diagram

4.12

$$
\begin{array}{ccc}
G_{f(x)} & \xrightarrow{\ \psi_{f(x)}\ } & (f_*F)_{f(x)} \\
\downarrow{a_x} & & \downarrow{b_x} \\
(f_*G)_x & \xrightarrow{\ \varphi_x\ } & F_x
\end{array}
$$

Recalling from 4.3 that a_x is an isomorphism, we see how
the stalks of φ are determined by those of ψ. Injectivity
of 4.11 thus follows from 2.2.i.

To prove surjectivity of 4.11 let $\psi: G \to f_*F$ be given.
For an open set U of X form the product

$$
\prod_{x \in U} b_x \psi_{f(x)} : \prod_{x \in U} G_{f(x)} \longrightarrow \prod_{x \in U} F_x
$$

It is left to the reader to see that this map will transform
$\Gamma(U, f^*G)$ into $\Gamma(U, \widetilde{F})$ where $\widetilde{F}$ denotes the sheaf associated
the presheaf F. In this way we obtain a morphism $f^*G \to \widetilde{F}$
which we compose with the inverse of the isomorphism $F \to \widetilde{F}$,
2.4 to obtain a morphism $\varphi: f^*G \to F$. It is now easy to see
that $\psi = f_*\varphi \circ a$.

Q.E.D.

Corollary 4.13. The additive functor $f_*: Sh(X) \to Sh(Y)$
transforms injectives into injectives.

Proof. Let I be an injective sheaf on X. The formula
$Hom(G, f_*I) = Hom(f^*G, I)$ and the fact that f^* is exact,
show that $G \mapsto Hom(G, f_*I)$ is an exact functor on $Sh(Y)$.

Q.E.D.

II.5 Continuous maps and cohomology

In this section we shall discuss the action of a continuous map $f: X \to Y$ on cohomology.

Let there be given a sheaf G on Y . We shall introduce a natural map

$$5.1 \qquad\qquad f^*: H^{\cdot}(Y,G) \longrightarrow H^{\cdot}(X,f^*G)$$

To describe this choose an injective resolution $G \to J^{\cdot}$ on Y and an injective resolution $f^*J^{\cdot} \to I^{\cdot}$ on X, I.6.1. Since f^* is exact this makes $I^{\cdot}$ an injective resolution of f^*G. The composite

$$\Gamma(Y,J) \xrightarrow{\ a\ } \Gamma(X,f^*J^{\cdot}) \longrightarrow \Gamma(X,I^{\cdot})$$

represents 5.1 on the chain level by definition.

__Scolium 5.2.__ Given a $\Gamma(Y,-)$-acyclic resolution $t: G \to T$ and a $\Gamma(X,-)$-acyclic resolution $s: f^*G \to S^{\cdot}$ and a morphism of complexes $\psi: T^{\cdot} \to f_*S^{\cdot}$ making the following diagram commutative

$$
\begin{array}{ccc}
G & \xrightarrow{\ \ t\ \ } & T^{\cdot} \\
\downarrow{\scriptstyle a} & & \downarrow{\scriptstyle \psi} \\
f_*f^*G & \xrightarrow{\ f_*s\ } & f_*S^{\cdot}
\end{array}
$$

then $f^*: H^{\cdot}(Y,G) \to H^{\cdot}(X,f^*G)$ is represented by the chain map

$$\Gamma(Y,T^{\cdot}) \xrightarrow{\ \psi\ } \Gamma(Y,f_*S^{\cdot}) = \Gamma(X,S^{\cdot})$$

    <u>Proof</u>. Choose an injective resolution $T^{\cdot} \to J^{\cdot}$ on Y and an injective resolution $f*J^{\cdot} \to I^{\cdot}$ on X, I.6.1. Consider the following commutative diagram

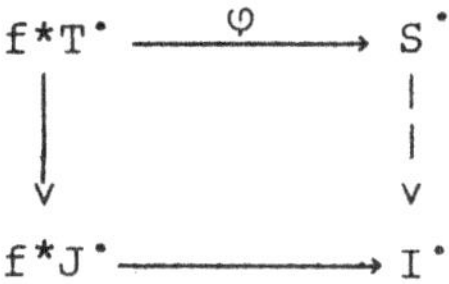

where φ corresponds to ψ by adjunction 4.7. Fill in the dotted arrow to make the square homotopy commutative. Finally consider the homotopy commutative diagram

$$\begin{array}{ccccc}
\Gamma(Y,T^{\cdot}) & \xrightarrow{\ a\ } & \Gamma(X,f*T^{\cdot}) & \xrightarrow{\ \varphi\ } & \Gamma(X,S^{\cdot}) \\
\downarrow & & \downarrow & & \downarrow \\
\Gamma(Y,J^{\cdot}) & \xrightarrow{\ a\ } & \Gamma(X,f*J^{\cdot}) & \longrightarrow & \Gamma(X,I^{\cdot})
\end{array}$$

and use the commutative diagram 4.10. Let us finally remark that the two extreme vertical arrows are quasi-isomorphisms I.7.5.

$$\text{Q.E.D.}$$

Closed subspaces

    Let $i: Z \to X$ denote the inclusion of a closed subspace Z of the topological space X. For a sheaf E on Z and a point x of X we have

$$(i_*E)_x = \varinjlim_{x \in U} \Gamma(U \cap Z, E)$$

From this results immediately that

$$5.3 \qquad (i_*E)_x = \begin{cases} E_x & \text{for } x \in Z \\ 0 & \text{for } x \in X-Z \end{cases}$$

As a consequence we see that the functor

$$i_*: \operatorname{Sh}(Z) \to \operatorname{Sh}(X) \quad \text{is } \underline{\text{exact}}$$

From 5.3 and 4.12 follows that adjunction yields an isomorphism $i^*i_*E \stackrel{\sim}{\to} E$. Moreover,

$$5.4 \qquad H^{\bullet}(X, i_*E) \stackrel{\sim}{\to} H^{\bullet}(Z, E)$$

$\underline{\text{Proof}}$. An injective resolution $E \to I^{\bullet}$ on Z transforms into an injective resolution $i_*E \to i_*I^{\bullet}$ on X. The composite

$$I^{\bullet}(X, i_*I^{\bullet}) \stackrel{a}{\longrightarrow} \Gamma(Z, i^*i_*I^{\bullet}) \longrightarrow \Gamma(Z, I^{\bullet})$$

yields the desired isomorphism.

$$\text{Q.E.D.}$$

$\underline{\text{Lemma 5.5}}$. Let $i: Z \to X$ denote the inclusion of a closed subspace and F a sheaf on X. The adjunction morphism $a: F \to i_*i^*F$ has the following mapping property: Any morphism $F \to G$, into a sheaf G, whose stalks are zero outside Z, admits a unique factorization through $a: F \to i_*i^*F$.

Proof. Let us first remark that the adjunction morphism
$a: F \to i_* i^* F$ is an epimorphism in $Sh(X)$ as it follows by
localization using 5.3. Let us consider a morphism $f: F \to G$.

This gives rise to a commutative diagram in $Sh(X)$
where a and b are
adjunction morphisms.

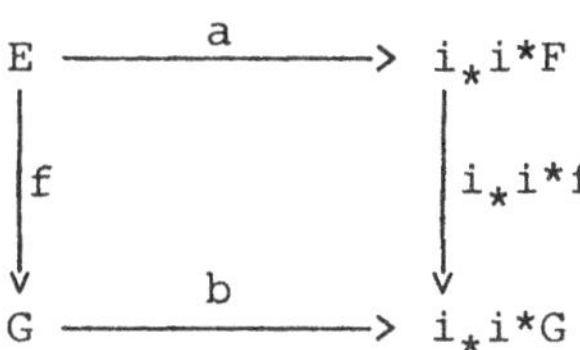

$$
\begin{array}{ccc}
E & \xrightarrow{\;a\;} & i_* i^* F \\
\downarrow f & & \downarrow i_* i^* f \\
G & \xrightarrow{\;b\;} & i_* i^* G
\end{array}
$$

It follows by localization that b is an isomorphism in $Sh(X)$.

For a sheaf F on X we shall often use the abuse of
notation $H^\cdot(Z,F)$ for $H^\cdot(Z,i^*F)$, where $i: Z \to X$ is the
inclusion of the subspace Z.

Mayer-Vietoris sequence 5.6. Let A and B denote
closed subspaces of the topological space X. A sheaf F on
X gives rise to a long exact sequence

$$\to H^p(A \cup B, F) \to H^p(A,F) \oplus H^p(B,F) \to H^p(A \cap B, F) \to$$

Proof. Let us put names to the inclusions $h: A \cup B \to X$,
$i: A \to X$, $j: B \to X$, $k: A \cup B \to X$. Consider the exact sequence

5.7 $\qquad 0 \longrightarrow h_* h^* F \longrightarrow i_* i^* F \oplus j_* j^* F \longrightarrow k_* k^* F \longrightarrow 0$

where the morphisms are sum and difference of adjunction
morphisms, taking into account the mapping principle 5.5.
This sequence is exact as one sees by localization. The long
cohomology sequence can be identified by means of 5.4.

$$\text{Q.E.D.}$$

Open subspaces

Let $j: U \to X$ denote the inclusion of an open subspace and F a sheaf on X. Let us remark that j*F has a particular simple description in this case: the sections over an open set $V \subseteq U$ is given by

$$5.8 \qquad \Gamma(V,j^*F) = \Gamma(V,F)$$

Let us consider a flabby resolution $F \to D^{\cdot}$ on X. This is transformed into a flabby resolution $j^*F \to j^*D^{\cdot}$ on U, which fits into a commutative diagram

As a consequence of Scolium 5.2 we can represent

$$j^*: \ H^{\cdot}(X,F) \longrightarrow H^{\cdot}(U,j^*F)$$

by the chain map

$$5.9 \qquad r_{UX}: \ \Gamma(X,D^{\cdot}) \longrightarrow \Gamma(U,D^{\cdot})$$

Mayer Vietoris sequence 5.10. Let U and V be open subsets of X. Any sheaf F on X induces a long exact sequence

$$\to H^n(U \cup V,F) \to H^n(U,F) \oplus H^n(V,F) \to H^n(U \cap V,F) \to$$

Proof. Consider the sequence

$$0 \to \Gamma(U \cup V,F) \to \Gamma(U,F) \oplus \Gamma(V,F) \to \Gamma(U \cap V,F) \to 0$$

where the two maps are sum and difference of restriction maps.
Note that this sequence is exact whenever F is a flabby sheaf
on X. In general replace F by a flabby resolution $D^{\cdot}$
and pass to cohomology using 5.9.

Q.E.D.

Application to $R^n f_*$

Let us return to a general continuous map $f: X \to Y$. The
additive functor

$$f_*: Sh(X) \to Sh(Y)$$

is <u>left exact and transforms injectives into injectives</u> as it
follows from the fact that f_* has a left adjoint which is
exact by 4.4 and 4.8. The n'th derived functor of this is denoted
$R^n f_*$, $n \in \mathbb{Z}$.

Proposition 5.11. Let F be a sheaf on X and $n \in \mathbb{Z}$.
The cohomology sheaf $R^n f_* F$ is the sheaf associated to the presheaf
$$V \longmapsto H^n(f^{-1}(V),F) \ .$$

Proof. Let $I^{\cdot}$ be an injective resolution of F on X.
We have
$$R^n f_* F = H^n f_* I^{\cdot}$$

It follows from 2.7 that $R^n f_* F$ is the sheaf associated to
the presheaf $V \longmapsto H^n \Gamma(V, f_* I^{\cdot})$. Using 5.9 we find that

$$H^n \Gamma(V, f_* I^{\cdot}) = H^n \Gamma(f^{-1}(V), I^{\cdot}) = H^n(f^{-1}(V),F)$$

which is the desired result.

Q.E.D.

II.6   Locally closed subspaces. $h_!, h^!$.

Given a sheaf F on a topological space X and a section s of F over the open subset U of X. We define $\underline{\text{Supp}(s)},$ $\underline{\text{the support of}}$ $\underline{s}$

$$\text{Supp}(s) = \{x \in U \mid s_x \neq 0\}$$

Notice that the subset $\text{Supp}(s)$ is closed relative to U.

We shall be concerned with a $\underline{\text{locally closed}}$ subspace W of X. This means that any point $v \in W$ has an open neighbourhood V in X such that $W \cap V$ is closed relative to V. The inclusion of W in X will be denoted $h: W \to X$.

$\underline{\text{Definition 6.1}}.$ For a sheaf E on W we let $h_! E$ denote the sheaf on X whose sections over the open set U of X is given by

$$\Gamma(U, h_! E) = \{s \in \Gamma(W \cap U, E) \mid \text{Supp}(s) \text{ is closed rel. to } U\}$$

The restriction maps of $h_! E$ are induced from $h_* E,$ of which it is a subsheaf. We record this as a $\underline{\text{canonical monomorphism}}$

$$6.2 \qquad\qquad h_! E \longrightarrow h_* E$$

From the description of the sections of $h_! E$ follows that

$$6.3 \qquad\qquad (h_! E)_x = \begin{cases} E_x & \text{for } x \in W \\ 0 & \text{for } x \in X-W \end{cases}$$

which in particular shows $\underline{\text{that}}$ $\underline{h_!}$ $\underline{\text{is exact}}.$

Proposition 6.4. The functor $h_!: Sh(W) \to Sh(X)$ is an equivalence between the category of sheaves on W and the full subcategory of $Sh(X)$ made up of the sheaves for which

$$F_x = 0 \quad \text{for all} \quad x \in X-W$$

The inverse functor is induced by $f*$.

Proof. For a sheaf E on W let us record the following simple identity among sheaves on W

6.5 $$h*h_!E = E$$

Next, consider a sheaf F on X whose stalks vanish outside W. A close examination reveals that the adjunction morphism $F \to h_*h*F$ may be factored through the monomorphism 6.2 as follows

$$F \to h_!h*F \to h_*h*F$$

to yield an isomorphism $F \cong h_!h*F$.

Q.E.D.

We shall construct a functor $h^!$

$$Sh(W) \underset{h_!}{\overset{h^!}{\longleftrightarrow}} Sh(X)$$

which satisfies the following identity

$$Hom(h_!E,F) = Hom(E,h^!F)$$

for sheaves E on W and sheaves F on X. Otherwise expressed

$\underline{\text{Proposition 6.6}}$. The functor $h_!: Sh(W) \to Sh(X)$ has a right adjoint $h^!: Sh(X) \to Sh(W)$.

$\underline{\text{Proof}}$. For a sheaf F on X we let F^W denote the sheaf on X whose sections over the open subset U are given by

$$\Gamma(U,F^W) = \{s \in \Gamma(U,F) \mid Supp(s) \subsetneq W\}$$

The stalks of F^W are zero at all points outside W, so if we put $h^!F = h*F^W$ we get according to the last line of 6.4 applied to F^W

$$6.7 \qquad \Gamma(U,h_!h^!F) = \{s \in \Gamma(U,F) \mid Supp(s) \subsetneq W\}$$

This formula provides a monomorphism $h_!h^!F \to F$. It follows from 6.7 that any morphism $G \to F$ where G is a sheaf on X whose stalks are zero outside W may be factored through the monomorphism $h_!h^!F \to F$. For a sheaf E on W we get accordingly

$$Hom(h_!E,F) = Hom(h_!E,h_!h^!F)$$

By the first part of 6.4 we have an isomorphism

$$h_!: Hom(E,h^!F) \longrightarrow Hom(h_!E,h_!h^!F)$$

We can now compose the two isomorphisms. $\hfill$ Q.E.D.

$\underline{\text{Proposition 6.8}}$. The functor $h^!: \mathrm{Sh}(X) \to \mathrm{Sh}(W)$ is left exact and transforms injective sheaves into injective sheaves.

$\underline{\text{Proof}}$. The left exactness is a formal consequence of the presence of a left adjoint $h_!$ to $h^!$. The second statement is a formal consequence of the exactness of $h_!$.

$$\text{Q.E.D.}$$

Open and closed subspaces

We shall make a close investigation of the special case of open and closed subsets.

$\underline{\text{Proposition 6.9}}$. i.) For the inclusion $i: Z \to X$ of a closed subspace we have $i_* = i_!$.

ii) For the inclusion $j: U \to X$ of an open subspace of X, we have $j^* = j^!$.

$\underline{\text{Proof}}$. i) For a sheaf E on Z we have a natural map 6.2 $i_!E \to i_*E$ which is an isomorphism as one checks by localization using 6.3 and 5.3.

ii) For a sheaf F on X we have according to the proof of 6.6 that $j^!F = j^*F^W$. Combined with the inclusion $F^W \to F$ this yields a $\underline{\text{monomorphism } j^!F \to j^*F}$. One checks by localization that this is an isomorphism.

$$\text{Q.E.D.}$$

$\underline{\text{Corollary 6.10}}$. The inclusion $j: U \to X$ of an open subspace transforms an injective sheaf I on X into an injective sheaf j^*I on U.

110

<u>Proof</u>. Combine 6.9.ii and 6.8.

Q.E.D.

In the case where U and Z are complementary subspaces of X, a sheaf F on X give rise to an exact sequence

6.11 $\qquad 0 \to j_! j^*F \to F \to i_* i^*F \to 0$

of sheaves on X, where the two morphism are adjunction morphisms. The exactness is seen by localization.

For the benefit of the reader we shall make a small tableaux over the basic functors

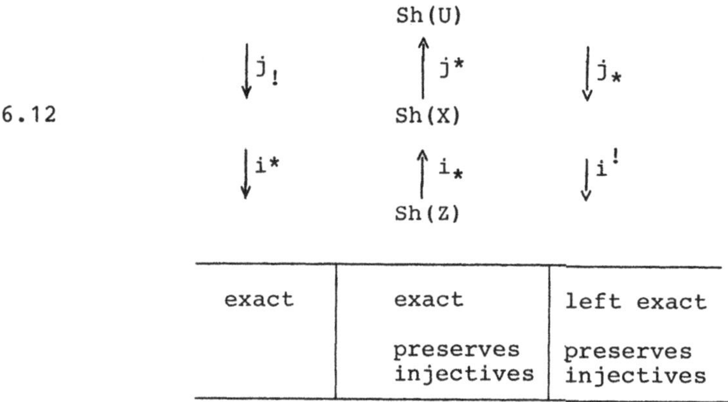

6.12

The adjoint relationship between the various functions can be read off the table. Identities:

$$j^*j_* = 1 \qquad j^*j_! = 1 \qquad i^!i_* = 1 \qquad i^*i_* = 1$$

Inverse image of locally closed subspaces

Let us consider a continuous map $f: X \to Y$ and a locally closed subspace B of Y. Let us notice that $A = f^{-1}(B)$ is a locally closed subspace of X. This can be seen by remarking that a subspace is locally closed if and only if it is the intersection between an open and a closed subspace. - Let $h: B \to Y$ and $g: A \to X$ denote the inclusions. We shall establish a natural isomorphism

6.13 $\qquad\qquad f^*h_! G = g_! f^*G$

$$\begin{array}{ccc} X & \xleftarrow{\ \ g\ \ } & A \\ f \downarrow & & f \downarrow \\ Y & \xleftarrow{\ \ h\ \ } & B \end{array}$$

for all sheaves G on B.

<u>Construction</u>. To be precise, let $\varphi: A \to B$ denote the map satisfying $h\varphi = fg$. We start with the adjunction morphism $G \to \varphi_*\varphi^*G$ from which we deduce a morphism $h_*G \to f_*g_*\varphi^*G$ or by adjunction a morphism

$$\theta: f^*h_*G \to g_*\varphi^*G$$

We have monomorphisms $f^*h_! G \to f^*h_*G$ and $g_!\varphi^*G \to g_*\varphi^*G$, compare 6.2. One can now check by localization that θ will induce on isomorphisms between these two subsheaves.

$$\text{Q.E.D.}$$

From 6.13 we deduce by adjunction a natural isomorphism

6.14 $\qquad\qquad h^! f^*F = f_* g^! F$

for all sheaves F on X.

112

<u>Fundamental diagrams</u>

For a locally closed subset W of X and a sheaf F on X we put

$$6.15 \qquad\qquad F_W = h_! h^* F$$

where h: W → X denotes the inclusion of W in X.

Closed subsets S ⊆ Z of X with complements O ⊇ U give rise to an exact commutative diagram

$$
6.16 \qquad
\begin{array}{ccccccccc}
0 & \longrightarrow & F_U & \longrightarrow & F & \longrightarrow & F_Z & \longrightarrow & 0 \\
 & & \downarrow & & \downarrow{\scriptstyle 1} & & \downarrow & & \\
0 & \longrightarrow & F_O & \longrightarrow & F & \longrightarrow & F_S & \longrightarrow & 0
\end{array}
$$

For closed subsets A and B of X with complements U and V respectively we have exact commutative diagrams

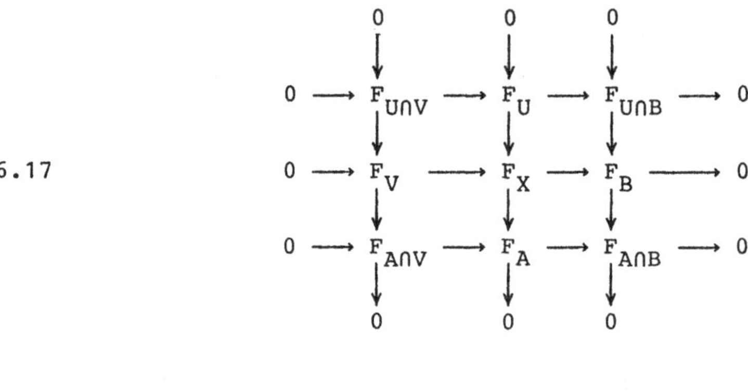

$$
6.17 \qquad
\begin{array}{ccccccccc}
 & & 0 & & 0 & & 0 & & \\
 & & \downarrow & & \downarrow & & \downarrow & & \\
0 & \longrightarrow & F_{U \cap V} & \longrightarrow & F_U & \longrightarrow & F_{U \cap B} & \longrightarrow & 0 \\
 & & \downarrow & & \downarrow & & \downarrow & & \\
0 & \longrightarrow & F_V & \longrightarrow & F_X & \longrightarrow & F_B & \longrightarrow & 0 \\
 & & \downarrow & & \downarrow & & \downarrow & & \\
0 & \longrightarrow & F_{A \cap V} & \longrightarrow & F_A & \longrightarrow & F_{A \cap B} & \longrightarrow & 0 \\
 & & \downarrow & & \downarrow & & \downarrow & & \\
 & & 0 & & 0 & & 0 & &
\end{array}
$$

$$
6.18 \qquad
\begin{array}{ccccccccc}
 & & 0 & & 0 & & 0 & & \\
 & & \downarrow & & \downarrow & & \downarrow & & \\
0 & \longrightarrow & F_{U \cap V} & \longrightarrow & F_U \oplus F_V & \longrightarrow & F_{U \cup V} & \longrightarrow & 0 \\
 & & \downarrow & & \downarrow & & \downarrow & & \\
0 & \longrightarrow & F & \xrightarrow{\;\binom{1}{1}\;} & F \oplus F & \xrightarrow{(1,-1)} & F & \longrightarrow & 0 \\
 & & \downarrow & & \downarrow & & \downarrow & & \\
0 & \longrightarrow & F_{A \cup B} & \longrightarrow & F_A \oplus F_B & \longrightarrow & F_{A \cap B} & \longrightarrow & 0 \\
 & & \downarrow & & \downarrow & & \downarrow & & \\
 & & 0 & & 0 & & 0 & &
\end{array}
$$

II.7   Cup product

Throughout this section we shall work with a fixed
commutative ring k. By a <u>k-sheaf</u> F on the topoloigcal space
X we understand a sheaf F together with a k-module structure
on F(U) for each open set U of X, making the restriction
maps k-linear. By a <u>morphism</u> f: F → G of k-sheaves we
understand a morphism of sheaves such that f(U): F(U) ⟶ G(U)
is k-linear for all open subsets U of X. The category of
k-sheaves on X will be denoted

$$7.1 \qquad\qquad Sh(X,k)$$

The Hom functor in this category is denoted Hom_k.

Let D be a k-module. The sheaf $\underset{\sim}{D}$ introduced in 1.10
is in a natural way a k-sheaf. For any k-sheaf F on X we
have a natural isomorphism

$$7.2 \qquad\qquad Hom_k(\underset{\sim}{D},F) = Hom_k(D,\Gamma(X,F))$$

<u>Proof</u>. Let p: X → Pt denote the projection of X onto
a point. The formula 7.2 may be interpreted as expressing that
p_* is a right adjoint to p*, II.4.7.

$$Q.E.D.$$

<u>Proposition 7.3</u>. The category Sh(X,k) is an abelian
category with enough injectives. Any injective object in
Sh(X,k) is a flabby sheaf.

114

<u>Proof</u>. The previous chapters II.1-6 could very well have been worked out for k-sheaves. We shall freely refer to these sections as if this had been done.

In the present context we need to supplement 3.1 with the fact that "the category of k-modules has enough injectives" in order to prove the first part of the proposition.

The second part follows by a straightforward modification of the proof of 3.5. Alternatively we can consider the inclusion $j: U \to X$ of an open subspace and the corresponding adjunction morphism $j_! j^* \underset{\sim}{k} \to \underset{\sim}{k}$, which is a monomorphism. An injective object E in $Sh(X,k)$ will transform this into a surjective map

$$\mathrm{Hom}_k(\underset{\sim}{k},E) \longrightarrow \mathrm{Hom}_k(j_! j^* \underset{\sim}{k},E)$$

The first term is $\Gamma(X,E)$ as it follows from 7.2. The second term can be evaluated by means of 6.6 and 6.9.

$$\mathrm{Hom}(j_! j^* \underset{\sim}{k},E) \ = \ \mathrm{Hom}(j^* \underset{\sim}{k},j^*E) \ = \ \mathrm{Hom}(\underset{\sim}{k},j^*E) \ = \ \Gamma(U,E)$$

Looking a bit closer one sees that the surjective map above may be identified with $r_{UX}: \Gamma(X,E) \to \Gamma(U,E)$.

Q.E.D.

The same ideas will prove that

<u>Proposition 7.4</u>. Any k-sheaf F has a resolution of the form

$$\to P_n \to P_{n-1} \to \ldots \to P_1 \to P_o \to F \to 0$$

where P_s is a direct sum of sheaves of the form $j_! \underset{\sim}{k}$
where $j: U \to X$ is the inclusion of an open subspace of X.

    <u>Proof</u>. It suffices to establish the existence of an epi-
morphism $P \to F$ where P is a sheaf of type described anove.
For the inclusion $j: U \to X$ of an open subset we have the
adjunction formula 6.6 and 6.9

$$\mathrm{Hom}(j_! \underset{\sim}{k}, F) = \mathrm{Hom}(\underset{\sim}{k}, j*F) = \Gamma(U,F)$$

It follows that $s \in \Gamma(U,F)$ determines a morphism of sheaves
$j_! \underset{\sim}{k} \to F$, which transforms $1 \in \Gamma(U, j_! \underset{\sim}{k})$ into $s \in \Gamma(U,F)$. The
result follows by varying s and U.

Q.E.D.

    For any k-sheaf F on X and $p \in \mathbb{Z}$ we have

7.5
$$\boxed{\mathrm{Ext}_k^{\,\cdot}(\underset{\sim}{k}, F) = H^{\cdot}(X,F)}$$

    <u>Proof</u>. Let $F \to I^{\cdot}$ denote an injetive resolution of F
in $\mathrm{Sh}(X,k)$. Thus

$$\mathrm{Ext}_k^{\,\cdot}(\underset{\sim}{k}, F) = H^{\cdot}\mathrm{Hom}_k(\underset{\sim}{k}, I^{\cdot})$$

By 7.3 and 3.5 $F \to I^{\cdot}$ is a $\Gamma(X,-)$-acyclic resolution of
F in $\mathrm{Sh}(X)$ thus by the acyclicity theorem I.7.5

$$H^{\cdot}(X,F) = H^{\cdot}\Gamma(X,I^{\cdot})$$

From formula 7.2 we get with $D = k$

$$\Gamma(X,I^{\cdot}) = \mathrm{Hom}_k(\underset{\sim}{k}, I^{\cdot})$$

and the result follows.
Q.E.D.

The formula 7.5 constitutes the basis for introducing cup products in sheaf cohomology. Let us first notice that this in case $F = \underset{\sim}{k}$ gives

$$7.6 \qquad \mathrm{Ext}_k(\underset{\sim}{k},\underset{\sim}{k}) = H^\cdot(X,\underset{\sim}{k})$$

which makes $H^\cdot(X,\underset{\sim}{k})$ into an associative ring and provides $H^\cdot(X,F)$ with the structure of a right $H^\cdot(X,\underset{\sim}{k})$-module, I.8. This is called the <u>cup product</u> and is denoted

$$7.7 \qquad \xi \cup \alpha \in H^{p+q}(X,F) \qquad ; \quad \xi \in H^p(X,F),\ \alpha \in H^q(X,\underset{\sim}{k})$$

Let $f: X \to Y$ be a continuous map and G a sheaf on Y. We have the formula

$$7.8 \qquad f^*(\eta \cup \beta) = f^*\eta \cup f^*\beta \qquad ; \quad \eta \in H^p(Y,G),\ \beta \in H^q(Y,k)$$

<u>Verification</u>. This formula is best described in terms of derived categories. Let us start with three functors

$$\mathrm{Sh}(Y,k) \xrightarrow{\ f^*\ } \mathrm{Sh}(X,k)$$

$$\Gamma(Y,-) \searrow \qquad \swarrow \Gamma(X,-)$$

$$k\text{-}mod$$

and a natural transformation of additive functors 4.2

$$a: \Gamma(Y,-) \longrightarrow \Gamma(X,-) \circ f^*$$

This gives us three derived functors I.7.9

$$D^+(Y,k) \xrightarrow{\ Rf_* \ } D^+(X,k)$$

7.9 $R\Gamma(Y,-)$ $R\Gamma(X,-)$

$$D^+(k)$$

and a natural transformation (compose I.7.11 and I.7.12)

7.10 $$R\Gamma(Y,-) \longrightarrow R\Gamma(X,-) \circ Rf_*$$

In this picture the formula 7.8 becomes clear recalling that
f_* is exact.

Once we have put up this machinery we might as well apply
it to triangles to get by means of I.7.10

Proposition 7.11. A short exact sequence of sheaves on Y

$$0 \longrightarrow E \longrightarrow F \longrightarrow G \longrightarrow 0$$

will induce a commutative ladder

$$\begin{array}{ccccccccc}
\longrightarrow & H^p(Y,E) & \longrightarrow & H^p(Y,F) & \longrightarrow & H^p(Y,G) & \longrightarrow & H^{p+1}(Y,E) & \longrightarrow \\
& \downarrow & & \downarrow & & \downarrow & & \downarrow & \\
\longrightarrow & H^p(X,f_*E) & \longrightarrow & H^p(X,f_*F) & \longrightarrow & H^p(X,f_*G) & \longrightarrow & H^{p+1}(X,f_*E) & \longrightarrow
\end{array}$$

II.8 Tensor product of sheaves

Throughout this section k denotes a fixed commutative ring. In this context it is important to notice that <u>tensor-product commutes with direct limits</u>.

<u>Lemma 8.1</u>. Let (F_a, f_{ba}) denote direct system of k-modules over the directed set I . For any k-module E there is a canonical isomorphism

$$\varinjlim E \otimes_k F_a \xrightarrow{\sim} E \otimes_k \varinjlim F_a$$

<u>Proof</u>. Let $f_a: F_a \to \varinjlim F_a$ denote the canonical map. Consider the system of maps

$$1 \otimes f_a: E \otimes_k F_a \to E \otimes_k \varinjlim F_a \quad ; \ a \in I$$

of abelian groups. We shall check that this satisfies the universal mapping property II.0.2. - To do so we shall appeal to the description of $\mathrm{Hom}_{\mathbb{Z}}(E \otimes_k F, G)$ as the set of bilinear maps $m: E \times F \to G$ which satisfies

$$m(re,f) = m(e,rf) \quad ; \quad r \in k, \ e \in E, \ f \in F$$

It is now easy to conlude by means of II.0.3.

Q.E.D.

Definition 8.2. Let F and G be k-sheaves on the topological space X. The tensor product $F \otimes_k G$ is the sheaf associated to the presheaf

$$U \longmapsto F(U) \otimes_k G(U)$$

As a consequence of 2.4 and 8.1 we find that

$$8.3 \qquad (F \otimes_k G)_x = F_x \otimes_k G_x \qquad ; \quad x \in X$$

from which we conclude that tensor product of sheaves has exactness properties similar to those of ordinary tensor product.

Tensor product of complexes

Given two complexes $R^{\cdot}$ and $S^{\cdot}$ of k-sheaves. The tensor product $R^{\cdot} \otimes_k S^{\cdot}$ is the complex given by

$$[R^{\cdot} \otimes_k S^{\cdot}]^n = \bigoplus_{p+q=n} R^p \otimes S^q$$

and differential is given by

$$8.4 \qquad \partial^{p+q}(r^p \otimes s^q) = \partial^p(r^p) \otimes s^q + (-1)^p r^p \otimes \partial^q(s^q)$$

This is in accordance with the general rules of I.11.

Commutativity 8.5. Let $R^{\cdot}$ and $S^{\cdot}$ be complexes of k-sheaves on X. The formula

$$\sigma(r^p \otimes s^q) = (-1)^{pq} r^q \otimes s^p$$

120

defines a natural commutativity constraint

$$\sigma\colon R^{\bullet} \otimes_k S^{\bullet} \to S^{\bullet} \otimes_k R^{\bullet}$$

Translation 8.6. Let $A^{\bullet}$ and $B^{\bullet}$ denote complexes of k-sheaves, $m,n \in \mathbb{Z}$. The map

$$a^{m+p} \otimes b^{n+q} \longmapsto (-1)^{pn}\, a^{m+p} \otimes b^{n+q}$$

induces a natural isomorphism

$$\tau_{m,n}\colon A^{\bullet}[m] \otimes B^{\bullet}[n] \overset{\sim}{\to} A^{\bullet} \otimes B^{\bullet}[m+n]$$

which is related to the commutatitivy constraint by the following commutative diagram

8.7
$$\begin{array}{ccc}
A^{\bullet}[m]\otimes B^{\bullet}[n] & \xrightarrow{\ \tau_{m,n}\ } & A^{\bullet}\otimes B^{\bullet}[m+n] \\
\downarrow{\scriptstyle \sigma} & & \downarrow{\scriptstyle (-1)^{mn}\sigma[m+n]} \\
B^{\bullet}[n]\otimes A^{\bullet}[m] & \xrightarrow{\ \tau_{n,m}\ } & B^{\bullet}\otimes A^{\bullet}[m+n]
\end{array}$$

Associativity. The usual associatitivy constraint for the tensor product extends without difficulties to complexes. Notice the following commutative diagram

8.8
$$\begin{array}{ccc}
A^{\bullet}[m] \otimes B^{\bullet}[n] \otimes C^{\bullet}[r] & \xrightarrow{\ 1\otimes\tau_{n,r}\ } & A^{\bullet}[m]\otimes(B^{\bullet}\otimes C^{\bullet})[n+r] \\
\downarrow{\scriptstyle \tau_{m,n}\otimes 1} & & \downarrow{\scriptstyle \tau_{m,n+r}} \\
(A^{\bullet}\otimes B^{\bullet})[m+n]\otimes C^{\bullet}[r] & \xrightarrow{\ \tau_{m+n,r}\ } & (A^{\bullet}\otimes B^{\bullet}\otimes C^{\bullet})[m+n+r]
\end{array}$$

<u>Homotopy 8.9</u>. Given homotopic morphisms $f,g: A^\bullet \to B^\bullet$ and homotopic morphisms $u,v: P^\bullet \to Q^\bullet$. Then

$$f \otimes u, \quad g \otimes v \quad : A^\bullet \otimes P^\bullet \to B^\bullet \otimes Q^\bullet$$

are homotopic morphisms.

<u>Proof</u>. Use the general construction I.11.6.

Q.E.D.

<u>Triangles 8.10</u>. Given a triangle in $K^+(X,k)$

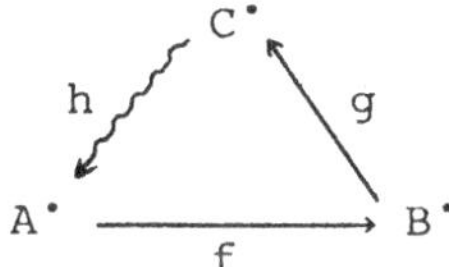

Then for any complex of sheaves $P^\bullet$,

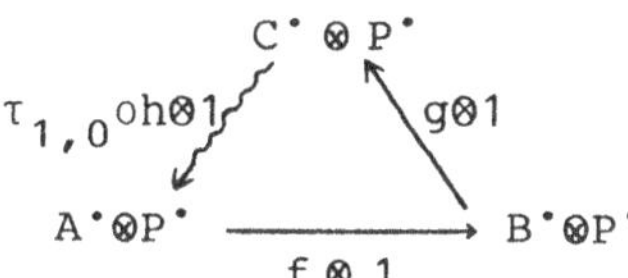

is a triangle.

<u>Proof</u>. We may assume that the sequence

$$0 \to A^\bullet \to B^\bullet \to C^\bullet \to 0$$

is a chainwise split exact sequence which comes equipped with a chainwise splitting $s: C \to B$ such that $fh = \partial s - s\partial$. This

122

gives rise to a chainwise splitting $S = s \otimes 1$ of the sequence

$$0 \longrightarrow A^{\cdot} \otimes P^{\cdot} \xrightarrow{f \otimes 1} B^{\cdot} \otimes P^{\cdot} \xrightarrow{g \otimes 1} C^{\cdot} \otimes P^{\cdot} \longrightarrow 0$$

$$(\partial S - S\partial)(x^p \otimes y^q) = \partial(sx^p \otimes y^q) - S(\partial x^p \otimes y^q + (-1)^p x^p \otimes \partial y^q) =$$

$$\partial sx^p \otimes y^q + (-1)^p sx^p \otimes \partial y^q - s\partial x^p \otimes y^q - (-1)^p sx^p \otimes y^q =$$

$$((\partial s - s\partial)x^p) \otimes y^q = (f \otimes 1)(h \otimes 1)(x^p \otimes y^q)$$

Q.E.D.

Let us call special attention to the isomorphism

$$8.11 \qquad\qquad \tau_{0,n} : M^{\cdot} \otimes \underset{\sim}{k}[n] \longrightarrow M^{\cdot}[n]$$

explicitly given by $\tau_{0,n}(x^p \otimes 1[n]) = (-1)^{np} x^p$. The effect of
tensoring the triangle 8.10 from the right with $\underset{\sim}{k}[n]$ is the
triangle

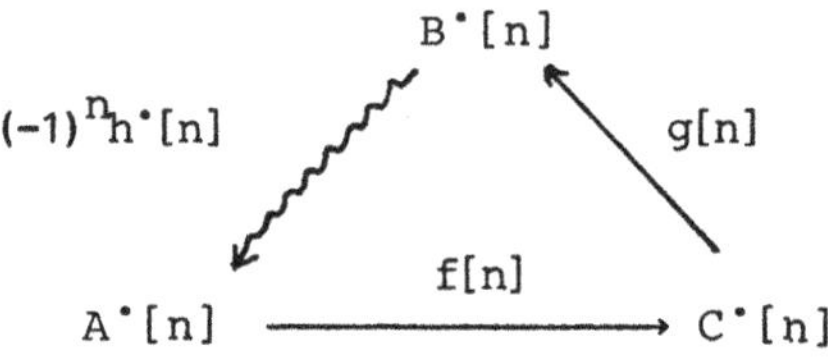

The tensor product makes $K^+(X,k)$ into a category with an internal tensor product satisfying the pentagon and hexagon axiom
of Mac Lane (1) 158(5), 180(3).

II.9 Local cohomology

In this section we consider the inclusion $i: A \to X$ of closed sets of an arbitrary topological space. The object of study is the behaviour of a sheaf in a neighbourhood of A.

Definition 9.1. For a sheaf F on X we define

$$\Gamma_A(X,F) = \{s \in \Gamma(X,F) \mid \text{Supp}(s) \subseteq A\}$$

the sections of F with support in A.

This is a left exact functor. The p'th derived functor of $\Gamma_A(X,-)$ evaluated on a sheaf F is denoted $H_A^{\,p}(X,F)$ and is called the p'th local cohomology group with support in A and coefficients F.

Proposition 9.2. Let U denote the complement of the closed subset A of X. A sheaf F on X gives rise to a long exact sequence

$$\to H_A^{\,p}(X,F) \to H^p(X,F) \to H^p(U,F) \to$$

Proof. Consider the tautological exact sequence

$$0 \to \Gamma_A(X,F) \to \Gamma(X,F) \to \Gamma(U,F)$$

and note that r_{UX} is surjective in case F is a flabby sheaf. Thus an injective resolution $F \to I^{\bullet}$ gives rise to a short exact sequence of complexes

124

$$0 \longrightarrow \Gamma_A(X,I^{\cdot}) \longrightarrow \Gamma(X,I^{\cdot}) \longrightarrow \Gamma(U,I^{\cdot}) \longrightarrow 0$$

which gives rise to a cohomology sequence.

Q.E.D.

<u>Corollary 9.3</u>. For a flabby sheaf F on X

$$H_A^{\,p}(X,F) = 0 \qquad ; \; p > 0$$

Suppose we have given two closed subsets A and B of X. For a sheaf F on X we deduce the following commutative exact diagram where U and V denotes the complements of A and B

9.4

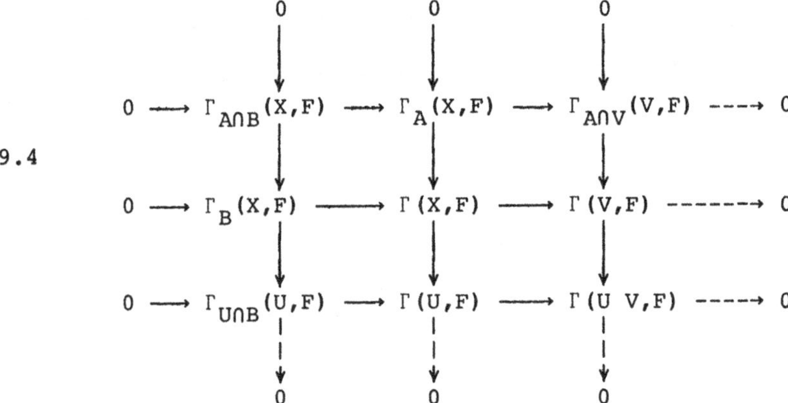

In case F is flabby this diagram gives a 3×3-diagram consisting of 6-exact sequences as it follows by inspection. If we apply the sequence above to a flabby resolution of F we get six long exact sequences related in various ways. Let us record the one resulting from the 1. row

<u>Excision exact sequence 9.5.</u>

$$\to H_{A-V}^{\,p}(X,F) \to H_A^{\,p}(X,F) \to H_{A\cap V}^{\,p}(V,F) \to$$

here A is a closed subset of X and V is an open subset. In case V contains A this degenerates into an

<u>Excision isomorphism 9.6.</u>

$$H_A^{\,p}(X,F) \overset{\sim}{\to} H_A^{\,p}(V,F) \qquad ; \qquad p \in \mathbb{Z}$$

<u>Functoriality 9.7.</u> Let A be a closed subset of X and B a closed subset of Y. A continuous map $f: X \to Y$ with $f(X-A) \subseteq Y-B$ will for any sheaf G on Y induce a commutative ladder, where $V = Y - B$ and $U = X - A$

$$\begin{array}{ccccccccc}
\longrightarrow & H_B^{\,p}(Y,G) & \longrightarrow & H^p(Y,G) & \longrightarrow & H^p(V,G) & \longrightarrow & H_B^{\,p+1}(Y,G) & \to \\
& \downarrow & & \downarrow & & \downarrow & & \downarrow & \\
\longrightarrow & H_A^{\,p}(X,f{\star}G) & \to & H^p(X,f{\star}G) & \to & H^p(U,f{\star}G) & \to & H_A^{\,p+1}(X,f{\star}G) & \to
\end{array}$$

<u>Proof</u>. Let $G \to J^{\cdot}$ and $f{\star}J^{\cdot} \to I^{\cdot}$ be injective resolutions. This gives rise to a commutative exact diagram of complexes

$$\begin{array}{ccccccccc}
0 & \longrightarrow & \Gamma_B(Y,J^{\cdot}) & \longrightarrow & \Gamma(Y,J^{\cdot}) & \longrightarrow & \Gamma(V,J^{\cdot}) & \longrightarrow & 0 \\
& & \downarrow & & \downarrow & & \downarrow & & \\
0 & \longrightarrow & \Gamma_A(X,I^{\cdot}) & \longrightarrow & \Gamma(X,I^{\cdot}) & \longrightarrow & \Gamma(U,I^{\cdot}) & \longrightarrow & 0
\end{array}$$

from which the result follows.

Q.E.D.

Interpretation in terms of Ext

Let us now pass to the framwork of k-sheaves, where k is a fixed commutative ring. As before, we consider the inclusion $i: A \to X$ of a closed subspace and the inclusion $j: U \to X$ of the complement. Recall the exact sequence 6.11

$$0 \longrightarrow j_! \underset{\sim}{k} \longrightarrow \underset{\sim}{k} \longrightarrow i_* \underset{\sim}{k} \longrightarrow 0$$

For a k-sheaf F this gives an exact sequence

9.8 $$\to \text{Ext}^p(i_* \underset{\sim}{k}, F) \to \text{Ext}^p(\underset{\sim}{k}, F) \to \text{Ext}^p(j_! \underset{\sim}{k}, F) \to$$

which we may identify with the sequence 9.2 as it follows from 7.5 and the proof of 7.3. - The formula

$$H^{\cdot}(X, k) = \text{Ext}^{\cdot}(\underset{\sim}{k}, \underset{\sim}{k})$$

allows us to introduce cup product in local cohomology according to the general principles of I.8

9.9 $$\alpha \cup \beta \in H_A^p(X, \underset{\sim}{k}) \quad ; \quad \alpha \in H^p(X, \underset{\sim}{k}), \beta \in H_A^q(X, \underset{\sim}{k})$$

Let us notice that the boudary operator of the sequence

$$\to H^q(X, \underset{\sim}{k}) \to H^q(U, \underset{\sim}{k}) \overset{\partial}{\to} H_A^{q+1}(X, \underset{\sim}{k}) \to$$

satisfies the formula, compare I.8.5

9.10 $$\partial(\alpha \cup \beta) = (-1)^p \alpha \cup \partial\beta \quad ; \alpha \in H^p(X, \underset{\sim}{k}), \quad \beta \in H^q(U, \underset{\sim}{k})$$

The same type of formula is valid for the excision sequence 9.5.
To see this, remark that the diagram 9.4 can be derived from
6.17. - Let us take the opportunity to mention that the Mayer-
Vietoris sequence 5.10 can be derived from 6.18 again with a
formula like 9.10 as a consequence. In a similar way we derive
from 6.18 a

<u>Mayer-Vietoris sequence 9.11</u>.

$$\longrightarrow H_A^q(X,k) \oplus H_B^q(X,k) \longrightarrow H_{A\cup B}^q(X,k) \overset{\partial}{\longrightarrow} H_{A\cap B}^{q+1}(X,k) \longrightarrow$$

again satisfying a relation like 9.10.

<u>The extraordinary cup product</u>

Let $i: Z \to X$ denote the inclusion of a closed subspace.
The adjunction morphism $\underset{\sim}{k} \to i_*\underset{\sim}{k}$ induces an isomorphism

$$9.12 \qquad \operatorname{Ext}^\cdot(i_*\underset{\sim}{k}, i_*\underset{\sim}{k}) \cong H^\cdot(Z, \underset{\sim}{k})$$

<u>Proof</u>. Let $\underset{\sim}{k} \to I^\cdot$ denote an injective resolution in
$Sh(Z,k)$. We have by adjunction 4.8

$$\Gamma(Z, I^\cdot) = \operatorname{Hom}(i^*\underset{\sim}{k}, I^\cdot) = \operatorname{Hom}(\underset{\sim}{k}, i_*I^\cdot) = \operatorname{Hom}(i_*\underset{\sim}{k}, i_*I^\cdot)$$

where we have used Lemma 5.5. Conclusion by the fact that $i_*I^\cdot$
is an injective resolution of $i_*\underset{\sim}{k}$ in $Sh(X,k)$.

Q.E.D.

If we combine formula 9.12 with the fact that

$$9.13 \qquad \text{Ext}^{\cdot}(i_{*}\underset{\sim}{k},F) = H_{Z}^{\cdot}(X,F)$$

we can introduce the <u>extraordinary cup product</u>

$$9.14 \qquad \beta \cup \gamma \in H_{Z}^{q+r}(X,F) \qquad ; \ \beta \in H_{Z}^{q}(X,F), \ \gamma \in H^{r}(Z,k)$$

The two cup products are interrelated by the following formulas.
We let $r: H_{Z}^{\cdot}(X,k) \to H^{\cdot}(X,k)$ denote the restriction map

$$9.15 \qquad \alpha \cup (\beta \cup \gamma) = (\alpha \cup \beta) \cup \gamma$$

$$\alpha \in H^{p}(X,k), \ \beta \in H_{Z}^{q}(X,k), \ \gamma \in H^{r}(Z,k)$$

$$9.16 \qquad r(\alpha) \cup \beta = \alpha \cup i^{*}r(\beta)$$

$$\alpha \in H_{Z}^{p}(X,k), \ \beta \in H_{Z}^{q}(X,k)$$

$$9.17 \qquad \alpha \cup \beta = (-1)^{pq}\beta \cup i^{*}(\alpha)$$

$$\alpha \in H^{p}(X,k), \ \beta \in H_{Z}^{q}(X,k)$$

The first formula is obvious, the second formula will be verified below while the third formula will be proved in the next section.

<u>Verification of 9.16</u>. We shall work in the category $D^{+}(X,k)$. Let $I^{\cdot}$ be an injective resolution of $\underset{\sim}{k}$, $J^{\cdot}$ an injective resolution of $i_{*}\underset{\sim}{k}$ and let $\pi: I^{\cdot} \to J^{\cdot}$ represent the adjunction morphism $\underset{\sim}{k} \to i_{*}\underset{\sim}{k}$. According to 9.12 we have that

$$[\pi,1]: \ [J^{\cdot},J^{\cdot}[p]] \ \longrightarrow \ [I^{\cdot},J[p]] \qquad ; \quad p \in \mathbb{Z}$$

is an isomorphism. The map $\ i^*: H^p(X,k) \to H^p(Z,k)\ $ can be described as the composite

$$[I^{\cdot},I^{\cdot}[p]] \ \xrightarrow{\ [1,\pi]\ } \ [I^{\cdot},J^{\cdot}[p]] \ \xrightarrow{\ [\pi,1]^{-1}\ } \ [J^{\cdot},J^{\cdot}[p]]$$

while the restriction $\ r: H_Z^{\ p}(X,k) \to H^p(X,k)\ $ is represented

$$[\pi,1]: \ [J^{\cdot},I^{\cdot}[p]] \ \longrightarrow \ [I^{\cdot},I^{\cdot}[p]]$$

Consider $\ \alpha \in [J^{\cdot},I^{\cdot}[p]]\ $ and $\ \beta \in [J^{\cdot},I^{\cdot}[q]]$. We have $\ r(\alpha) = \alpha\pi$, $r(\beta) = \beta\pi\ $ and $\ i^*r(\alpha) = \pi\alpha$. This gives

$$\beta \cup i^*r(\alpha) \ = \ \beta[p](\pi\alpha) \ = \ (\beta\pi[p])\alpha \ = \ r(\beta) \cup \alpha$$

which is the desired result.

Q.E.D.

II.10 Cross products

In this section we shall perform a considerable extension of the cup product introduced in the previous section. We shall work with a fixed commutative ring k. Given a topological space X, two closed subspaces A and B and k-sheaves E and F on X. We shall construct a <u>cup product</u>

$$10.1 \qquad \alpha \cup \beta \in H_{A \cap B}^{p+q}(X, E \otimes F) \qquad ; \alpha \in H_A^p(X,E), \beta \in H_B^q(X,F)$$

Let us list some of its properties. First of all the following diagram is <u>commutative</u>, here U denotes the complement of A

$$10.2$$

$$\begin{array}{ccccccc}
H_A^p(X,E) & \longrightarrow & H^p(X,E) & \longrightarrow & H^p(U,E) & \overset{\partial}{\longrightarrow} & H_A^{p+1}(X,E) \\
\downarrow{\scriptstyle \cup \beta} & & \downarrow{\scriptstyle \cup \beta} & & \downarrow{\scriptstyle \cup \beta} & & \downarrow{\scriptstyle \cup \beta} \\
H_{A \cap B}^p(X,E \otimes F) & \to & H_B^p(X,E \otimes F) & \to & H_{U \cap B}^p(X,E \otimes F) & \overset{\partial}{\to} & H_{A \cap B}^{p+1}(X,E \otimes F)
\end{array}$$

The cup product is <u>anticommutative</u>

$$10.3 \qquad \alpha \cup \beta = (-1)^{pq} \beta \cup \alpha \qquad ; \quad \alpha \in H_A^p(X,E), \quad \beta \in H_B^q(X,F)$$

In the sense that the canonical symmetry $E \otimes F \overset{\sim}{\to} F \otimes E$ transforms $\alpha \cup \beta$ into $(-1)^{pq} \beta \cup \alpha$. - Given a continuous map $f: W \to X$ and closed subsets C and D of W with $f(W-C) \subseteq X-A$ and $f(W-D) \subseteq X-B$, then

$$10.4 \qquad f^*(\alpha \cup \beta) = f^*\alpha \cup f^*\beta \qquad ; \quad \alpha \in H_A^p(X,k), \quad \beta \in H_B^q(B,k)$$

$$10.5 \qquad \alpha \cup (\beta \cup \gamma) = (\alpha \cup \beta) \cup \gamma$$

$$\alpha \in H_A^p(X,E), \quad \beta \in H_B^q(X,F), \quad \gamma \in H_C^r(X,G)$$

Given a sequence of k-sheaves

$$0 \longrightarrow E \longrightarrow F \longrightarrow G \longrightarrow 0$$

which is <u>pointwise split exact</u>, i.e. whose localization at each point of X is a split exact sequence of k-modules. For any k-sheaf D on X we have

10.6 $\qquad \partial(\alpha \cup \beta) = (\partial\alpha) \cup \beta \qquad ; \ \alpha \in H_A^p(X,G), \ \beta \in H_B^q(X,D)$

Cross products

Let A be a closed subset of the topolocial space X and B a closed subset of the topological space Y. For a sheaf F on X and a sheaf G on Y we consider

$$p^*: H_A^{\,\cdot}(X,F) \to H_{A\times Y}^{\,\cdot}(X\times Y, p^*F) \qquad q^*: H_B^{\,\cdot}(Y,G) \to H_{X\times B}^{\,\cdot}(X\times Y, q^*G)$$

We can now define a cohomology class, the <u>cross product</u>

$$\alpha \times \beta \in H_{A\times B}^{p+q}(X\times Y, p^*F \otimes q^*G) \qquad ; \qquad \alpha \in H_A^p(X,F), \quad \beta \in H_B^q(Y,G)$$

by the following formula involving the cup product

10.7 $\qquad\qquad \alpha \times \beta = p^*\alpha \cup q^*\beta$

In case $Y = X$ we let $\Delta: X \to X \times X$ denote the diagonal map, and

132

we deduce from 10.4 that

$$10.8 \qquad \alpha \cup \beta = \Delta^*(\alpha \times \beta) \qquad ; \qquad \alpha \in H_A^p(X,F), \ \beta \in H_B^q(Y,G)$$

We shall now construct the cup product. The basic idea is to construct a category with an internal tensor product in which we can represent cohomology classes as arrows and cup product as tensor product of arrows.

An auxilary category

Let X denote a toplogical space and k a commutative ring. The category $K^+(X,k)$ comes equipped with an internal tensor product obeying certain rules II.8. This is certainly not the case for $D^+(X,k)$ being the homotopy category of injective sheaves. Alternatively $D^+(X,k)$ can be obtained from $K^+(X,k)$ by inverting all quasi-isomorphisms XI.2.6. This is a perfect description when k is a field since the system of quasi-isomorphisms is stable under $\otimes$. In general we shall invert a smaller class of quasi-isomorphisms namely the <u>pointwise homotopy equivalences</u>, i.e. morphisms $f: R^\cdot \to S^\cdot$ in $K^+(X,k)$ such that for each point $x \in X$, $f_x: R_x^\cdot \to S_x^\cdot$ is a homotopy equivalence of complexes of k-modules. The class of pointwise homotopy equivalences is stable under tensor product, moreover

<u>Lemma 10.9.</u> For any bounded below complex of k-sheaves $F^\cdot$ there exists a pointwise homotopy equivalence $F^\cdot \to S^\cdot$ where $S^\cdot$ is bounded below complex of flabby k-sheaves.

$\underline{\text{Proof}}$. Recall from 3.6 the Godement construction which to a sheaf F assigns the Godement resolution $C^\cdot F$. The construction is functorial, thus we may consider $C^\cdot F^\cdot$ as a double complex I.11. Let us consider the double complex $F^{\cdot\cdot}$ given by

$$F^{pq} = \begin{cases} F^q & \text{for} \quad p = 0 \\ 0 & \text{for} \quad p \neq 0 \end{cases}$$

The Godement construction furnishes a morphism $F^{\cdot\cdot} \to C^\cdot(F^\cdot)$ of double complexes which induces a morphism of single complexes

$$F^\cdot \longrightarrow \text{Tot}^\cdot(C^\cdot(F^\cdot))$$

This is a pointwise homotopy equivalence: In fact for $x \in X$ a homotopy inverse $C^\cdot(F^\cdot)_x \to F^{\cdot\cdot}_x$ is furnished by the canonical local retractions, compare the remarks after 3.10. Finally use the formula I.11.6 to construct the needed homotopies.

Q.E.D.

The class P of pointwise homotopy equivalences satisfies the condition FR 1-5 of Verdier compare XI.1,2,6. The localized category will be denoted

$$G^+(X,k) = P^{-1}K^+(X,k)$$

By the fact that P is stable under tensor product, the category $G^+(X,k)$ comes equipped with an internal tensor product. The following two lemmas reveal the role $G^+(X,k)$ is going to play.

$\underline{\text{Lemma } 10.10}$. Let D be a k-sheaf with the property that $\text{Ext}_k^p(D,S) = 0$ for all $p \geq 1$ and all flabby sheaves S. Then for any k-sheaf F and any $p \in \mathbb{Z}$

$$\text{Ext}_k^p(D,F) = \text{Hom}(D,F[p])$$

where Hom is calculated in $G^+(X,k)$.

$\underline{\text{Proof}}$. Throughout the proof the functor Hom is calculated in $G^+(X,k)$. Let us first prove that for a complex of flabby sheaves $S^{\cdot}$ in $K^+(X,k)$ then

$$[D,S^{\cdot}] = \text{Hom}(D,S^{\cdot})$$

Suppose first that $f \in [D,S^{\cdot}]$ represents the morphism zero $D \to S^{\cdot}$ in $G^+(X,k)$. This means that we can find $s: S^{\cdot} \to T^{\cdot}$ in P such that $sf = 0$ in $K^+(X,k)$. By Lemma 10.9 we may assume that $T^{\cdot}$ is a complex of flabby sheaves. The quasi-isomorphism $s: S^{\cdot} \to T^{\cdot}$ is a morphism between complexes of $\text{Hom}(D,-)$ acyclic objects. Thus we conclude from the acyclicity theorem I.7.5 that f is zero.

To see that our map is surjective notice that the general element of $\text{Hom}(D,S^{\cdot})$ is a fraction $D \xrightarrow{f} T \xleftarrow{s} S$ where s is in P. Again by 10.9 we may assume that T consists of flabby sheaves. We conclude from the acylicity theorem that f may be factored through s.

To prove the stated result choose $s: F \to S^{\cdot}$ in P where $S^{\cdot}$ is a complex of flabby sheaves and choose an injective resolution $S^{\cdot} \to I^{\cdot}$. We have

$$\text{Hom}(D,F[p]) \xrightarrow{\sim} \text{Hom}(D,S^{\bullet}[p]) = [D,S^{\bullet}[p]] = [D,I^{\bullet}[p]]$$

The last group is $\text{Ext}_k^p(D,F)$.

Q.E.D.

Lemma 10.11. Let $h: W \to X$ denote the inclusion of a locally closed subspace. Then

$$\text{Ext}^p(h_* \underset{\sim}{k}, S) = 0 \qquad , \; p \geq 1$$

for any <u>flabby</u> k-sheaf S.

Proof. Let us write $W = A \cap V$ where A is a closed subset of and V an open subset of X. If we let U denote the complement of A in X we have an exact sequence of k-sheaves, compare the first column of 6.17.

$$0 \longrightarrow \underset{\sim}{k}_{U \cap V} \longrightarrow \underset{\sim}{k}_V \longrightarrow \underset{\sim}{k}_W \longrightarrow 0$$

from which we derive the isomorphism, compare 9.4

$$\text{Ext}_k^{\bullet}(h_* \underset{\sim}{k}, F) = H_{A \cap V}^{\bullet}(V,F)$$

The result follows from 9.3.

Q.E.D.

Let us record an important canonical isomorphism. For locally closed subsets A and B of X we have

$$10.12 \qquad\qquad \underset{\sim}{k}_{A \cap B} = \underset{\sim}{k}_A \otimes \underset{\sim}{k}_A$$

136

Let us now consider closed subsets A and B of X and k-sheaves
E and F. Cohomology classes $\alpha \in H_A^p(X,E)$ and $\beta \in H_B^q(X,F)$ may
be viewed as morphisms in $G^+(X,k)$

$$\alpha: \underset{\sim}{k}_A \to E[p] \qquad \beta: \underset{\sim}{k}_B \to F[q]$$

We then define the <u>cup product</u> by the formula

$$10.13 \qquad \alpha \cup \beta = \tau_{p,q} \circ \alpha \otimes \beta$$

taking the isomorphism 10.12 into account.

We shall now proceed to establish the formulas 10.1-6. The
basic principle is simply to write up a suitable commutative diagram in the derived category $G^+(X,k)$.

<u>Proof of 10.3</u>. According to 8.7 we have the following commuta-
tive diagram

$$\begin{array}{ccccccc}
\underset{\sim}{k}_{A\cap B} & \xrightarrow{\sim} & \underset{\sim}{k}_A \otimes \underset{\sim}{k}_B & \longrightarrow & E[p] \otimes F[q] & \xrightarrow{\tau_{p,q}} & E \otimes F[p+q] \\
\downarrow{\scriptstyle 1} & & \downarrow{\scriptstyle \sigma} & & \downarrow{\scriptstyle \sigma} \quad {\scriptstyle (-1)^{pq}}\downarrow & & \downarrow{\scriptstyle \sigma[p+q]} \\
\underset{\sim}{k}_{B\cap A} & \longrightarrow & \underset{\sim}{k}_B \otimes \underset{\sim}{k}_A & \longrightarrow & F[q] \otimes E[p] & \xrightarrow{\tau_{q,p}} & F \otimes E[p+q]
\end{array}$$

<u>The cup product extends 9.9</u> as it follows from the following
commutative diagram

$$\begin{array}{ccccc}
\underset{\sim}{k} \otimes \underset{\sim}{k}_A & \xrightarrow{1\otimes\beta} & \underset{\sim}{k} \otimes \underset{\sim}{k}[q] & \xrightarrow{\alpha\otimes 1} & \underset{\sim}{k}[p] \otimes \underset{\sim}{k}[q] \\
\downarrow & & \downarrow{\scriptstyle \tau_{0,q}} & & \downarrow{\scriptstyle \tau_{p,q}} \\
\underset{\sim}{k}_A & \xrightarrow{\beta} & \underset{\sim}{k}[q] & \xrightarrow{\alpha[q]} & \underset{\sim}{k}[p+q]
\end{array}$$

Proof of formula 9.17. Let us first establish the following formula, where $i: A \to X$ denotes the inclusion

$$\beta \cup i^*\alpha = \beta \cup \alpha \quad ; \alpha \in H^p(X,k), \beta \in H_A^q(X,k)$$

here the left hand side is defined in 9.14 and the right hand side is defined in 10.13. This can be read off the commutative diagram

$$\begin{array}{ccccc}
\underset{\sim}{k}_A \otimes k & \xrightarrow{1\otimes\alpha} & \underset{\sim}{k}_A \otimes k[p] & \xrightarrow{\beta\otimes 1} & \underset{\sim}{k}[q] \otimes \underset{\sim}{k}[p] \\
\downarrow & & \downarrow{}^{\tau_{0,p}} & & \downarrow{}^{\tau_{p,q}} \\
\underset{\sim}{k}_A & \xrightarrow{i_*i^*\alpha} & \underset{\sim}{k}_A[p] & \xrightarrow{\beta[p]} & \underset{\sim}{k}[p+q]
\end{array}$$

Formula 9.17 is now a consequence of 10.3.

Proof of 10.6. The pointwise, split exact sequence

$$0 \to D \to F \to G \to 0$$

gives rise to a triangle in $G^+(X,k)$. Let $\partial: G \to E[1]$ denote the third side of the triangle. We leave it to the reader to check that the accompanying diagram is commutative.

$$\begin{array}{ccc}
\underset{\sim}{k}_A \otimes \underset{\sim}{k}_B & \xrightarrow{\quad 1 \quad} & \underset{\sim}{k}_A \otimes \underset{\sim}{k}_B \\
\downarrow{}^{\alpha\otimes 1} & & \downarrow{}^{\partial[p]\alpha\otimes 1} \\
G[p] \otimes \underset{\sim}{k}_B & \xrightarrow{\partial[p]\otimes 1} & E[p+1] \otimes \underset{\sim}{k}_B \\
\downarrow{}^{1\otimes\beta} & & \downarrow{}^{1\otimes\beta} \\
G[p] \otimes D[q] & \xrightarrow{\partial[p]\otimes 1} & E[p+1] \otimes D[q] \\
\downarrow{}^{\tau_{p,q}} & & \downarrow{}^{\tau_{p+1,q}} \\
G \otimes D[p+q] & \xrightarrow{\partial\otimes 1[p+q]} & E \otimes D[p+q+1]
\end{array}$$

138

<u>Proof of 10.2.</u> The short exact sequence

$$0 \longrightarrow \underset{\sim}{k}_U \longrightarrow \underset{\sim}{k} \longrightarrow \underset{\sim}{k}_A \longrightarrow 0$$

is pointwise split exact. Let $\partial: \underset{\sim}{k}_A \to \underset{\sim}{k}_U[1]$ denote the characteristic arrow. If we tensor the sequence above with $\underset{\sim}{k}_B$ we get the sequence

$$a \longrightarrow \underset{\sim}{k}_{U \cap B} \longrightarrow \underset{\sim}{k}_B \longrightarrow \underset{\sim}{k}_{A \cap B} \longrightarrow 0$$

The characteristic arrow of this sequence may be identified with $\partial \otimes 1: \underset{\sim}{k}_A \otimes \underset{\sim}{k}_B \to \underset{\sim}{k}_U[1] \otimes \underset{\sim}{k}_B$.

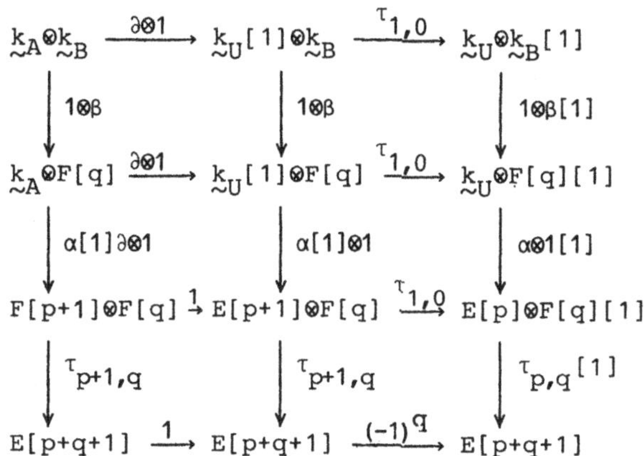

From the commutative diagram above we deduce

$$(\partial \alpha) \cup \beta = (-1)^q \partial(\alpha \cup \beta) \qquad ; \ \alpha \in H_A^p(X,E), \ \beta \in H_B^q(Y,F)$$

Multiply this formula with $(-1)^{p+1}$ to get the desired result, compare I.8.5.

$$\text{Q.E.D.}$$

$\underline{\text{Proposition 10.14}}$. Let $\underset{\sim}{k} \to C^{\cdot}$ be an acyclic resolution in $Sh(X,k)$ which is a pointwise homotopy equivalence. Given a morphism of complexes $\wedge: C^{\cdot} \otimes C^{\cdot} \to C^{\cdot}$ which makes the accompanying diagram commutative. Given

$$
\begin{array}{ccc}
\underset{\sim}{k} \otimes \underset{\sim}{k} & \longrightarrow & \underset{\sim}{k} \\
\downarrow & & \downarrow \\
C^{\cdot} \otimes C^{\cdot} & \xrightarrow{\wedge} & C^{\cdot}
\end{array}
$$

$$a \in \Gamma(X, C^p) \quad \text{with} \quad \partial a = 0 \quad \text{representing} \quad \alpha \in H^p(X,k)$$
$$b \in \Gamma(X, C^q) \quad \text{with} \quad \partial b = 0 \quad \text{representing} \quad \beta \in H^q(X,k)$$

then $a \wedge b$ represents $\alpha \cup \beta$

$\underline{\text{Proof}}$. Consider the following commutative diagram in $G^+(X,k)$

$$
\begin{array}{ccccccc}
\underset{\sim}{k} & \xrightarrow{\alpha \otimes \beta} & \underset{\sim}{k}[p] \otimes \underset{\sim}{k}[q] & \xrightarrow{\tau_{p,q}} & \underset{\sim}{k} \otimes \underset{\sim}{k}[p+q] & \longrightarrow & \underset{\sim}{k}[p+q] \\
\downarrow & & \downarrow & & \downarrow & & \downarrow \\
\underset{\sim}{k} & \xrightarrow{a \otimes b} & C^{\cdot}[p] \otimes C^{\cdot}[q] & \xrightarrow{\tau_{p,q}} & C^{\cdot} \otimes C^{\cdot}[p+q] & \xrightarrow{\wedge[p+q]} & C^{\cdot}[p+q]
\end{array}
$$

Notice $\tau_{p,q}(a \otimes b) = (-1)^{0q} a \otimes b = a \otimes b$ and the result follows.

Q.E.D.

II.11 Flat sheaves

Let k denote a fixed commutative ring. A k-module F is called __flat__ if any exact sequence

$$0 \to R \to S \to T \to 0$$

of k-modules is transformed into an exact sequence

$$0 \longrightarrow R{\otimes}F \longrightarrow S{\otimes}F \longrightarrow T{\otimes}F \longrightarrow 0$$

Let us notice that a projective k-module is flat, and that any direct limit of flat modules is flat as it follows from 8.1. Let us mention a theorem of D.Lazard to effect that any flat module can be realized as a direct limit of finitely generated free modules, Bourbaki (2) A.X.14. The reader may use this to simplify some of the proofs in these notes.

__Lemma 11.1__. Given an exact sequence

$$0 \longrightarrow P_n \longrightarrow P_{n-1} \longrightarrow \cdots P_0 \longrightarrow F \longrightarrow 0$$

of k-modules. If $F, P_0, \ldots, P_{n-1}$ are flat modules then P_n is a flat module.

__Proof__. By splitting the sequence into short exact sequences one sees that it suffices to treat the case $n = 1$. This case is easy to treat by means of I.1.3.

Q.E.D.

<u>Theorem 11.2</u>. Let F. be a bounded above complex of flat k-modules and R. a bounded above complex of k-modules with H.(R.) = 0. Then H.(F.⊗R.) = 0.

<u>Proof</u>. We shall change to the "upper dot" notation in order not to mess up the truncation notation I.5.7. - It suffices to prove the following statement for all $n \in \mathbb{Z}$:

"Any complex $R^{\cdot}$ with $H^{\cdot}(R^{\cdot}) = 0$ and $R^i = 0$ for $i > n$ has $H^0(R^{\cdot}\otimes F^{\cdot}) = 0$ ". - This is done by increasing induction on n, noticing that the statement is obvious for $n \ll 0$. In order to prove "$n \Rightarrow n+1$" consider the exact sequence

$$0 \longrightarrow \tau_{\leq n}R^{\cdot} \longrightarrow R^{\cdot} \longrightarrow \tau_{\geq n}R^{\cdot} \longrightarrow 0$$

The resulting short exact sequence of complexes

$$0 \longrightarrow (\tau_{\leq n}R^{\cdot})\otimes F^{\cdot} \longrightarrow R^{\cdot}\otimes F^{\cdot} \longrightarrow (\tau_{\geq n}R^{\cdot})\otimes F^{\cdot} \longrightarrow 0$$

gives an exact sequence

$$H^0((\tau_{\leq n}R^{\cdot})\otimes F^{\cdot}) \longrightarrow H^0(R^{\cdot}\otimes F^{\cdot}) \longrightarrow H^0(\tau_{\geq n}R^{\cdot})\otimes F^{\cdot})$$

from which the result follows.

Q.E.D.

<u>Corollary 11.3</u>. Let R. → S. be a quasi-isomorphism of bounded above complexes of k-modules and F. a bounded above complex of flat k-modules. Then

$$R.\otimes F. \longrightarrow S.\otimes F.$$

142

is a quasi-isomorphism.

Proof. Consider the mapping cone of $R. \to S.$ and combine 11.2 with 8.10.

Q.E.D.

Corollary 11.4. Let $F.$ be a bounded above complex of flat modules with $H.(F.) = 0$. Then for any bounded above complex $R.$ we have $H.(F.\otimes R.) = 0$.

Proof. Choose a projective resolution $P. \to R.$ and notice that $P.\otimes F. \to R.\otimes F.$ is a quasi-isomorphism 11.3 and apply 11.2 to the complex $R.\otimes F.$

Q.E.D.

Let us briefly discuss the functors Tor_i. Given k-modules M and N. Choose projective resolutions $P. \to M$ and $Q. \to N$ and consider the following two quasi-isomorphisms, 11.3

$$P.\otimes N \Leftarrow P.\otimes Q. \Rightarrow M\otimes Q.$$

This makes sense to the following definition

11.5 $\qquad\qquad \mathrm{Tor}_i(M,N) = H_i(P.\otimes N) = H_i(M\otimes Q.) = H_i(P.\otimes Q.)$

It follows from 11.4 that we can use <u>resolutions $P.$</u> <u>and $Q.$ of flat modules</u> in the definition 11.5.

Let us now consider a topological space X. We shall generalize the discussion above to $K^-(X,k)$, the homotopy category of bounded below complexes of k-sheaves on X.

Definition 11.6. A k-sheaf F on X is called _flat_ if for every point x of X the stalk F_x is a flat k-module.

Proposition 11.7. Let $R. \to S.$ be a quasi-isomorphism in $K^-(X,k)$. For any complex $F.$ of flat k-sheaves in $K^-(X,k)$

$$R.\otimes F. \longrightarrow S.\otimes F.$$

is a quasi-isomorphism.

Proof. Follows from 11.3 by localization 8.3.

Q.E.D.

Theorem 11.8. For any $R.$ in $K^-(X,k)$ there exists a quasi-isomorphism $F. \to R.$ where $F.$ is a bounded above complex of flat k-sheaves (a _flat resolution of $R.$_).

Proof. For any k-sheaf R there exists an epimorphism $F \to R$ where F is a flat k-sheaf, 7.4. In general, follow a procedure dual to the one used in the proof of I.6.1 see also I.9.

Q.E.D.

$\underline{\text{Proposition 11.9}}$. Given quasi-isomorphisms E. → R. and
F. → R. in $K^-(X,k)$ where P. and Q. are complexes of
flat sheaves.Then there exists a
commutative diagram consisting of
quasi-isomorphisms in $K^-(X,k)$
where D. is a complex of flat
sheaves.

$$
\begin{array}{ccc}
D. & \longrightarrow & E. \\
\downarrow & & \downarrow \\
F. & \longrightarrow & P.
\end{array}
$$

$\underline{\text{Proof}}$. According to the proof of XI.2.4 we can construct
such a diagram consisting of quasi-isomorphisms, but no conditions on D. . Combine this with 11.8 to get the result.

$$\text{Q.E.D.}$$

$\underline{\text{Definition 11.10}}$. Let P. and Q. be complexes in $K^-(X,k)$.
Choose flat resolutions E.↠P. and F. → G. . We put

$$
P. \overset{L}{\otimes} G. = E. \otimes F.
$$

and define the Tor-sheaves

$$
Tor_i^{\,k}(P.,Q.) = H_i(E. \otimes_k F.)
$$

The constructions above may be interpreted in terms of the
derived category $D^-(X,k)$ introduced in XI.2.6°. According to
11.8 the derived category $D^-(X,k)$ is equivalent to the category
defined from the homotopy category of bounded above complexes of
flat sheaves by inverting all quasi-isomorphisms, see XI.6. –
This makes it evident that $D^-(X,k)$ comes equipped with an $\underline{\text{internal}}$
$\underline{\text{tensor product}}$, which allow us to interprete the formulas from
II.8 in $D^-(X,k)$.

II.12 $Hom(E,F)$

Let X denote a topological space and k a commutative ring. We shall work with the category $Sh(X,k)$. - Two sheaves E and F give rise to a presheaf

$$U \mapsto Hom(j^*E, j^*F) \quad ; \; j: U \to X$$

which is easily seen to be <u>a sheaf</u>. This will be denoted $Hom(E,F)$. Thus with the notation above

12.1 $$\Gamma(U, Hom(E,F)) = Hom(j^*E, j^*F)$$

Given sheaves E,F,G on X we have a fundamental isomorphism

12.2 $$Hom(E \otimes F, G) = Hom(E, Hom(F,G))$$

<u>Proof</u>. It suffices to prove a formula of the form

$$Hom(E \otimes F, G) = Hom(E, Hom(F,G))$$

where E,F and G are presheaves on X. This follows from

$$Hom(\Gamma(U,E) \otimes \Gamma(U,F), \Gamma(U,G)) = Hom(\Gamma(U,E), Hom(\Gamma(U,F), \Gamma(U,G)))$$

which we consider well known from algebra.

Q.E.D.

<u>Corollary 12.3</u>. If F is a flat k-sheaf and G an injective k-sheaf, then $Hom(F,G)$ is an injective k-sheaf.

In case $F^{\cdot}$ and $G^{\cdot}$ are complexes we define $Hom^{\cdot}(F^{\cdot}, G^{\cdot})$ by

12.4 $$Hom^n(F^{\cdot}, G^{\cdot}) = \prod_{p \in \mathbb{Z}} Hom(F^n, G^{n+p})$$

the differential being similar to that of I.4.3.

III. Cohomology with Compact Support

<u>III.1</u>   <u>Locally compact spaces</u>

Let X denote a locally compact space, i.e. a Hausdorff topological space in which every point has a compact neighbourhood.
Let us prove two simple facts about locally compact spaces.

1.1   Any neighbourhood of a point $x \in X$ contains a

         compact neighbourhood of x.

<u>Proof</u>. It suffices to prove this for an open neighbourhood
V of x which is contained in a compact subset K. Let us
consider the compact set K-V. For each point $y \in K-V$ choose
disjoint open neighbourhoods V_y of y and U^y of x. The
covering (V_y) of K-V can be refined to a finite covering $(V_y)_{y \in S}$
of K-V. The compact set $K-U_{x \in S}V_y$ is contained in V and
contains $\bigcap_{y \in S} U^y \cap V$.

                                                          Q.E.D.

1.2   For any open set U containing a compact set K,

         there exists a compact set $L \subseteq U$ with $K \subseteq \overset{\bullet}{L}$.

<u>Proof</u>. For each $x \in K$ choose a compact neighbourhood N_x
of x contained in U. The covering $(\overset{\bullet}{N}_x)_{x \in K}$ of K can be

refined to a finite covering $(\dot{N}_x)_{x \in S}$ of K. Put $L = \bigcup_{x \in S} N_x$.

Q.E.D.

<u>Definition 1.3</u>. For a sheaf F on the locally compact space X put

$$\Gamma_c(X,F) = \{s \in \Gamma(X,F) \mid \mathrm{Supp}(s) \text{ is compact}\}$$

the sections of F over X with <u>compact support</u>.

This defines a left exact functor

$$\Gamma_c(X,-): \quad \mathrm{Sh}(X) \longrightarrow Ab$$

The i'th derived functor of this evaluated on the sheaf F will be denoted $H_c^i(X,F)$ and is called the i'th <u>cohomology group with compact support with coefficients in F</u>.

Consider a continuous map $f: X \longrightarrow Y$ between locally compact spaces which is <u>proper</u>, i.e. such that the inverse image of any compact subset of Y is a compact subset of X. For a sheaf G on Y, the adjunction map $a: \Gamma(Y,G) \to \Gamma(X,f*G)$ will transform a section of G with compact support into a section of f*G with compact support and thus induce a natural transformation

1.4 $$a: \Gamma_c(Y,G) \longrightarrow \Gamma_c(X,f*G)$$

Since f* is exact we can extend this to

1.5 $$f*: H_c^{\cdot}(Y,G) \longrightarrow H_c^{\cdot}(X,f*G)$$

by choosing an injective resolution $G \to J^\cdot$ and an injective resolution $f^*J^\cdot \to I^\cdot$. The composite

$$\Gamma_c(Y,J^\cdot) \overset{a}{\to} \Gamma_c(X,f^*J^\cdot) \to \Gamma_c(X,I^\cdot)$$

represents 1.5 on the chain level, noticing that $I^\cdot$ is an injective resolution of f^*G.

Proceeding as in II.7.9 we can derive from an exact sequence of sheaves on Y

$$0 \longrightarrow E \longrightarrow F \longrightarrow G \longrightarrow 0$$

a commutative ladder of abelian groups

1.6
$$\begin{array}{ccccccccc}
\longrightarrow & H_c^p(Y,E) & \longrightarrow & H_c^p(Y,F) & \longrightarrow & H_c^p(Y,G) & \longrightarrow & H_c^{p+1}(Y,E) & \longrightarrow \\
& \downarrow & & \downarrow & & \downarrow & & \downarrow & \\
\longrightarrow & H_c^p(X,f^*E) & \longrightarrow & H_c^p(X,f^*F) & \longrightarrow & H_c^p(X,f^*G) & \longrightarrow & H_c^{p+1}(X,f^*E) & \longrightarrow
\end{array}$$

Closed subspaces

Consider in particular the inclusion $i: Z \longrightarrow X$ of a closed subspace. A sheaf E on Z gives rise to a canonical isomorphism, compare II.5.4

1.7
$$H_c^\cdot(Z,E) = H_c^\cdot(X,i_*E)$$

For closed subspaces A and B of X the exact sequence II.5.7 yields a

Mayer-Vietoris sequence 1.8

$$\longrightarrow H_c^p(A \cup B,F) \longrightarrow H_c^p(A,F) \oplus H_c^p(B,F) \longrightarrow H_c^p(A \cap B,F)$$

III.2 Soft sheaves

Let X denote a locally compact space. We shall introduce
an important class of $\Gamma_c(X,-)$-acyclic sheaves on X.

Definition 2.1. A sheaf S on X is called <u>soft</u> if whenever there is given data

1)   a compact subset K of X

2)   an open subset U of X containing K

3)   a section s of S over U

then there exists a section t of S over X, such that s
and t have the same restriction to some open neighbourhood of
K contained in U.

The following proposition gives rise to an alternative
formulation of softness.

Theorem 2.2. Let $i: Z \to X$ denote the inclusion of a compact
subset into a locally compact space X. For any sheaf F on X,
there is a canonical isomorphism

$$\varinjlim \Gamma(U,F) \xrightarrow{\sim} \Gamma(Z,i^*F)$$

the limit is taken over all open $U \supseteq Z$.

Proof. Notice that the canonical map

$$\varinjlim \Gamma(U,F) \to \Gamma(Z,i^*F)$$

is injective for trivial reasons. To prove that the map is surjective consider $s \in \Gamma(Z,i^*F)$. For each point z of Z choose

a compact neighbourhood K^z of z in Z and an open subset
U^z of X containing K^z and $s^z \in \Gamma(U^z, F)$ with $(s^z)_x = s_x$
for all $x \in K^z$. Since Z is compact, finitely many K^z's
will cover Z. This leaves us with the following problem:

Given compact subsets K and L of X, sections s
and t of F over open subsets containing K and L respectively, whose stalks agree on K ∩ L. Find a section of F
over an open subset of X containing K ∪ L which agrees with
s in a neighbourhood of K and t in a neighbourhood of L.

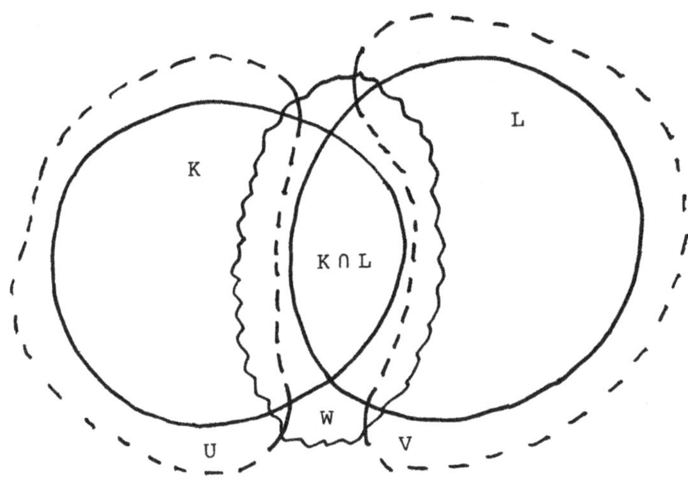

Choose an open neighbourhood W of K ∩ L in X such that s
and t have the same restriction to W. Choose disjoint open
subsets U and V of X containing L-W and K-W respectively, 1.2. Finally use the sheaf axiom to construct a section of
F over V ∪ U ∪ W which extends s and t.

Q.E.D.

This theorem shows that <u>a sheaf S on X is soft if</u> <u>and only if for any compact subset K of X, the restriction</u> <u>map $\Gamma(X,S) \to \Gamma(K,S)$ is surjective.</u> In the rest of this section we shall utilize this formulation which is technically easier to work with once we make two remarks on sections over closed subsets.

<u>Gluing sections 2.3</u>. Let Z and W be two closed subsets of a topolocial space X and F a sheaf on X. Given $s \in \Gamma(Z,F)$ and $t \in \Gamma(W,F)$ which have the same restrictions to $W \cap Z$ then there exists one and only one section of F over $Z \cup W$ with restriction s to Z and restriction t to W, as it follows from II.5.6.

<u>Extension by zero 2.4</u>. Let Z be a closed subset of the topological space X and F a sheaf on X. Any section s of F over Z with restriction 0 to ∂Z can be extended to a section of F over X with restriction 0 to the complement of Z in X.

To see this let us notice that $\overset{\bullet}{Z}$ and X-Supp(s) form an open covering of X, which we can use to glue the restriction of s to $\overset{\bullet}{Z}$ and the zero section over X-Supp(s) to a section of F over X.

Let us mention that a subspace W of a locally compact space X is locally compact if and only if W is a <u>locally closed</u> subspace of X, compare II.6.

$\underline{\text{Corollary 2.5}}$. Let h: W → X denote the inclusion of a
locally closed subspace W of X. A soft sheaf S on X
will induce a soft sheaf h*S   on W.

$\underline{\text{Corollary 2.6}}$. Let i: Z → X denote the inclusion of
a closed subspace. A soft sheaf S on X has surjective restriction map

$$\Gamma_c(X,S) \longrightarrow \Gamma_c(Z,S)$$

$\underline{\text{Proof}}$. Given $s \in \Gamma_c(Z,i*S)$.   Choose a compact subset L
of X whose interior $\overset{\bullet}{L}$ contains Supp(s)

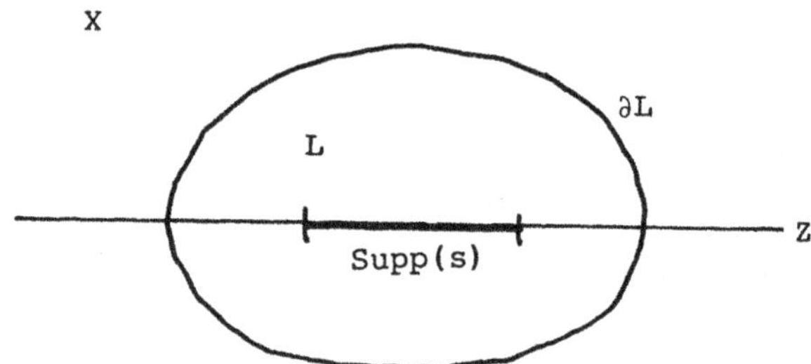

Let $\bar{s}$ denote the section of S over ∂L ∪ (Z ∩ L) with restriction
0 to ∂L and whose restriction to L ∩ Z coincides with the
restriction of s to L ∪ Z, 2.3. Extend s to a section t
of S over L, and finally extend t by zero outside L, 2.4.

Q.E.D.

$\underline{\text{Theorem 2.7}}$. Any soft sheaf S on the locally compact space
X is $\Gamma_c(X,-)$-acyclic, i.e.

$$H_c^n(X,S) = 0 \quad \text{for} \quad n \geq 1$$

$\underline{\text{Proof}}$. Let there be given a short exact sequence of sheaves on X

$$0 \longrightarrow E \xrightarrow{\ e\ } F \xrightarrow{\ f\ } G \longrightarrow 0$$

with E soft. Let us prove that the projection

$$\Gamma_c(X,F) \longrightarrow \Gamma_c(X,G)$$

is surjective. - Assume first that X is compact. So, let there be given a section s of G over X. For each point x of X we can choose a compact neighbourhood K_x of x and a section t_x of F over K_x which projects onto the restriction of s to K_x. We can refine the covering $(\mathring{K}_x)_{x \in X}$ to a finite covering. - This leads us to the following construction:

Given compact subsets K and L of X, $u \in \Gamma(K,F)$ and $v \in \Gamma(L,F)$ which projects onto s.   The section

$$r_{L \cap K, K}(u) - r_{L \cap K, L}(v)$$

represents a section of E over $L \cap K$ and can as such be extended to a section $w \in \Gamma(L,E)$. We can now consider the sections $u \in \Gamma(K,F)$ and $v + e(w) \in \Gamma(L,F)$ and glue these together to get a section of F over $K \cup L$ which projects onto s. - We can now apply this construction to our finite covering to obtain a section of F over X which projects onto $s \in \Gamma(X,G)$.

In case X is no longer compact we can choose a compact set K such that $\mathring{K}$ contains Supp(s). Choose a section t of F over K which projects onto the restriction of s to K.

The restriction of t to ∂K represents a section of E over ∂K which we extend to a section of E over K, and subtract from t. Thus we may assume that the restriction of t to ∂K is zero. We can now extend t by zero outside K, to obtain a global section of F which projects onto s.

Let us remark that these considerations proves that if we add the assumption that F is soft, then G is soft.

Let us now return to our soft sheaf S and choose an exact sequence

$$0 \longrightarrow S \longrightarrow I \longrightarrow T \longrightarrow 0$$

with I injective. Using that I is flabby we conclude first that I is soft and next that T is soft. We can now conclude from the first part that $H_c^1(X,S) = 0$.   The general case follows by an induction similar to the one applied in the proof of II.3.5.

$$\text{Q.E.D.}$$

Proposition 2.8. Let S denote a sheaf on a locally compact space X. If every point of X has an open neighbourhood U such that the restriction of S to U is a soft sheaf on U, then S is soft.

Proof. Let us first treat the case where X is compact. By assumption we can find a finite number of compact subsets $X_1,\ldots,X_n$ of X such that the restriction of S to each of these is soft. To prove that S is soft consider a compact subset K of X and $s \in \Gamma(K,S)$. Put $X^p = K \cup X_1 \cup X_p$, $p = 0,\ldots,n$. Suppose we have already extended $s = s^0$ for a section s^p of S over X^p. By assumption we can extend the restriction of s^p

to $X^p \cap X_{p+1}$ to a section σ^{p+1} of S over X_{p+1}. By 2.3 we can find a section s^{p+1} of S over X^{p+1} which extends s^p and σ^{p+1}. We can now proceed by induction

Let us now consider the general case. We know that the restriction of S to any compact subset of X is soft. Let us attempt to extend a section s of S over the compact subset K to a global section. Choose a compact subset L of X with $K \subseteq \overset{\bullet}{L}$. Extend the section over $K \cup \partial L$ which is s on K and zero on ∂L to a section of S over L. Use 2.4 to extend that section to a global section which is zero outside L.

$$Q.E.D.$$

Sheaves of modules

Let X denote a locally compact space equipped with a sheaf of commutative rings A. It is understood that the sections over any open set is a commutative ring with 1 and that the restriction maps respect the ring structure and maps 1 to 1. - Let E be a sheaf of A-modules on X, i.e. $\Gamma(U,E)$ carries a structure of unitary $\Gamma(U,A)$-modules and the restriction maps satisfy the following condition

$$r_{VU}(ae) = r_{VU}(a)r_{VU}(e) \qquad ; \ a \in \Gamma(V,A), \ e \in \Gamma(V,E)$$

Proposition 2.9. Let A be a soft sheaf of commutative rings. Any sheaf E of A-modules is soft.

Proof. Let K be a compact subset of X and L a compact neighbourhood of K. There exists $\sigma \in \Gamma(X,A)$ such that $\sigma = 1$ in an open neighbourhood of K and such that the restriction of σ to the complements of $\overset{\bullet}{L}$ is zero.

156

To see this let us consider the section on $K \cup \partial L$ which
is 1 on K and 0 on ∂L. Extend this to a section τ of
A over L and let σ be the extension of τ by zero outside L,
2.4.

Let us now show that E is soft; so let s be a section
of E over an open neighbourhood U of K. Let L be a compact
neighbourhood of K contained in U and σ a global section
of A as above. If we let τ denote the restriction of σ to
L, then the section τs of E over L is zero on ∂L,
and therefore extendable to all of X.

Q.E.D.

<u>Local operators 2.10</u>. Let S and T be sheaves on the
locally compact space X and $D: \Gamma(X,S) \to \Gamma(X,T)$ a linear map
which satisfies the following condition

$$\text{Supp } D(f) \subseteq \text{Supp}(f) \quad \text{for all} \quad f \in \Gamma_c(X,S).$$

If f is soft then there exists a unique morphism of sheaves
$d: S \to T$ which induces D on the level of global sections.

<u>Proof</u>. For any point of X the restriction maps
$\Gamma_c(X,S) \to S_x$ is surjective. Thus the uniqueness follows from
II.2.2. - Given an open subset U of F and $s \in \Gamma(U,S)$. Cover
U with open subsets U_i such that there exists $t_i \in \Gamma_c(X,S)$
whose restriction to U_i equals that of s. Notice that $D(s_j)$
is independent of s_j since D is a local operator, i.e.
satisfies the support condition. Using that T is a sheaf we
can glue the $D(s_j)$'s together to get a section of T over U.
The remaining details are left to the reader.

Q.E.D.

III.3 Soft sheaves on $\mathbb{R}^n$

A basic sheaf on $\mathbb{R}^n$ is the sheaf C^∞ of <u>smooth</u>, i.e. infinitely often differentiable, $\mathbb{R}$-valued functions. We shall prove that this sheaf is soft on the basis of the following classical.

<u>Lemma 3.1.</u> There exists a smooth, positive real function β on $\mathbb{R}$ with compact support and $\beta(0) \neq 0$.

<u>Proof.</u> For $t > 0$ consider the function $\exp(-1/t)$. By a simple induction we find that

$$\frac{d^k}{dt^k} \exp(-1/t) = P_k(1/t) \exp(-1/t) \qquad ; \; k \in \mathbb{N}$$

where P_k is a polynomial of degree $2k$. Whence

$$\lim_{t \downarrow 0} \frac{d^k}{dt^k} \exp(-1/t) = 0$$

As a consequence we can define a smooth function α by

$$\alpha(t) = \begin{cases} \exp(-1/t) & \text{for } t > 0 \\ 0 & \text{for } t \leq 0 \end{cases}$$

For β take the function $\beta(t) = \alpha(1-t)\alpha(1+t)$.

Q.E.D.

<u>Theorem 3.2.</u> The sheaf C^∞ of smooth functions on $\mathbb{R}^n$ is soft.

158

<u>Proof</u>. Let us first remark that for any $a \in \mathbb{R}^n$ and any neighbourhood U of a there exists a positive smooth function b on $\mathbb{R}^n$ with support in U and $b(a) \neq 0$. With the notation of 3.1, such functions can be supplied by the formula

$$b(x) = \prod_{i=1}^{n} \beta(m(x_i - a_i))$$

where m is a sufficiently large integer.

For a compact set K and a compact neighbourhood L of K there exists a positive smooth function with support in L and with strictly positive values on K: For each $a \in K$ choose a positive smooth function φ_a on $\mathbb{R}^n$ with support in L and $\varphi_a(a) \neq 0$. Let U_a be the set of points $x \in \mathbb{R}^n$ where $\varphi_a(x) > 0$. Cover K with finitely many U_a's and use the corresponding sum $\Sigma \varphi_a$.

For a compact set K and a compact neighbourhood N of K there exists a smooth function on $\mathbb{R}^n$ which is zero outside N and which is constant 1 in a neighbourhood of K: Choose a compact neighbourhood L of K with $L \subseteq \mathring{N}$ and choose a positive smooth function φ on $\mathbb{R}^n$ which is strictly positive on L and which is zero outside N. Choose another positive smooth function ψ which is strictly positive on $N - \mathring{L}$ and which is zero in a neighbourhood of K. This can be done since there exists a compact neighbourhood of $N - \mathring{L}$ which does not meet K, K and $L - \mathring{N}$ being disjoint compacts. The function $\varphi + \psi$ is strictly positive on N, thus the function $\varphi(\varphi + \psi)^{-1}$ is defined in an open neighbourhood of N, is zero outside N and takes the value 1 in a neighbourhood of K.

The sheaf C^∞ is soft: Let s be a section of C^∞ over an open neighbourhood U of the compact set K. Choose a compact neighbourhood N of K contained in U and a smooth function on $\mathbb{R}^n$ which is zero outside N and which is 1 in a neighbourhood of K. The function φs can be extended to all of $\mathbb{R}^n$.

Q.E.D.

Corollary 3.3. The sheaf C of continuous real functions on $\mathbb{R}^n$ is soft. More generally the sheaf $C^{(p)}$ of p-times continuously differentiable real functions on $\mathbb{R}^n$ is soft.

Proof. Notice that $C^{(p)}$ is a sheaf of modules over the soft sheaf C^∞ and apply 2.9.

Q.E.D.

Example 3.4. The fundamental example of a locally compact space is the real line $\mathbb{R}$. The constant sheaf $\underset{\sim}{\mathbb{R}}$ has the "Calculus resolution" which is soft

$$0 \longrightarrow \underset{\sim}{\mathbb{R}} \longrightarrow C^{(1)} \xrightarrow{\ D\ } C \longrightarrow 0$$

where D is ordinary differentiation. To calculate cohomology we consider the exact sequence

$$0 \longrightarrow \Gamma_c(\mathbb{R}, C^{(1)}) \xrightarrow{\ D\ } \Gamma_c(\mathbb{R}, C) \xrightarrow{\ \int\ } \mathbb{R} \longrightarrow 0$$

where $\int$ is given by the formula

$$\int f = \int_{-\infty}^{+\infty} f(x)\,dx$$

160

Consequently we find

3.5
$$H_c^{\,i}(\mathbb{R},\underset{\sim}{\mathbb{R}}) = \begin{cases} 0 & \text{for } i \neq 1 \\ \mathbb{R} & \text{for } i = 1 \end{cases}$$

Let us notice another corollary to 3.2.

Corollary 3.6. Let W be a locally closed subspace of $\mathbb{R}^n$.
The sheaf of continuous real valued functions on W is a soft
sheaf on W.

Proof. Let $h: W \to \mathbb{R}^n$ denote the inclusion of W in $\mathbb{R}^n$
and let C resp. D be the sheaf of continuous functions on
$\mathbb{R}^n$ resp. W. Notice that we have a canonical morphism

$$C \longrightarrow h_* D$$

which we can describe as follows: Let U be an open subset of
$\mathbb{R}^n$ and $f \in \Gamma(U,C)$. We can compose f with the restriction of
h to $h^{-1}(U) = W \cap U$ to obtain a continuous function on $W \cap U$
i.e. an element of $\Gamma(U, h_* D) = \Gamma(h^{-1}(U), D)$. By adjunction we
derive a morphism of sheaves of rings $h*C \to D$. Thus we have
equipped D with the structure of a $h*C$-module. The sheaf
$h*C$ is soft according to 2.5. Conclusion by 2.9.

Q.E.D.

Example 3.7. On the unit interval [0,1] we have an exact
sequence of sheaves analogous to the one in 3.4

$$0 \longrightarrow \underset{\sim}{\mathbb{R}} \longrightarrow C^{(1)} \xrightarrow{\ D\ } C \longrightarrow 0$$

where $C^{(1)}$ and C respectively denotes the sheaf continuously differentiable functions on $[0,1]$ resp. continuous functions. The sheaf $C^{(1)}$ is soft since it may be viewed as a sheaf of modules over the restriction of C^∞ to $[0,1]$, compare the proof of 3.6. The sequence above furnishes a soft resolution of $\underset{\sim}{\mathbb{R}}$ and we find

$$3.8 \qquad\qquad H^i([0,1],\underset{\sim}{\mathbb{R}}) = 0 \qquad ; \quad i \geq 1.$$

Using general topology we can generalize 3.6 to an arbitrary locally compact space.

<u>Urysohn 3.9</u>. The sheaf of $\mathbb{R}$-valued continuous functions on a locally compact space X is soft.

<u>Proof</u>. Let us first prove that given a compact subset K of X and a compact neighbourhood N of K then there exists a continuous function on X, which is constant 1 in a neighbourhood of K and constant 0 outside N. - For this, choose compact sets L and M with

$$K \subseteq \overset{\bullet}{L} \subseteq L \subseteq \overset{\bullet}{M} \subseteq M \subseteq \overset{\bullet}{N} \subseteq N$$

In the compact space N we have to disjoint compact subsets L and $N-\overset{\bullet}{M}$. According to Urysohn's lemma we can find a continuous function f on N with value 1 on L and value zero on $N-\overset{\bullet}{M}$. We can now use the open covering $X-M,\overset{\bullet}{N}$ to construct a continuous function which agrees with f on $\overset{\bullet}{N}$ and is 0 outside M. We can now conclude the proof as in 3.2 or 2.9.

Q.E.D.

III.4   The exponential sequence

The multiplicative group of complex numbers with absolute value 1 is denoted S^1. This group fits into an exact sequence

4.1
$$0 \longrightarrow \mathbb{Z} \xrightarrow{2\pi} \mathbb{R} \xrightarrow{\text{expi}} S^1 \longrightarrow 0$$

where the map expi is given by

4.2
$$\text{expi } v = \cos v + i \sin v$$

The map expi has <u>local sections</u>, in the sense that for every $z \in S^1$ there exists an open neighbourhood U of z in S^1 and a continuous map s: U → $\mathbb{R}$, such that expi ∘ s = 1: Over the part of S^1 which lies in the open right halfplane we have the following section

4.3
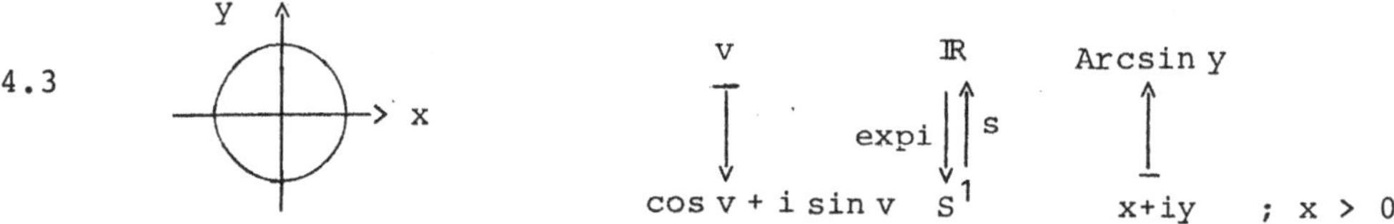

By translation in the group S^1 we see that expi   has local sections everywhere. From the presence of local sections we conclude that expi   is an <u>open map</u>, i.e. transforms open sets into open sets.

Let X be an arbitrary topological space. The exponential sequence above gives rise to a short exact sequence of sheaves

4.4
$$0 \longrightarrow \underset{\sim}{\mathbb{Z}} \longrightarrow C(\mathbb{R}) \xrightarrow{\text{expi}} C(S^1) \longrightarrow 0$$

where $C(\mathbb{R})$ and $C(S^1)$ denotes the sheaf of continuous functions on X with value in $\mathbb{R}$ and S^1 respectively. In this section we shall exploit this sequence in three simple cases.

The real line $\mathbb{R}$

We shall base our investigations on the following classical result.

Lemma 4.5. Any continuous map $f: \mathbb{R} \to S^1$ can be factored $f = \mathrm{expi} \circ F$, where $F: \mathbb{R} \to \mathbb{R}$ is a continuous map.

Proof. Let us first solve the analogous problem for $f: [0,1] \to S^1$. For $t \in [0,1]$ choose an open interval I_t with center in t such that the restriction of f to $I_t \cap [0,1]$ admits a factorization through expi. Let J_t denote the open interval which makes up the middle third of I_t. According to Borel-Heine we can find a finite subset S of $[0,1]$ such that $(J_t)_{t \in S}$ covers $[0,1]$. Choose an integer n such that $1/n$ is smaller than the length of each of the intervals J_t, $t \in S$. The restriction of f to an interval of the form $[p/n, p+1/n]$ can be written $f = \exp \circ F_p$, where $F_p: [p/n, p+1/n] \to \mathbb{R}$ is continuous. Notice that $F_p(p/n) - F_{p-1}(p/n)$ is an integral mul-

164

tiple of 2π. We can now successively fit the F_p's together
by adding integral multiples of 2π to get the desired factorization of f.

Returning to the original problem, we can for each $n \in \mathbb{Z}$
choose a lifting $F^n\colon [n,n+1] \to S^1$ of the restriction of f
to that interval and finally fit these together.

Q.E.D.

<u>Theorem 4.6</u>. The exponential sequence 4.4 is a soft resolution
of the constant sheaf $\underset{\sim}{\mathbb{Z}}$ on $\mathbb{R}$.

<u>Proof</u>. Let us prove that for any open subset U of $\mathbb{R}$ the
following sequence

$$\Gamma(U,C(\mathbb{R})) \longrightarrow \Gamma(U,C(S^1)) \to 0$$

is exact. Remark first that this is true in case U is an
open interval as it follows from 4.5. The general case follows
from the fact that U is union of disjoint open intervals.

Let us finally prove that $C(S^1)$ is soft. So let there
be given a compact subset K of $\mathbb{R}$ and a section t of
$C(S^1)$ over an open neighbourhood U of K. Lift t to a
section s of $C(\mathbb{R})$ over U. After shrinking U we may
extend s to a global section r of $C(\mathbb{R})$, since $C(\mathbb{R})$ is
soft. The section $\expi \circ r$ will deliver the looked for global
section of $C(S^1)$.

Q.E.D.

On the basis of the resolution 4.4 we will show that

$$4.7 \qquad H_c^{\cdot}(\mathbb{R}; \mathbb{Z}) = 0, \mathbb{Z}, 0, \ldots$$

To do so we introduce the <u>winding number</u>, $I(s)$ of $s \in \Gamma_c(\mathbb{R}, C(S^1))$ To this end choose a continuous map $t: \mathbb{R} \to \mathbb{R}$ such that $s = \text{expi} \circ t$, 4.5. Since s has compact support it follows that $t(x)$ is constant for large values of x, resp. small values of x. Put

$$4.8 \qquad I(s) = \frac{1}{2\pi} (t(+\infty) - t(-\infty))$$

It is left to the reader to verify exactness of the sequence

$$0 \longrightarrow \Gamma_c(\mathbb{R}, C(\mathbb{R})) \xrightarrow{\text{expi}} \Gamma_c(\mathbb{R}, C(S^1)) \xrightarrow{I} \mathbb{Z} \longrightarrow 0$$

Let us calculate the action of a <u>proper</u> continuous map $f: \mathbb{R} \to \mathbb{R}$ on $H_c^1(\mathbb{R}; \mathbb{Z})$. Remark first that proper in this context means that $|f(x)| \longrightarrow +\infty$ for $x \to +\infty$ and $x \to -\infty$. Taking into account that $f^*\underset{\sim}{\mathbb{Z}} = \underset{\sim}{\mathbb{Z}}$ we can identify the adjunction morphism with the morphism

$$\underset{\sim}{\mathbb{Z}} \xrightarrow{\quad \circ f \quad} f_*\underset{\sim}{\mathbb{Z}}$$

which to an open set V and a locally constant function $s: V \to \mathbb{Z}$ associates the locally constant function $s \circ f: f^{-1}(V) \to \mathbb{Z}$. The morphism $\circ f$ can be lifted to the exponential resolution as follows, compare Scolium II.5.2

$$0 \longrightarrow \underset{\sim}{\mathbb{Z}} \xrightarrow{2\pi} C(\mathbb{R}) \xrightarrow{\text{expi}} C(S^1) \longrightarrow 0$$

$$\downarrow \circ f \qquad\qquad \downarrow \circ f \qquad\qquad \downarrow \circ f$$

$$0 \longrightarrow f_* \underset{\sim}{\mathbb{Z}} \xrightarrow{2\pi} f_* C(\mathbb{R}) \xrightarrow{\text{expi}} f_* C(S^1) \to 0$$

We shall evaluate the action f^* of f on $H_c^1(\mathbb{R}, \mathbb{Z})$ in terms of the symbol $[f]$, which we now define

4.9

$f(+\infty)$	$+\infty$	$+\infty$	$-\infty$	$-\infty$
$f(-\infty)$	$-\infty$	$+\infty$	$+\infty$	$-\infty$
$[f]$	1	0	-1	0

With the notation of 4.8 let us prove the formula

$$I(s \circ f) = [f]\, I(s) \qquad ; \quad s \in \Gamma_c(\mathbb{R}, C(S^1))$$

Choose $t: \mathbb{R} \longrightarrow \mathbb{R}$ such that $s = \text{expi}\, t$. This gives $s \circ f = \text{expi}\, t \circ f$ and consequently, 4.8

$$I(s \circ f) = \frac{1}{2\pi}\, (t(f(+\infty)) - t(f(-\infty)))$$

which ensures us the validity of the formula, i.e. that

$$\begin{array}{ccc}
\Gamma_c(\mathbb{R}, C(S^1)) & \xrightarrow{\quad I \quad} & \mathbb{Z} \\
\downarrow{\scriptstyle \circ f} & & \downarrow{\scriptstyle \cdot [f]} \\
\Gamma_c(\mathbb{R}, C(S^1)) & \xrightarrow{\quad I \quad} & \mathbb{Z}
\end{array}$$

is a commutative diagram. Thus we have proved that

$$4.10 \qquad f^*\alpha = [f]\alpha \qquad ; \qquad \alpha \in H_c^1(\mathbb{R}, \mathbb{Z})$$

for a proper continuous map $\ f: \mathbb{R} \to \mathbb{R}$.

We shall now use the exponential resolution to make a closer study of the sequence 3.4

$$0 \to \Gamma_c(\mathbb{R}, C^{(1)}) \to \Gamma_c(\mathbb{R}, C) \to H_c^1(\mathbb{R}, \mathbb{R}) \longrightarrow 0$$

which tells us that a continuous function $\ f: \mathbb{R} \longrightarrow \mathbb{R}\ $ with compact support represents a cohomology class.

<u>Proposition 4.11</u>. A continuous real function φ on $\mathbb{R}$ with compact support represents an integral cohomology class[*] if and only if

$$\int_{-\infty}^{+\infty} \varphi(x)\, dx \quad \text{is an integer.}$$

<u>Proof</u>. Consider the following exact, commutative diagram of sheaves on $\mathbb{R}$

$$
\begin{array}{ccccccccc}
0 & \longrightarrow & \mathbb{R} & \longrightarrow & C^{(1)}(\mathbb{R}) & \xrightarrow{\ D\ } & C(\mathbb{R}) & \longrightarrow & 0 \\
 & & \downarrow & & \downarrow & & \downarrow & & \\
0 & \longrightarrow & \mathbb{C} & \longrightarrow & C^{(1)}(\mathbb{C}) & \xrightarrow{\ D\ } & C(\mathbb{C}) & \longrightarrow & 0
\end{array}
$$

where $C(\mathbb{C})\ $ resp. $\ C^{(1)}(\mathbb{C})\ $ denotes the sheaf of $\mathbb{C}$-valued continuous functions resp. continuously differentiable functions.

---

[*] Quite generally let X denote a locally compact space. A cohomology class $\theta \in H_c^n(X,\mathbb{R})\ $ is called <u>integral</u>, if it belongs to the image of

$$H_c^n(X,\mathbb{Z}) \longrightarrow H_c^n(X,\mathbb{R})$$

induced from the inclusion of sheaves $\mathbb{Z} \to \mathbb{R}$.

The unlabelled arrows are inclusions. It follows that it suffices to prove a result similar to 4.11 with $\mathbb{R}$ replaced by $\mathbb{C}$. - Next consider the following diagram of sheaves on $\mathbb{R}$

$$
\begin{array}{ccccccccc}
0 & \longrightarrow & \underset{\sim}{\mathbb{Z}} & \xrightarrow{2\pi} & C^{(1)}(\mathbb{R}) & \xrightarrow{\exp i} & C(S^1) & \longrightarrow & 0 \\
& & \downarrow & & \downarrow{\scriptstyle \frac{1}{2\pi}} & & \downarrow{\scriptstyle \frac{1}{2\pi i}\,\mathrm{Dlog}} & & \quad *) \\
0 & \longrightarrow & \underset{\sim}{\mathbb{C}} & \longrightarrow & C^{(1)}(\mathbb{C}) & \xrightarrow{D} & C(\mathbb{C}) & \longrightarrow & 0
\end{array}
$$

where the unlabelled arrows are simple inclusions. This induces a commutative diagram

$$
\begin{array}{ccc}
\Gamma_c(\mathbb{R},\, C(S^1)) & \xrightarrow{\quad I \quad} & \mathbb{Z} \\
\downarrow{\scriptstyle \frac{1}{2\pi i}\,\mathrm{Dlog}} & & \downarrow \\
\Gamma_c(\mathbb{R},\, C(\mathbb{C})) & \xrightarrow{\quad \int \quad} & \mathbb{C}
\end{array}
$$

as the following computation shows: An element f of $\Gamma_c(\mathbb{R},\, C(S^1))$ may be represented by $\exp i \circ g$ where $g \in \Gamma(\mathbb{R},\, C(\mathbb{R}))$,

$$
\int_{-\infty}^{+\infty} \mathrm{Dlog}(\exp i\, g(x))\, dx = i \int_{-\infty}^{+\infty} g'(x)\, dx =
$$

$$
i(\lim_{x \to +\infty} g(x) - \lim_{x \to -\infty} g(x)) = 2\pi i\, I(f)
$$

where we have used the formula 4.8.

Q.E.D.

---

*) The evaluation of Dlog on f is $\dfrac{f'}{f}$.

The unit interval [0,1]

Let us consider the exponential sequence of sheaves on the unit interval [0,1]

$$0 \longrightarrow \mathbb{Z} \xrightarrow{2\pi} C(\mathbb{R}) \xrightarrow{\text{expi}} C(S^1) \longrightarrow 0$$

This is a <u>soft resolution</u> of the constant sheaf $\mathbb{Z}$ on [0,1]. To prove that $C(S^1)$ is soft we can use the remark that a continuous map $V \to S^1$, where V is a relative open subset of [0,1], can be extended to a continuous map on an open neighbourhood of V in $\mathbb{R}$. - From the proof of 4.5 follows that the sequence

$$0 \to \mathbb{Z} \xrightarrow{2\pi} \Gamma([0,1], C(\mathbb{R})) \xrightarrow{\text{expi}} \Gamma([0,1], C(S^1)) \longrightarrow 0$$

is exact. Thus we have found that

4.12 $\qquad\qquad H^i([0,1], \mathbb{Z}) = 0 \qquad ; i \geq 1$

A result of fundamental importance.

The 1-sphere S^1

Let us once more consider the exponential sequence of sheaves

$$0 \longrightarrow \mathbb{Z} \xrightarrow{2\pi} C(\mathbb{R}) \xrightarrow{\text{expi}} C(S^1) \longrightarrow 0$$

This is a <u>soft resolution</u> of the constant sheaf $\mathbb{Z}$ on S^1 as it follows from the local character of softness, 2.8.

In order to calculate cohomology by means of the exponential sequence we shall introduce the notion of _degree_ of a continuous map $f: S^1 \to S^1$. To do so let us remark that it follows from 4.5 that we can find a commutative diagram of continuous maps

4.13

$$
\begin{array}{ccc}
\mathbb{R} & \xrightarrow{\ F\ } & \mathbb{R} \\
\downarrow{\scriptstyle \text{expi}} & & \downarrow{\scriptstyle \text{expi}} \\
S^1 & \xrightarrow{\ f\ } & S^1
\end{array}
$$

The map F is unique up to an integral multiple of 2π . Thus it makes sense to put

4.14

$$
\boxed{\ \deg(f) \;=\; \frac{1}{2\pi}(F(2\pi) - F(0))\ }
$$

The following sequence of abelian groups is exact

$$
0 \longrightarrow \mathbb{Z} \xrightarrow{\ 2\pi\ } \Gamma(S^1, C(\mathbb{R})) \xrightarrow{\ \text{expi}\ } \Gamma(S^1, C(S^1)) \xrightarrow{\ \deg\ } \mathbb{Z} \longrightarrow 0
$$

Proof. To see that the degree map is surjective, let us notice that $m \in \mathbb{Z}$ gives rise to a map $z \longmapsto z^m$ of degree m . Conversely, let $f: S^1 \to S^1$ be a map of degree zero with the notation from 4.13, remark that the expression $F(x+2\pi) - F(x)$ is independent of x . Thus we find that F is periodic with period 2π , which shows that f can be factored through expi.

Q.E.D.

The product hk of two continuous maps $h,k: S^1 \to S^1$ is given by $z \longmapsto h(z)k(z)$. We have

4.15

$$
\deg(hk) = \deg(h) + \deg(k)
$$

as it follows rather easily from the definition.

The notion of degree introduced in 4.14 is multiplicative in the sense that two continuous maps $f,g: S^1 \to S^1$ satisfy

$$4.16 \qquad \deg(g \circ f) = \deg(g)\deg(f)$$

<u>Proof</u>. Choose continuous maps $F,G: \mathbb{R} \to \mathbb{R}$ such that

$$f \circ \text{expi} = \text{expi} \circ F, \quad g \circ \text{expi} = \text{expi} \circ G$$

From these formulas follows $\quad g \circ f \circ \text{expi} = \text{expi} \circ G \circ F$

According to the definition 4.14 we find

$$\deg(g \circ f) = \frac{1}{2\pi}[G(F(2\pi)) - G(F(0))] =$$

$$\frac{1}{2\pi}[G(F(0) + 2\pi \deg(f)) - G(F(0))]$$

We leave it to the reader to prove the formula

$$G(x+2\pi n) - G(x) = 2\pi n \deg(g) \quad ; \ x \in \mathbb{R}, \quad n \in \mathbb{Z}$$

Finally, use this with $x = F(0)$ and $n = \deg(f)$ to get 4.16.

$$\text{Q.E.D.}$$

<u>Proposition 4.17</u>. The action of a continuous map $f: S^1 \to S^1$ on $H^1(S^1, \mathbb{Z})$ is given by

$$f^*\alpha = \deg(f)\alpha \qquad ; \ \alpha \in H^1(S^1, \mathbb{Z})$$

<u>Proof</u>. Lift f to the exponential resolution as follows

$$0 \longrightarrow \mathbb{Z} \xrightarrow{2\pi} C(\mathbb{R}) \xrightarrow{\text{expi}} C(S^1) \longrightarrow 0$$

$$\Big\downarrow \text{of} \qquad\qquad \Big\downarrow \text{of} \qquad\qquad \Big\downarrow \text{of}$$

$$0 \longrightarrow f_*\mathbb{Z} \xrightarrow{2\pi} f_*C(\mathbb{R}) \xrightarrow{\text{expi}} f_*C(S^1) \longrightarrow 0$$

It follows from 4.16 that the following diagram

$$\begin{array}{ccc} \Gamma(S^1,C(S^1)) & \xrightarrow{\text{deg}} & \mathbb{Z} \\ \Big\downarrow \text{of} & & \Big\downarrow \text{deg}(f) \\ \Gamma(S^1,C(S^1)) & \xrightarrow{\text{deg}} & \mathbb{Z} \end{array}$$

is commutative, and the result follows.

$$\text{Q.E.D.}$$

<u>Proposition 4.18</u>. Any continuous endomorphism of the topological group S^1 has the form $z \longmapsto z^n$; $n \in \mathbb{Z}$.

<u>Proof</u>. Let $f: S^1 \to S^1$ be an endomorphism and choose $F: \mathbb{R} \to \mathbb{R}$ as in 4.13. By adding a constant to F we may assume that $F(0) = 0$. Since f is an endomorphism we have

$$\text{expi} (F(x+y) - F(x) - F(y)) = 1 \quad ; \ x,y \in \mathbb{R}$$

or otherwise expressed $F(x+y) - F(x) - F(y) \in 2\pi\,\mathbb{Z}$

Using that $\mathbb{R}$ is connected we get that F is a continuous endomorphism of $\mathbb{R}$. Consequently F must have the form $x \longmapsto ax$. It follows that a is an integral multiple of 2π.

$$\text{Q.E.D.}$$

III.5   Cohomology of direct limits

Let us consider a locally compact space X and a direct system of sheaves $(F_\lambda, f_{\mu\lambda})$ over a directed set I.

Theorem 5.1. The canonical map, $p \in \mathbb{N}$

$$\varinjlim H_c^P(X,F_\lambda) \underset{\sim}{\to} H_c^P(X, \varinjlim F_\lambda)$$

is an isomorphism.

Proof. Put $F = \lim F_\lambda$ and let $f_\lambda: F_\lambda \to F$ denote the canonical map. Let us at once notice that for the inclusion $i: Z \to X$ of a (closed) subspace we have

$$i^*F = \varinjlim i^*F_\lambda$$

as it follows by localization I.2.8. - Let us prove that the canonical map

$$5.2 \qquad\qquad \varinjlim \Gamma_c(X,F_\lambda) \to \Gamma_c(X,F)$$

is an isomorphism. Let us first prove injectivity in case X is compact. So let there be given $s \in \Gamma(X,F_\lambda)$ which maps to zero in $\Gamma(X,F)$. For each $x \in X$ we can find $\lambda(x) \geq \lambda$ and an open neighbourhood U_x of x such that $f_{\lambda(x),\lambda}(s)$ has restriction zero to U_x. Let us choose a finite set of points S of X such that $(U_x)_{x \in S}$ covers X and choose $\mu \in I$, such that $\lambda(x) \leq \mu$ for all $x \in S$. It follows that $f_{\mu\lambda}(s) = 0$.

174

Let us prove surjectivity of 5.2 in case X is compact.
So let there be given $s \in \Gamma(X,F)$. For each $x \in X$ choose
a compact neighbourhood K_x of x in X, $\lambda(x) \in I$ and
$s^x \in \Gamma(K_x, F_{\lambda(x)})$ such that $f_{\lambda(x)}(s^x)$ equals the restriction
of s to K_x. Since X is compact we can choose a finite
subset T of X such that $(K_x)_{x \in T}$ covers X. Consider
two points, $x,y \in T$: according to our first result, the
canonical map

$$\varinjlim \ \Gamma(K_x \cap K_y, F_\lambda) \rightarrow \Gamma(K_x \cap K_y, F)$$

is injective. It follows that we can choose $\mu \in I$ with $\mu \geq \lambda(x)$
for all $x \in T$, and such that $f_{\mu,\lambda(x)}(s^x)$ and $f_{\mu,\lambda(y)}(s^y)$
have the same restriction to $K_x \cap K_y$ for all $(x,y) \in T \times T$. The
section $t \in \Gamma(X,F_\mu)$ with restriction $f_{\mu,\lambda(x)}(s^x)$ to K_x for
all $x \in T$ will now be mapped onto $s \in \Gamma(X,F)$.

Let us now prove injectivity of 5.2 in general. Given
$s \in \Gamma_c(X,F_\lambda)$ which is mapped to zero in $\Gamma_c(X,F)$. Using the
previous result for $K = \mathrm{Supp}(s)$ we find $\mu \geq \lambda$ such that
$f_{\mu\lambda}(s) = 0$ in $\Gamma(K,F_\mu)$. From this we conclude by localization
that $f_{\mu\lambda}(s) = 0$ in $\Gamma(X,F_\mu)$.

Let us prove that 5.2 is surjective. So let there be given
$s \in \Gamma_c(X,F)$. Choose a compact set K containing $\mathrm{Supp}(s)$ in its
interior. Choose $\lambda \in I$ and $t \in \Gamma(K,F_\lambda)$ such that $f_\lambda(t)$
equals the restriction of s to K. Since the restriction of
$f_\lambda(t)$ to ∂K is zero, we can find $\mu \geq \lambda$ such that $f_{\mu\lambda}(t)$
has restriction zero to ∂K. We can now extend $f_{\mu\lambda}(t)$ by
zero outside K to obtain a global section of F with compact
support which is transformed into $s \in \Gamma(X,F)$ by f_λ.

Let us remark that $\underline{if}$ F_λ $\underline{\text{is soft for all}\ \lambda \in I,}$ then $\varinjlim F_\lambda$ $\underline{\text{is soft}}$. This follows from the first part and the remarks following 2.4.

Let us finally consider the Godement resolution $C^\cdot F_\lambda$ of F_λ, II.3.6. This gives us a direct system of complexes of flabby sheaves on X. The direct limit $\varinjlim C^\cdot F_\lambda$ is a soft resolution of F by our previous remark. Thus

$$H_c^{\;p}(X,F) \;=\; H^p\Gamma_c(X,\varinjlim C^\cdot F_\lambda) \;=\; H^p\varinjlim \Gamma_c(X,C^\cdot F_\lambda)$$

$$=\; \varinjlim H^p\Gamma_c(X,C^\cdot F_\lambda) \;=\; \varinjlim H_c^{\;p}(X,F_\lambda)$$

and our journey ends.

$$\text{Q.E.D.}$$

$\underline{\text{Corollary 5.3}}$. Let X denote a compact space which is $\underline{\text{acyclic}}$, i.e. for which

$$\mathbb{Z} = H^0(X,\mathbb{Z}) \quad \text{and} \quad H^i(X,\mathbb{Z}) = 0 \quad \text{for}\ i \geq 1$$

Then for any abelian group k

$$k \;\tilde{\rightarrow}\; H^0(X,k) \quad \text{and} \quad H^i(X,k) = 0 \quad \text{for}\ i \geq 1$$

$\underline{\text{Proof}}$. The abelian group k is direct limit of its finitely generated subgroups. Thus by 5.1 it suffices to treat the case where k is finitely generated. - Choose a finitely generated free resolution $0 \longrightarrow \mathbb{Z}^m \longrightarrow \mathbb{Z}^n \longrightarrow k \longrightarrow 0$ The long exact sequence on cohomology gives at once $H^i(X,k) = 0$ for $i \geq 1$. The assumption $\mathbb{Z} = H^0(X,\mathbb{Z})$ implies that X is connected, and consequently that any locally constant function with values in k is constant.

$$\text{Q.E.D.}$$

III.6   Proper base change and proper homotopy

Let $f: X \to Y$ be a <u>proper</u> map between locally compact spaces. Recall that this means that the inverse image of a compact subset of Y is a compact subset of X . In particular the <u>fiber</u> $f^{-1}(y)$ over a point $y \in Y$ is compact.

6.1   A proper continuous map $f: X \to Y$ between
          locally compact spaces is <u>closed</u>, i.e. a closed
          subset Z of X is transformed into a closed
          subset $f(Z)$ of Y.

<u>Proof</u>. Let y be a point of Y which does not belong to $f(Z)$. Choose a compact neighbourhood N of y in Y and consider the compact subset $Z \cap f^{-1}(N)$ of X . This is mapped onto

$$f(Z \cap f^{-1}(N)) = f(Z) \cap N$$

This shows that $f(Z) \cap N$ is compact. Thus $N - f(Z)$ is a neighbourhood of y in Y which does not meet $f(Z)$.

Q.E.D.

<u>Theorem 6.2</u>. Let $f: X \to Y$ be a proper map between locally compact spaces, and let F be a sheaf on X . For $y \in Y$ and $n \in \mathbb{N}$ the restriction map

$$(R^n f_* F)_y \to H^n(f^{-1}(y), F)$$

is an isomorphism.

<u>Proof</u>. Recall from II.5.11 that $R^n f_* F$ is the sheaf on Y associated to the presheaf

$$V \longmapsto H^n(f^{-1}(V), F)$$

As a consequence we have

$$(R^n f_* F)_y = \varinjlim_V H^n(f^{-1}(V), F)$$

where the limit is taken over all open neighbourhoods V of y in Y. By lemma 6.3 below we have an isomorphism

$$\varinjlim H^n(U, F) \xrightarrow{\sim} H^n(f^{-1}(y), F)$$

where the limit is taken over all open neighbourhoods U of $f^{-1}(y)$ in X. From the fact that f is closed 6.1, follows that any such U contains an open subset of the form $f^{-1}(V)$ where V is an open neighbourhood of y in Y:

$$U \supseteq f^{-1}(Y - f(X-U))$$

Thus we may identify the two direct limits above and the result follows.

Q.E.D.

<u>Lemma 6.3</u>. Let K be a compact subset of the locally compact space X. For any sheaf F on X there is a canonical isomorphism

$$\varinjlim H^n(U, F) \xrightarrow{\sim} H^n(K, F)$$

178

where the limit is taken over all open neighbourhoods U of K in X.

Proof. Let $i: K \to X$ denote the inclusion. Recall from 2.2 that the formula is correct for $n = 0$, i.e. that

$$\varinjlim \Gamma(U,F) \xrightarrow{\sim} \Gamma(K,i*F)$$

Let us remark that $i*$ transforms soft sheaves into soft sheaves. In particular an injective resolution $I^\cdot$ of F transforms into a soft resolution $i*I^\cdot$ of $i*F$. Thus

$$H^n(K,i*F) \ = \ H^n(K,i*I^\cdot) \ = \ H^n(\varinjlim \Gamma(U,I^\cdot)) \ =$$

$$\varinjlim H^n\Gamma(U,I^\cdot) \ = \ \varinjlim H^n(U,F)$$

where the last step is justified by II.5.9.

$$\text{Q.E.D.}$$

Vietoris-Begle mapping theorem 6.4. Let $f: X \to Y$ be a proper map between locally compact spaces with acyclic fibers, 5.3. For any sheaf G on Y the induced map

$$f*: H_c^\cdot(Y,G) \longrightarrow H_c^\cdot(X,f*G)$$

is an isomorphism.

Proof. Let us first prove that the adjunction map $a: G \to f_*f^*G$ is an isomorphism. This will be done by localization, so consider a point y of Y. Let us first remark that the canonical map

$$(f_* f^* G)_y \to H^0(f^{-1}(y), G_y)$$

is an isomorphism since f is proper 6.2. Thus we may identify a_y with the canonical map

$$G_y \longrightarrow H^0(f^{-1}(y), G_y)$$

which is an isomorphism according to 5.3.

Let $f^* G \to I^{\cdot}$ be an injective resolution. Let us prove that $H^i f_* I^{\cdot} = 0$ for $i \geq 1$. This is done by localization so consider a point y of Y. According to 6.1 we have

$$(H^i f_* I^{\cdot})_y = R^i f_* (f^* G)_y = H^i(f^{-1}(y), G_y)$$

The last group is zero by 5.3. - Using that f_* is left exact and transforms injectives to injectives we conclude that $f_* f^* G \to f_* I^{\cdot}$ is an injective resolution of $f_* f^* G$. Combine this with the fact that the adjunction morphism $G \to f_* f^* G$ is an isomorphism and conclude by an adaption of Scolium II.5.2 to cohomology with compact support.

$$\text{Q.E.D.}$$

Example 6.5. For the n-cube $[0,1]^n$ we have

$$H^i([0,1]^n, \underset{\sim}{k}) = 0 \qquad ; \quad i \geq 1$$

for any constant sheaf $\underset{\sim}{k}$. This follows from 4.12 and 6.4 using

180

the projection

$$[0,1]^n \to [0,1]^{n-1} \quad ; \quad (x_1, x_2, \ldots, x_n) \longmapsto (x_2, \ldots, x_n)$$

Let $f,g\colon X \to Y$ be proper continuous maps between locally compact spaces. A <u>proper homotopy</u> from f to g is a proper continuous map $F\colon X \times [0,1] \to Y$ with

$$F(x,0) = f(x), \quad F(x,1) = g(x) \quad ; \quad x \in X$$

<u>Proposition 6.6</u>. Two properly homotopic maps $f,g\colon X \to Y$ between locally compact spaces induces the same map

$$f^*, g^*\colon \quad H_c^{\cdot}(Y,\underset{\sim}{k}) \to H_{\cdot c}^{\cdot}(X,\underset{\sim}{k})$$

for any abelian group k.

<u>Proof</u>. Let $F\colon X \times [0,1] \to Y$ be a proper homotopy from f to g. Let $p\colon X \times [0,1] \longrightarrow X$ denote the projection and for $t \in [0,1]$ let $i_t\colon X \to X \times [0,1]$ be given by $i_t(x) = (x,t)$, $x \in X$. By Vietoris-Begle, $p^*\colon H_c^{\cdot}(X,\underset{\sim}{k}) \to H_c^{\cdot}(X \times [0,1],\underset{\sim}{k})$ is an isomorphism. From the formula $p \circ i_t = 1$ we deduce $i_t^* \circ p^* = 1$. This shows that

$$i_t^*\colon H_c^{\cdot}(X \times [0,1],\underset{\sim}{k}) \to H_c^{\cdot}(X,\underset{\sim}{k})$$

is independent of $t \in [0,1]$. Notice that $f = F \circ i_0$ and $g = f \circ i_1$ consequently

$$F^* = i_0^* \circ F^* = i_1^* \circ F^* = g^*$$

as required.

Q.E.D.

The Vietoris-Begle theorem can be refined in various ways. Here is one such refinement.

Proposition 6.7. Let $f: X \to Y$ be a proper continuous map between locally compact spaces. Assume that the fibers all satisfy the condition

$$H^0(f^{-1}(y),\mathbb{Z}) = \mathbb{Z} \quad \text{and} \quad H^i(f^{-1}(y),\mathbb{Z}) = 0 \quad \text{for} \quad i = 1,\ldots,n.$$

Then for any sheaf G on Y the induced map

$$f^*: H_c^p(Y,G) \to H_c^p(X,f^*G)$$

is an isomorphism for $p = 0,\ldots,n-1$ and a monomorphism for $p = n$.

Proof. Let $f^*G \to I^{\cdot}$ and $G \to J^{\cdot}$ be injective resolutions on X and Y respectively. Choose a morphism of complexes $\psi: J^{\cdot} \to f_*I^{\cdot}$ making the following diagram commutative

$$\begin{array}{ccc} G & \longrightarrow & J^{\cdot} \\ {\scriptstyle a}\downarrow & & \downarrow{\scriptstyle \psi} \\ f_*f^*G & \longrightarrow & f_*I^{\cdot} \end{array}$$

It follows from the proof of 6.4 that $H^p\psi$ is an isomorphism for $p = 0,\ldots,n$. Consider the mapping cone triangle

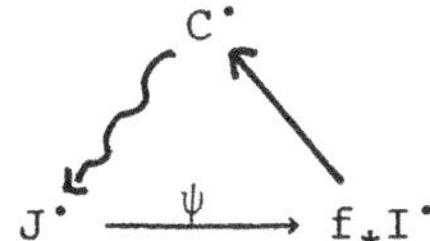

182

We can conclude that $H^p C^\cdot = 0$ for $p = 0,\ldots,n-1$. Since $C^\cdot$ is a bounded below complex of injectives this implies that $C^\cdot \to \tau_{\geq n} C^\cdot$ is a homotopy equivalence, I.5.8. - The transformed triangle

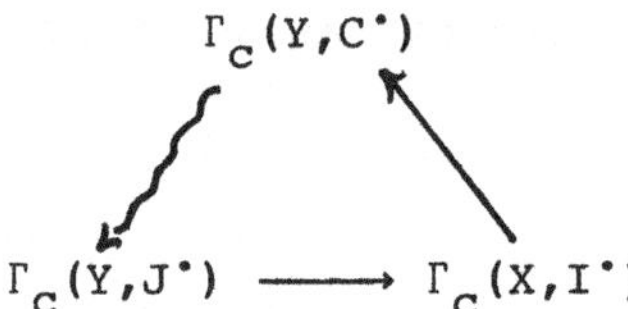

and the resulting long exact sequence

$$\to H^{n-1}(\Gamma_c(Y,C^\cdot)) \to H_c^{\ n}(Y,G) \xrightarrow{f^*} H_c^{\ n}(X,f^*G) \to$$

allows us to make the conclusion.

Q.E.D.

III.7   Locally closed subspaces

Let $h: W \to X$ denote the inclusion of the locally closed
subspace W of the locally compact space X.   For a sheaf E
on W we have the fundamental formula

$$7.1 \qquad \Gamma_c(X, h_! E) = \Gamma_c(W, E)$$

as it follows from the definition II.6.1.

Proposition 7.2. The inclusion of a locally closed subspace
$h: W \to X$ will transform a soft sheaf S on W into a soft
sheaf $h_! S$ on X.

Proof. Let K be a compact subset of X.   We are going to
prove that $\Gamma_c(X, h_! S) \to \Gamma(K, h_! S)$ is surjective. Let us introduce some notation and recall the formula II.6.13

$$
\begin{array}{ccc}
W \cap K & \xrightarrow{\quad q \quad} & K \\
\downarrow{\scriptstyle j} & & \downarrow{\scriptstyle i} \\
W & \xrightarrow{\quad h \quad} & X
\end{array}
\qquad
\begin{array}{l}
\text{II.6.13} \\[1em]
i^* h_! S = q_! j^* S
\end{array}
$$

$$\Gamma(K, h_! S) = \Gamma(K, q_! j^* S) = \Gamma_c(W \cap K, S)$$

Again by 7.1 $\Gamma_c(X, h_! S) = \Gamma_c(W, S)$.   From 2.6 we conclude that
$\Gamma_c(W, S) \to \Gamma_c(W \cap K, S)$ is surjective.

Q.E.D.

$\underline{\text{Corollary 7.3}}$. For any sheaf E on W we have a canonical isomorphism

$$H_c^{\cdot}(W,E) = H_c^{\cdot}(X,h_!E)$$

$\underline{\text{Proof}}$. Consider an injective resolution $E \to I^{\cdot}$ on W. Recall that an injective sheaf is soft, since it is a flabby sheaf. Accordingly $h_!E \to h_!I^{\cdot}$ is a $\Gamma_c(X,-)$-acyclic resolution. Using the acyclicity Theorem I.7.5 and formula 7.1 we get

$$H_c^{\cdot}(X,h_!S) = H^{\cdot}\Gamma_c(X,h_!I^{\cdot}) = H^{\cdot}\Gamma_c(W,I^{\cdot}) = H_c^{\cdot}(W,F)$$

$$\text{Q.E.D.}$$

Open and closed subspaces

Let us now turn to the case where $j: U \to X$ is the inclusion of an open subset of the locally compact space X. For a sheaf F on X, the adjunction morphism $j_!j^*F \to F$ will induce a morphism

$$H_c^{\cdot}(X,j_!j^*F) \to H_c^{\cdot}(X,F)$$

which we can compose with the isomorphism 7.3 to get a natural map of fundamental importance "extension by zero"

7.4 $$j_!: H_c^{\cdot}(U,F) \to H_c^{\cdot}(X,F)$$

<u>Meyer Vietoris sequence 7.5</u>. Two open subsets U and V of X give rise to a long exact sequence

$$\to H_c^P(U \cap V, F) \to H_c^P(U, F) \oplus H_c^P(V, F) \to H_c^P(U \cup V, F)$$

for any sheaf F on X.

<u>Proof</u>. Let us put names to the inclusions

$$h:\ U \cap V \to X,\quad i:\ U \to X,\quad j:\ V \to X,\quad k:\ U \cup V \to X.$$

Consider the exact sequence of sheaves on X

$$0 \to h_! h^* F \to i_! i^* F \oplus j_! j^* F \to k_! k^* F \to 0$$

where the morphisms are sum and difference between adjunction morphisms. The exactness is established by localization. The resulting cohomology sequence can be identified by means of 7.3.

$$\text{Q.E.D.}$$

Let $i:\ Z \to X$ denote the inclusion of the complement $Z = X - U$ in X. We can now use the fundamental sequence II 6.11

$$0 \longrightarrow j_! j^* F \longrightarrow F \longrightarrow i_* i^* F \longrightarrow 0$$

to derive a long exact sequence

$$7.6 \qquad\qquad \longrightarrow H_c^P(U, F) \longrightarrow H_c^P(X, F) \longrightarrow H_c^P(Z, F) \longrightarrow$$

<u>Variations over 7.6</u>. From the diagrams at the end of section II.6 can be deduced a number of useful exact sequences. We give some examples.

Given closed subspaces $S \subseteq Z$ of X with complements O and U respectively. Then we deduce from II.6.16 an exact commutative ladder of abelian groups

$$7.7 \quad \begin{array}{ccccccccc} \longrightarrow & H_c^{P-1}(Z,F) & \longrightarrow & H_c^P(U,F) & \longrightarrow & H_c^P(X,F) & \longrightarrow & H_c & \longrightarrow \\ & \downarrow i^* & & \downarrow j_! & & \downarrow 1 & & \downarrow i^* \\ \longrightarrow & H_c^{P-1}(S,F) & \longrightarrow & H_c^P(O,F) & \longrightarrow & H_c^P(X,F) & \longrightarrow & H_c^P(S,F) & \longrightarrow \end{array}$$

where $i: S \to Z$ and $j: U \to O$ denotes the inclusions.

Let there be given a proper, continuous map $f: X \to Y$, a closed subset B on Y with morphism $V = Y-B$. We put $A = f^{-1}B$ and $U = f^{-1}(V)$. A sheaf G on Y will give rise to a commutative ladder

$$7.8 \quad \begin{array}{ccccccccc} \longrightarrow & H_c^P(V,G) & \longrightarrow & H_c^P(X,G) & \longrightarrow & H_c^P(B,G) & \longrightarrow & H_c^{P+1}(V,G) & \longrightarrow 0 \\ & \downarrow & & \downarrow & & \downarrow & & \downarrow \\ \longrightarrow & H_c^P(U,f^*G) & \longrightarrow & H_c^P(X,f^*G) & \longrightarrow & H_c^P(A,f^*G) & \longrightarrow & H_c^{P+1}(U,f^*G) & \longrightarrow \end{array}$$

<u>Proof</u>. The exact sequence of sheaves on Y, in notation of II.6.15

$$0 \longrightarrow G_V \longrightarrow G \longrightarrow G_B \longrightarrow 0$$

transforms under f^* into the sequence

$$0 \longrightarrow f^*G_U \longrightarrow f^*G \longrightarrow f^*G_A \longrightarrow 0$$

as one sees by II.6.13. Conclusion by 1.6.

Q.E.D.

III.8   Cohomology of the n-sphere

In this section we shall make some geometric applications of cohomology with constant coefficients. We consider a fixed commutative ring k. The n-sphere

$$S^n = \{x \in \mathbb{R}^{n+1} \mid |x| = 1\}$$

has cohomology as follows

$$8.1 \qquad H^p(S^n, k) = \begin{cases} k & \text{for } p = 0, n \\ 0 & \text{otherwise} \end{cases}$$

Proof. Let us introduce the half spheres

$$S_+^n = \{x \in S^n \mid x_{n+1} \geq 0\} \qquad S_-^n = \{x \in S^n \mid x_{n+1} \leq 0\}$$

Each of these spaces are homeomorphic to $[0,1]^n$, thus they have vanishing higher cohomology groups by 6.5. Notice that $S_+^n \cap S_-^n = S^{n-1}$ and deduce the result from the Mayer-Vietoris sequence 1.8

$$\rightarrow H^p(S^n, k) \rightarrow H^p(S_+^n, k) \oplus H^p(S_-^n, k) \rightarrow H^p(S_+^n \cap S_-^n, k) \rightarrow$$

Q.E.D.

Brouwer's fixpoint theorem 8.2. Any continuous selfmap of the unitball $B^n = \{x \in \mathbb{R}^n \mid |x| \leq 1\}$ has a fixpoint.

$\underline{Proof}$. Let $f: B^n \to B^n$ be a continuous map without fix-points. We can then construct a continuous map $r: B^n \to S^{n-1}$ as follows

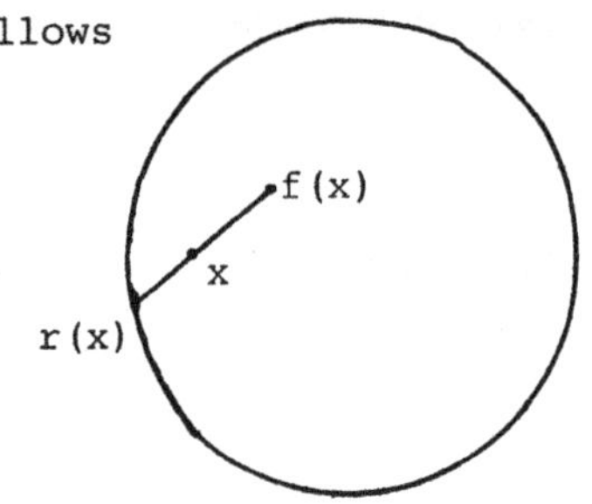

r(x) denotes the point where the oriented line from f(x) to x inter-sects S^{n-1}.

Let $i: S^{n-1} \to B^n$ denote the inclusion. Note, that $r \circ i = 1$. This implies that

$$i^*: H^n(B^n, \mathbb{Z}) \to H^n(S^{n-1}, \mathbb{Z})$$

is surjective, contradicting the calculation 8.2.

Q.E.D.

8.3 $\qquad H_c^p(\mathbb{R}^n, k) = \begin{cases} k & \text{for } p = n \\ 0 & \text{otherwise} \end{cases}$

$\underline{Proof}$. Let $N = (0,\ldots,0,1)$ denote the northpole of S^{n+1}. Stereographic projection from N will map $S^{n+1} - \{N\}$ homeomorphically onto $\mathbb{R}^n$. The result follows from the exact sequence

$$\to H_c^p(S^{n+1} - \{N\}, k) \to H_c^p(S^{n+1}, k) \to H_c^p(\{N\}, k) \to$$

and the formula 8.1.

Q.E.D.

<u>Cohomology of a closed halfspace</u> $\mathbb{R}_+^n = \{x \in \mathbb{R}^n \mid x_n \geq 0\}$

8.4 $$H_c^p(\mathbb{R}_+^n, k) = 0 \qquad \text{for all} \quad p \in \mathbb{N}$$

<u>Proof</u>. Use stereographic projection to see that $\mathbb{R}_+^n$ is homeomorphic to one half of the sphere S^{n+1} minus N , i.e. homeomorphic to $[0,1]^n - \{0\}$. The long exact sequence

$$H_c^p([0,1]^n - \{0\}, k) \to H^p([0,1]^n, k) \to H^p(\{0\}, k) \to$$

allow us to make the conclusion.

Q.E.D.

<u>The inclusion</u> $j: D \to \mathbb{R}^n$ <u>of an open disc</u> in $\mathbb{R}^n$ induces an isomorphism

8.5 $$j_! : H_c^{\,\cdot}(D, k) \xrightarrow{\sim} H_c^{\,\cdot}(\mathbb{R}^n, k)$$

<u>Proof</u>. Consider the exact sequence

$$\to H_c^p(D, k) \to H_c^p(\mathbb{R}^n, k) \to H_c^p(\mathbb{R}^n - D, k) \to$$

to see that it suffices to prove that $H_c^{\,\cdot}(\mathbb{R}^n - D, k) = 0$. This can be seen by the method used in the proof of 8.4.

Q.E.D.

<u>The degree of a proper map</u> $f: \mathbb{R}^n \to \mathbb{R}^n$ is by definition the integer <u>deg(f)</u> characterized by the formula

8.6 $$f^*\alpha = \deg(f)\alpha \qquad ; \; \alpha \in H_c^n(\mathbb{R}^n, \mathbb{Z}) \; .$$

$\underline{\text{Proposition 8.7.}}$ A proper map $f: \mathbb{R}^n \to \mathbb{R}^n$ whose degree is different from zero is surjective.

$\underline{\text{Proof.}}$ The image $Z = f(\mathbb{R}^n)$ is a closed subset of $\mathbb{R}^n$. Assume $Z \neq \mathbb{R}^n$ and let $i: Z \to \mathbb{R}^n$ denote the inclusion. Consider the exact sequence where j denotes the inclusion of $U = \mathbb{R}^n - Z$ in $\mathbb{R}^n$

$$H_c^n(U, \mathbb{Z}) \xrightarrow{j_!} H_c^n(\mathbb{R}^n, \mathbb{Z}) \xrightarrow{i^*} H_c^n(Z, \mathbb{Z})$$

The map $j_!$ is surjective as one sees by considering an open disc D contained in U and noticing that the inclusion $k: D \to \mathbb{R}^n$ induces an isomorphism

$$k_!: H_c^n(D, \mathbb{Z}) \longrightarrow H_c^n(\mathbb{R}^n, \mathbb{Z})$$

It follows that i^* is zero. Since $f: \mathbb{R}^n \to \mathbb{R}^n$ can be factored through i it follows that $f^* = 0$.

Q.E.D.

$\underline{\text{Proposition 8.8.}}$ A complex polynomial, $n \geq 1$

$$f(z) = z^n + a_1 z^{n-1} + \ldots + a_n$$

defines a proper map $f: \mathbb{C} \to \mathbb{C}$ of degree n.

$\underline{\text{Proof.}}$ Let us first show that f is properly homotopic to the map $z \mapsto z^n$. The formula

$$F(z,t) = z^n + ta_1 z^{n-1} + \ldots + ta_n$$

defines a continuous map $F: \mathbb{C} \times [0,1] \to \mathbb{C}$. To see that F
is proper, notice the inequality

$$|F(z,t)| \geq |z^n| \, (1 - |a_1 z^{-1}| \, \ldots \, -|a_n z^{-n}|)$$

and use this to find a constant r such that

$$|F(z,t)| \geq \tfrac{1}{2}|z^n| \qquad ; \quad t \in [0,1], \; |z| \geq r.$$

From this follows that $F: \mathbb{C} \times [0,1] \to \mathbb{C}$ is proper.

　　To see that $z \longmapsto z^n$ has degree n it suffices to calculate
the action on the open unit disc D^2. The exact sequence

$$H_c^1(B^2,\mathbb{Z}) \; \to \; H_c^1(S^1,\mathbb{Z}) \; \to \; H_c^2(D^2,\mathbb{Z}) \; \to \; H_c^2(B^2,\mathbb{Z})$$

shows that $H^1(S^1,\mathbb{Z}) \xrightarrow{\sim} H_c^2(D^2,\mathbb{Z})$. The action of $z \mapsto z^n$ on
$H^1(S^1,\mathbb{Z})$ is multiplication by n as it follows from 4.15.

Q.E.D.

__Fundamental theorem of algebra 8.9__. Any non constant complex
polynomial has a root in $\mathbb{C}$.

　　__Proof__. Combine 8.7. and 8.8.

Q.E.D.

__Proposition 8.10__. A transformation $\sigma \in Gl_n(\mathbb{R})$ has degree
$\mathrm{sign}(\det(\sigma))$, the sign of the determinant.

$\underline{\text{Proof}}$. The automorphism group of $H_c^{\,n}(\mathbb{R}^n,\mathbb{Z})$ is cyclic of order 2: multiplication with $+1$ or multiplication with -1. Using the fact that the commutator group of $\text{Gl}_n(\mathbb{R})$ is $\text{Sl}_n(\mathbb{R})$ we find that $\deg(\sigma)$ depends only on $\det(\sigma)$. Thus it suffices to treat the case of a diagonal matrix of the form

$$\sigma = \begin{Bmatrix} 1 & & 0 \\ & \ddots & \\ & & 1 \\ 0 & & r \end{Bmatrix}$$

This will be done by induction on the dimension. The case $n = 1$ has been treated in 4.10. The inductive step is accomplished by the Mayer-Vietoris sequence 1.8

$$\to H_c^{\,p}(\mathbb{R}^n,\mathbb{Z}) \to H_c^{\,p}(\mathbb{R}_+^{\,n},\mathbb{Z}) \oplus H_c^{\,p}(\mathbb{R}_-^{\,n},\mathbb{Z}) \to H_c^{\,p}(\mathbb{R}_+^{\,n} \cap \mathbb{R}_-^{\,n},\mathbb{Z}) \to$$

and the result 8.4.

$$\text{Q.E.D.}$$

$\underline{\text{Corollary 8.1}}$1. An orthogonal transformation $\sigma \in O_{n+1}(\mathbb{R})$ acts on $H^n(S^n,\mathbb{Z})$ as multiplication by $\det(\sigma)$, the determinant of σ.

$\underline{\text{Proof}}$. Let D^{n+1} denote the open unit disc in $\mathbb{R}^{n+1}$. It follows from 8.5 that $O_{n+1}(\mathbb{R})$ acts on $H_c^{\,n+1}(D^{n+1},\mathbb{Z})$ through the determinant. The exact sequence

$$\to H_c^{\,n}(B^{n+1},\mathbb{Z}) \to H_c^{\,n}(S^n,\mathbb{Z}) \to H_c^{\,n+1}(D^{n+1},\mathbb{Z}) \to H_c^{\,n+1}(B^{n+1},\mathbb{Z})$$

allows us to make the final conclusion.

$$\text{Q.E.D.}$$

By a _tangent vector field_ on a sphere S^p we understand a continuous map $v: S^p \to \mathbb{R}^{p+1}$ such that x and $v(x)$ are orthogonal for all $x \in S^p$. As an example, the sphere S^{2n-1} admits the tangent vector field

$$(-x_2, x_1, -x_4, x_3, \ldots, -x_{2n}, x_{2n-1}) \quad ; \quad x \in S^{2n-1}$$

Notice that each tangent vector is a unit vector.

The hairy ball theorem 8.12. Any continuous tangent vector field on an even dimensional sphere S^{2n} has a zero.

Proof. Assume to the contrary that $v: S^{2n} \to \mathbb{R}^{2n+1}$ represents a zero free tangent-vectorfield. By normalization we can assume that $v(x)$ is a unit vector for all $x \in S^{2n}$. Consider the homotopy $f: S^{2n} \times [0,1] \to S^{2n}$ given by

$$f(x,t) = \cos(\pi t)x + \sin(\pi t)v(x)$$

Notice, that $f(x,0) = x$ and $f(x,1) = -x$, $x \in S^{2n}$. This shows that the _antipodal_ map is homotopic to the identity, contradicting 8.11.

Q.E.D.

Topological manifolds

A topological space X is called a _topological n-manifold_, if it is a Hausdorff space, and each point x of X has an open neighbourhood homeomorphic to $\mathbb{R}^n$ or $\mathbb{R}^n_+$. In the first case x is called an _interior_ point, otherwise a _boundary point_. The set of boundary points forms the boundary ∂X of X.

In order to distinguish the two sets of points we shall introduce a presheaf $o\hbar$ on X. For the inclusion $j: V \to U$ of two open subsets of X consider

$$j_!: H_c^{\,n}(V,k) \longrightarrow H_c^{\,n}(U,k)$$

and the linear transposed map

$$8.13 \qquad \mathrm{Hom}_k(H_c^{\,n}(U,k),k) \to \mathrm{Hom}_k(H_c^{\,n}(V,k),k)$$

The presheaf $o\hbar$ on X defined by 8.13 is called the _orientation presheaf on_ X relative to k. From 8.4 and 8.5 we find

$$8.14 \qquad o\hbar_x \;\tilde{\to}\; \begin{cases} 0 & \text{if } x \in \partial X \\ k & \text{if } x \in X - \partial X \end{cases}$$

Proposition 8.15. A homemorphism between topological manifolds will transform boundary points into boundary points.

Proof. Follows from 8.14.

Q.E.D.

Example 8.16. A Möbius band M is not homeomorphic to an ordinary band $S^1 \times [0,1]$ since ∂M is connected while the ordinary band has a disconnected boundary.

III.9   Dimension of locally compact spaces

In this section we shall prove a number of vanishing theorems
for cohomology with compact support. A basic such result is

$$9.1 \qquad H_c^2(\mathbb{R},F) = 0 \qquad \text{for all sheaves } F \text{ on } \mathbb{R}$$

We shall derive this from the following general principle.

__Minimality principle 9.2.__ Let F be a sheaf on the locally
compact space X and let Z denote a family of closed subsets
of X stable under arbitrary intersections, in particular $X \in Z$.
Let $\alpha \in H_c^P(X,F)$ be a non vanishing cohomology class. For
$Z \in Z$ let $\alpha|_Z \in H_c^P(Z,F)$ denote the restriction of α to Z.
The ordered set

$$\{ Z \in Z \mid \alpha|_Z \neq 0 \}$$

contains minimal elements.

__Proof.__ Follows immediately from Zorn's lemma and lemma 9.3
below.
$$\text{Q.E.D.}$$

__Lemma 9.3.__ Let F denote a sheaf on the locally compact
space X and Z a downward directed family of subsets of X
with intersection W. We have

$$\varinjlim H_c^P(Z,F) \;\tilde{\rightarrow}\; H_c^P(W,F)$$

the limit taken over $Z \in Z$.

196

$\underline{\text{Proof}}$. For a closed subset Y of X, let $i: Y \to X$ denote the inclusion and put $F_Y = i_* i^* F$. For closed sets $Z \supseteq Y$, the adjunction morphism, II.5.5, will induce a morphism

$$F_Z \longrightarrow F_Y$$

and thereby create a $\underline{\text{direct system of sheaves on the set of}}$ $\underline{\text{closed subsets of}}$ X. With the notation above, we find by localization, that

$$\varinjlim F_Z \overset{\sim}{\to} F_W$$

Conclusion by 5.1 and 7.3.

Q.E.D.

$\underline{\text{Proof of 9.1}}$. Given $\alpha \in H_c^2(\mathbb{R}, F)$ with $\alpha \neq 0$. Choose a closed subset Z of $\mathbb{R}$ such that $\alpha|_Z \neq 0$ and which is minimal with that property. Let z be a point of Z which separates two points of Z and put

$$Z_- = \,]-\infty, z] \cap Z, \qquad Z_+ = [z, +\infty[\cap Z$$

Use the Mayer-Vietoris sequence 1.8

$$\to H_c^1(\{z\}, F) \to H_c^2(Z, F) \to H_c^2(Z_-, F) \oplus H_c^2(Z_+, F) \to$$

to obtain a contradiction.

Q.E.D.

<u>Definition 9.4</u>. Let X be a locally compact space. The
<u>dimension</u> of X, <u>dim X</u>, is the smallest integer n for which

$$H_c^{n+1}(X,F) = 0 \quad \text{for all sheaves } F \text{ on } X$$

Using imbedding into injective sheaves it is easy to deduce
that for any sheaf F on X

$$H_c^{i}(X,F) = 0 \quad \text{for } i > \dim X.$$

<u>Proposition 9.5</u>. For a locally compact space X we have

$$\dim X \times \mathbb{R} \leq \dim X + 1$$

<u>Proof</u>. Let X have finite dimension n, and let $\mathcal{Z}$ denote
the set of closed subsets of $X \times \mathbb{R}$ of the form $X \times Z$, where
Z is a closed subset of $\mathbb{R}$. For a sheaf F on $X \times \mathbb{R}$ and a non
trivial cohomology class

$$\alpha \in H_c^{n+2}(X \times \mathbb{R}, F) \qquad ; \alpha \neq 0$$

use the minimality principle 9.2 and a Mayer-Vietoris sequence
similar to the one used in the proof of 9.1 to deduce a contradiction.

Q.E.D.

<u>Example 9.6</u>. dim $\mathbb{R}^n$ = n.

$\underline{\text{Proposition 9.7}}$. A locally closed subspace W of the locally compact space X satisfies $\quad \dim W \leq \dim X$

$\underline{\text{Proof}}$. This is a consequence of 7.3.

$$\text{Q.E.D.}$$

$\underline{\text{Example 9.8}}$. For a proper closed subset Z of $\mathbb{R}^n$

$$H_c^n(Z,k) = 0$$

for any constant sheaf k. - Let $j: U \longrightarrow \mathbb{R}^n$ denote the inclusion of the complement and consider the exact sequence.

$$H_c^n(U,k) \xrightarrow{\ j_!\ } H_c^n(\mathbb{R}^n,k) \longrightarrow H_c^n(Z,k) \longrightarrow H_c^{n+1}(U,k)$$

The last group is zero by 9.7. The map $\quad j_!\quad$ is surjective as it follows by considering an open disc $D \subseteq U$ and using 8.5.

$\underline{\text{Proposition 9.9}}$. Let X be a locally compact space of dimension $\leq n$. Given an exact sequence

$$0 \to F \to S^0 \to S^1 \to \ldots \to S^{n-1} \to S^n \to 0$$

If $S^0, S^1, \ldots, S^{n-1}$ are soft sheaves, then S^n is soft.

$\underline{\text{Proof}}$. It follows from the long exact sequence III.7 and the material in III.2, that

A sheaf G on X is soft if and only if $H_c^1(U,G) = 0$ for all open subsets U of X.

By splitting the complex into short exact sequences one sees
that for any open set U

$$H_c^{1}(U,S^n) = H_c^{n+1}(U,F).$$

The last group is zero by 9.7.

Q.E.D.

<u>Definition 9.10</u>. For a point x of the locally compact
space X, we put

$$\dim_x X = \inf \dim U$$

where U varies over open neighbourhoods of x in X.

The concept of dimension is local in the sense made precise
by the formula

$$9.11 \qquad\qquad \dim X = \sup_{x \in X} \dim_x X$$

<u>Proof</u>. It follows from 9.7 that

$$\sup_x \dim_x X \le \dim X$$

Put $n = \sup_x \dim_x X$. For a given sheaf F on X choose a resolution

$$0 \to F \to S^0 \to S^1 \to \ldots \to S^{n-1} \to S^n \to 0$$

where $S^0,\ldots,S^{n-1}$ are soft sheaves. For each point x of X
choose an open neighbourhood U_x of x with $\dim U_x = \dim_x X$.
According to 9.9 the restriction of S^n to U_x is soft. It
follows from 2.8 that S^n is a soft sheaf on X. This proves
that $H_c^{n+1}(X,F) = 0$.

Q.E.D.

III.10   Wilder's finiteness theorem

In this section we shall use a Mayer-Vietoris technique to establish a basic finiteness result.

<u>Theorem 10.1.</u> Let X be a compact topological manifold with boundary and k a noetherian ring. Each cohomology module $H^i(X,k)$ is finitely generated over k.

We shall prove a theorem which is both more general and contains more information in the concrete case.

<u>Theorem 10.2.</u> Let X be a locally compact space, k a noetherian ring and F a k-sheaf on X. Suppose that for any $i \in \mathbb{N}$, any $x \in X$ and any compact neighbourhood N of x, there exists a compact neighbourhood $M \subseteq N$ of x such that the map $H^i(N,F) \to H^i(M,F)$ has finitely generated image. Then for any pair of compact subsets L and K of X with $L \supseteq K$ the restriction map

$$H^i(L,F) \longrightarrow H^i(K,F) \qquad\qquad ;\quad i \in \mathbb{Z}$$

has finitely generated image.

<u>Proof.</u> This is done by increasing induction on $i \in \mathbb{Z}$. To accomplish the inductive step, let us fix L and let L denote the set of compact subsets A for which there exists a compact subset C with $A \subseteq \overset{\bullet}{C} \subseteq C \subseteq \overset{\bullet}{L}$ such that $H^i(L,F) \to H^i(C,F)$ has finitely generated image. The set L has the following properties.

1) Every point x has a compact neighbourhood $M \in L$.

2) If $B \in L$ then any compact subset of B belongs to L.

3) If $A \in L$ and $R \in L$, then $A \cup R \in L$. Only 3) needs a proof which will be given below. It follows from Borel-Heine and 1), 2), 3) that every compact subset K of $L^{\cdot}$ belongs to L .

3) To see this choose for A compact sets B and C with $A \subseteq \dot{B} \subseteq B \subseteq \dot{C} \subseteq C \subseteq \dot{L}$ such that $H^i(L,F) \to H^i(C,F)$ has finite generated image. Choose similar data S and T for R. Consider the following commutative diagram constructed on the basis of Mayer-Vietoris sequences II.5.6

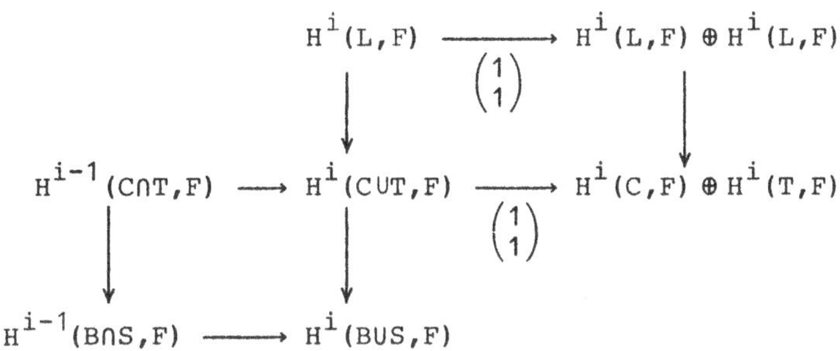

and conclude on the basis of the lemma below that the map $H^i(L,F) \to H^i(B \cup S,F)$ has finitely generated image.

Q.E.D.

<u>Lemma 10.3</u>. Given a commutative diagram of modules over a noetherian ring. Suppose that CDE is exact and that CE and BE have finitely generated image. Then the composite AE of AD and DE has finitely generated image.

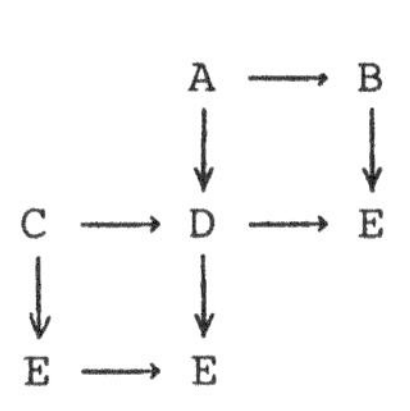

<u>Proof</u>. Left to the reader with the recommendation to reduce to the case where A = B and B = E.

Q.E.D.

IV. Cohomology and Analysis

<u>IV.1 Homotopy invariance of sheaf cohomology</u>

In this section we shall prove that sheaf cohomology with constant coefficient is a <u>homotopy invariant</u> of the space. Recall that continuous maps $f,g: X \to Y$ are said to be <u>homotopic</u> if there exists a continuous map $F: X \times [0,1] \to Y$ with

$$F(x,0) = f(x), \quad F(x,1) = g(x) \qquad ; \quad x \in X$$

<u>Theorem 1.1</u>. Homotopic continuous maps $f,g: X \to Y$ induce the same map

$$f*,g*: H^{\cdot}(Y,k) \longrightarrow H^{\cdot}(X,k)$$

for any abelian group k.

Let us first indicate how this is done in case X and Y are locally compact. The first thing to do is to prove the Vietoris-Begle mapping theorem 1.2 below for ordinary cohomology as opposed to cohomology with compact support, III.6.4. Next, adapt the proof of III.6.6 to ordinary cohomology. None of this offers any difficulties and is left to the reader.

<u>Vietoris-Begle mapping theorem 1.2</u>. Let $f: X \to Y$ be a proper map between locally compact spaces with acyclic fibers. For any sheaf G on Y the induced map

$$f^*: H^{\cdot}(Y,G) \longrightarrow H^{\cdot}(X,f*G)$$

is an isomorphism.

<u>Proof of 1.1</u>. Let us use an ad hoc terminology and say that a continuous map $f: X \to Y$ between topological spaces is proper if

1.3   i.   f is closed, i.e. the image of a closed set
               is closed.

          ii. Any two distinct points of X in the same fiber
               has disjoint neighbourhoods in X.

         iii. The fibers of f are compact.

For a proper map $f: X \to Y$ we have an isomorphism

1.4 $(R^p f_* F)_y = H^p(f^{-1}(y),F)$

for $y \in Y$ and F a sheaf on X. - This statement is proved exactly as III.6.2 with the following modification of III.6.3:

Let K be a compact subspace of X with the property that any two distinct points of K have disjoint neighbourhoods in X. Then for any sheaf F on X

$$H^n(K,F) = \varinjlim H^n(U,F)$$

as U varies over all open neighbourhoods of K in X.

To prove this, notice first that the case $n = 0$ is covered by the proof of III.2.2. This can be used to prove that a flabby sheaf on X restricts to a soft sheaf on K. The proof can be completed as the proof of III.6.3.

We can now generalize 1.2 to a continuous map $f: X \to Y$, proper in the sense introduced above. Finally, copy the proof of III.6.6, using Lemma 1.5 below.

Q.E.D.

<u>Lemma 1.5</u>. Let X be topological space and C a compact topological space. The projection $p: X \times C \to X$ is closed, i.e. the image of a closed set is a closed set.

<u>Proof</u>. Let Z denote a closed subset of $X \times C$ and x a point of X outside $p(Z)$. For each point $c \in C$ choose open neighbourhoods V_c of c and U^c of x such that $U^c \times V_c \cap Z = \emptyset$. Let S be a finite subset of C such that $(V_c)_{c \in S}$ cover C. The set $\bigcap_{c \in S} U^c$ is an open neighbourhood of x in X which does not meet $p(Z)$.

Q.E.D.

Application to fibre bundles

Let F denote a topological space. By a __fibre bundle with fiber__ F we understand a continuous map $f: X \to Y$ such that each point y of Y admits a neighbourhood V and a homeomorphism $g: f^{-1}(V) \overset{\sim}{\to} V \times F$ such that the following diagram is commutative

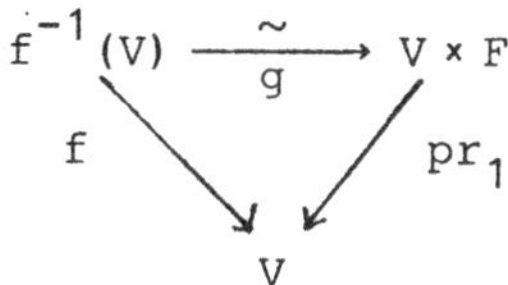

Theorem 1.6. Let $f: X \to Y$ be a fibre-bundle whose fibre F is homotopic to a compact space and let k denote an abelian group. For each point y of Y the canonical map

$$(R^p f_* \underset{\sim}{k})_y \longrightarrow H^p(f^{-1}(y), \underset{\sim}{k})$$

is an isomorphism.

Proof. The problem is local on Y so we may assume that $f: Y \times F \to Y$ is the projection onto the first factor. Let $g: F \to C$ be a homotopy equivalence, where C is compact and let $p: Y \times C \to Y$ denote the projection. According to 1.5, the map p satisfies the three conditions 1.3. Thus we can make the conclusion 1.4 for p. It remains to involve Theorem 1.1 in order to make the correct identifications.

Q.E.D.

IV.2 Locally compact space, countable at infinity

In this section we consider a locally compact space X which is <u>countable at infinity</u>, i.e. which is union of a countable family $(K_n)_{n \in \mathbb{N}}$ of compact subset. From this we can construct a filtration $(X^n)_{n \in \mathbb{N}}$ of X by compact subsets. We shall proceed inductively: $X^0 = K_0$, X^{n+1} is a compact neighbourhood of $X^n \cup K_n$. Notice,

$$2.1 \qquad X^n \subseteq \overset{\bullet}{X}{}^{n+1} \qquad ; \quad n \in \mathbb{N}$$

<u>Theorem 2.2</u>. Let X be a locally compact space, countable at infinity. A soft sheaf E on X satisfies

$$H^i(X,E) = 0 \qquad ; \quad i \geq 1$$

<u>Proof</u>. Consider a short exact sequence of sheaves on X

$$0 \longrightarrow E \overset{c}{\longrightarrow} F \overset{f}{\longrightarrow} G \longrightarrow 0$$

where E is soft. Let us prove that $\Gamma(X,F) \to \Gamma(X,G)$ is surjective. We shall use a filtration $(X^n)_{n \in \mathbb{N}}$ of X by compact subsets satisfying 2.1. Let there be given a section t of G over X. We shall proceed inductively: Suppose we have constructed a section s^n of F over X^n which projects onto the restriction t^n of t to X^n. Choose a lifting $r^{n+1} \in \Gamma(X^{n+1},F)$ of t^{n+1}. The difference between s^n and the restriction of r^{n+1} to X^n defines a section of E over X^n. Extend this to a section of E over X^{n+1} and add the result to r^{n+1}. This gives a section

s^{n+1} of F over X^{n+1} which extends s^n and projects onto t^{n+1}. - From the family of sections $(s^n)_{n \in \mathbb{N}}$ we can construct a section of F over X, namely the section s whose restriction to $\dot{X}^n$ agrees with the restriction of s^n to $\dot{X}^n$. It is clear that s projects onto t.

If we apply this result to an exact sequence with F injective, we get $H^1(X,E) = 0$. Let us notice that the sheaf G is soft by the proof of III.2.7. The cohomology sequence

$$\longrightarrow H^{p-1}(X,G) \longrightarrow H^p(X,E) \longrightarrow H^p(X,F)$$

will now allow us to conclude the proof by induction.

Q.E.D.

<u>Corollary 2.3</u>. Let X denote a locally compact space countable at infinity and E a soft sheaf on X. For any closed subspace Z of X the restriction

$$\Gamma(X,E) \longrightarrow \Gamma(Z,E)$$

is surjective.

<u>Proof</u>. Put $U = X-Z$ and let $i: Z \to X$ and $j: U \to X$ denote the inclusions. The exact sequence

$$0 \longrightarrow j_! j^*E \longrightarrow E \longrightarrow i_* i^*E \longrightarrow 0$$

consists of soft sheaves, III.2.5 and III.7.2. The exact sequence

208

$$0 \to \Gamma(X,j_!j^*E) \to \Gamma(X,E) \to \Gamma(Z,i^*E) \to H^1(X,j_!j^*E)$$

and 2.2. allows us to make the conclusion.

$$Q.E.D.$$

$\underline{\text{Proposition 2.4}}$. Let X denote a locally compact space countable at infinity. For any sheaf F on X and any closed subspace Z of X we have

$$\varinjlim H^p(U,F) \overset{\sim}{\to} H^p(Z,F) \qquad ; \ p \in \mathbb{N}$$

$\underline{\text{Proof}}$. Let us consider the case $p = 0$. The map

$$\varinjlim \Gamma(U,F) \to \Gamma(Z,F)$$

is injective for general reasons and surjective by 2.3. Consider an exact sequence

$$0 \longrightarrow F \longrightarrow I \longrightarrow J$$

where I and J are injective sheaves. This induces a commutative exact diagram

$$
\begin{array}{ccccccc}
0 & \longrightarrow & \varinjlim \Gamma(U,F) & \longrightarrow & \varinjlim \Gamma(U,I) & \longrightarrow & \varinjlim \Gamma(U,J) \\
& & \downarrow & & \downarrow & & \downarrow \\
0 & \longrightarrow & \Gamma(Z,F) & \longrightarrow & \Gamma(Z,I) & \longrightarrow & \Gamma(Z,J)
\end{array}
$$

The two vertical arrows to the right are isomorphisms by our previous remarks. From this we conclude that the last vertical arrow is an isomorphism.

In the general case, choose an injective resolution $F \to I^{\cdot}$. The resolution $i*F \to i*I^{\cdot}$ is a soft resolution and consequently a $\Gamma(Z,-)$-acyclic resolution as it follows from the fact that Z is countable at infinity and 2.2. Consequently

$$\varinjlim H^p(U,F) = H^p \varinjlim \Gamma(U,I^{\cdot}) = H^p\Gamma(Z,i*I^{\cdot}) = H^p(Z,F)$$

$$Q.E.D.$$

Let me list three useful remarks from pointset topology, which will facilitate the applications of the material of this section.

2.5   A locally compact space with a countable basis is countable at infinity.

2.6   A subspace of a space with a countable basis has itself a countable basis.

2.7   A metric space has a countable basis if and only if it contains a countable dense subset.

<u>Remark 2.8</u>. The results of this section can be generalized to a locally compact space which is paracompact. This can be done by using the fact that such a space is the topological sum of locally compact spaces each of which is countable at infinity, Bourbaki (2) §9, no. 10, Thm. 5. The general reference for sheaves on paracompact spaces is Godement (1).

IV.3 Complex Logarithms

In this section we shall make some sheaf theoretical interpretations of a number of elementary problems in ordinary complex analysis.

Argument of complex number

Let U denote a connected open subset of the complex plane. The exponential sequence

$$0 \longrightarrow \underset{\sim}{\mathbb{Z}} \xrightarrow{2\pi} C(\mathbb{R}) \xrightarrow{\text{expi}} C(S^1) \longrightarrow 0$$

gives an exact sequence

$$0 \longrightarrow \mathbb{Z} \xrightarrow{2\pi} \Gamma(U, C(\mathbb{R})) \xrightarrow{\text{expi}} \Gamma(U, C(S^1)) \longrightarrow H^1(U, \mathbb{Z}) \longrightarrow 0$$

where we have used that $H^1(U, C(\mathbb{R})) = 0$, 2.2 and III.3.3.

Let us suppose that U does not contain the origin, in which case $C(S^1)$ has a canonical section over U, namely $z \longmapsto z|z|^{-1}$. Any lifting of this section to $C(\mathbb{R})$ is called an _argument_ and is denoted _arg_. Thus

$$z = |z| \exp(i \arg z) \qquad ; z \in U$$

Collecting this together we have proved that

3.1   If $H^1(U, \mathbb{Z}) = 0$ then we can find a continuous argument function on U.

Logarithms of complex numbers

The multiplicative group of complex numbers $\mathbb{C}^*$ fits into an exact sequence

$$0 \longrightarrow \mathbb{Z} \xrightarrow{2\pi i} \mathbb{C} \xrightarrow{\exp} \mathbb{C}^* \longrightarrow 0$$

of topological groups. The exponential map has local sections in virtue of the decomposition

$$
\begin{array}{ccccc}
x+iy & \mathbb{C} & \longrightarrow & \mathbb{C}^* & z \\
\Big\downarrow & \Big\downarrow & & \Big\downarrow & \Big\downarrow \\
(x,y) & \mathbb{R}\oplus\mathbb{R} & \longrightarrow & \mathbb{R}^+\oplus S^1 & (|z|,z|z|^{-1}) \\
& (x,y) & \longmapsto & (e^x,\exp i\, y) &
\end{array}
$$

and the fact that $\exp i: \mathbb{R} \longrightarrow S^1$ has local sections. There results an exact sequence of sheaves on $\mathbb{C}$

$$3.2 \qquad 0 \longrightarrow \underset{\sim}{\mathbb{Z}} \xrightarrow{2\pi i} C(\mathbb{C}) \xrightarrow{\exp} C(\mathbb{C}^*) \longrightarrow 0$$

For a connected open subset U of $\mathbb{C}$ we have an exact sequence

$$0 \longrightarrow \mathbb{Z} \xrightarrow{2\pi i} \Gamma(U,C(\mathbb{C})) \xrightarrow{\exp} \Gamma(U,C(\mathbb{C}^*)) \longrightarrow H^1(U,\mathbb{Z}) \longrightarrow 0$$

where we have used $H^1(U,C(\mathbb{C})) = 0$.

If U does not contain the origin, the inclusion of U in $\mathbb{C}^*$ represents an element of $\Gamma(U,C(\mathbb{C}^*))$. Any lifting of this to a section of $C(\mathbb{C})$ over U will be called a <u>logarithm</u> on U. Thus a logarithm on U is a continuous map $\log: U \to \mathbb{C}$, such that

$$z = \exp(\log z) \qquad\qquad ; \ z \in U$$

Using the commutative diagram above we see that logarithm and argument are related by the formula

3.3
$$\log z = \log |z| + i \arg z \qquad ; \ z \in U$$

It follows from the lemma below that $\underline{\log z}$ $\underline{\text{is an analytic}}$ $\underline{\text{function on}}$ $\underline{U}$. From the relation $\exp(\log z) = z$ we find at once that

3.4
$$\frac{d}{dz} \log z = \frac{1}{z} \qquad\qquad ; \ z \in U$$

$\underline{\text{Lemma 3.5.}}$ Any continuous section of

$$\exp : \mathbb{C} \to \mathbb{C}^*$$

over any open subset U of $\mathbb{C}^*$ is analytic.

$\underline{\text{Proof.}}$ By a translation argument in the group $\mathbb{C}^*$ it suffices to construct one analytic section over some open neighbourhood of 1 in $\mathbb{C}^*$. - The power series

$$\ell(z) = \sum_{n=1}^{+\infty} (-1)^{n+1} \frac{z^n}{n} \qquad ; \ |z| < 1$$

has radius of convergence 1 and satisfies

$$\ell'(z) = \frac{1}{z+1} \qquad\qquad ; \ |z| < 1$$

Note, $\ell(0) = 0$. Let us prove that

$$\exp(\ell(z-1)) = z \qquad ; \ |z-1| < 1$$

Notice that the two functions have the same logarithmic derivative and both takes the value 1 for $z = 1$.

Q.E.D.

Corollary 3.6. Let $\mathcal{O}$ denote the sheaf of analytic functions on $\mathbb{C}$ and $\mathcal{O}*$ the sheaf of zerofree analytic functions on $\mathbb{C}$. Then we have an exact sequence of sheaves

$$0 \longrightarrow \underset{\sim}{\mathbb{Z}} \xrightarrow{\ 2\pi i\ } \mathcal{O} \xrightarrow{\ \exp\ } \mathcal{O}* \longrightarrow 0$$

n'th roots of unity 3.7. For an integer $n \geq 2$ let μ_n denote the n'th roots of unity. Raising to the n'th power gives a sequence of multiplicative sheaves on the complex plane

$$0 \longrightarrow \underset{\sim}{\mu}_n \longrightarrow \mathcal{O}* \xrightarrow{\ n\ } \mathcal{O}* \longrightarrow 0$$

which is exact as it follows from 3.6. For a connected open set U of $\mathbb{C}$ we get the exact sequence

$$0 \longrightarrow \mu_n \longrightarrow \Gamma(U, \mathcal{O}*) \xrightarrow{\ n\ } \Gamma(U, \mathcal{O}*) \longrightarrow H^1(U, \mu_n)$$

Let there be given a fixed $f \in \Gamma(U, \mathcal{O}*)$ where U is an open subset of $\mathbb{C}$ with $H^1(U, \mu_n) = 0$. We want to analyse the space

$$Z = \{(x, y) \in U \times \mathbb{C} \mid y^n = f(x)\}$$

Let us prove that this is homeomorphic to $U \times \mu_n$: Choose $g \in \Gamma(U, 0^*)$ with

$$g(z)^n = f(z) \qquad ; \quad z \in U$$

We can construct two continuous maps

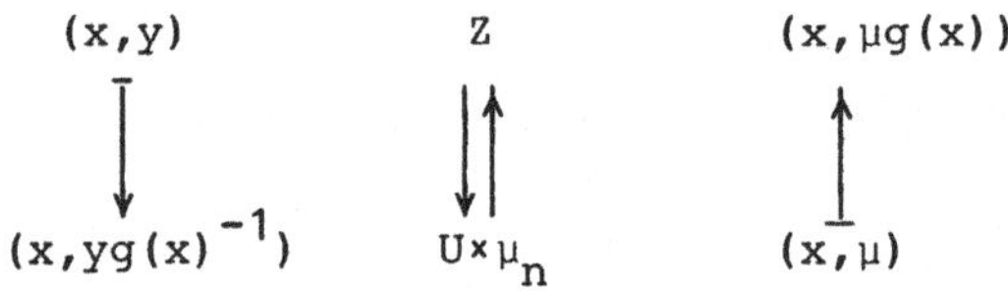

which are easily seen to be inverse to each other.

Let us drop the assumptions on U. The above result shows that $Z \to U$ is a covering space with μ_n as covering group.

The pointed complex plane $\mathbb{C}^*$

Let us recall that an analytic function f on $\mathbb{C}^*$ can be expanded in a Laurent series

$$f(z) = \sum_{n=-\infty}^{+\infty} a_n z^n \qquad ; \quad z \in \mathbb{C}^*$$

and that the residue of f is defined by

$$3.8 \qquad \qquad \mathrm{Res}(f, 0) = a_{-1}$$

From this follows easily that we have an exact sequence

$$0 \longrightarrow \mathbb{C} \longrightarrow \Gamma(\mathbb{C}^*, 0) \xrightarrow{\ D\ } \Gamma(\mathbb{C}^*, 0) \xrightarrow{\ \mathrm{Res}\ } \mathbb{C} \longrightarrow 0$$

Another basic feature is that a continuous map $f: \mathbb{C}^* \to \mathbb{C}^*$ has a _degree_, $\deg(f)$ given by

$$f^*\alpha = \deg(f)\alpha \qquad ; \quad \alpha \in H^1(\mathbb{C}^*,\mathbb{Z})$$

Using the material from III.4 we deduce an exact sequence

$$0 \longrightarrow \mathbb{Z} \xrightarrow{2\pi i} \Gamma(\mathbb{C}^*,C(\mathbb{C})) \xrightarrow{\exp} \Gamma(\mathbb{C}^*,C(\mathbb{C}^*)) \xrightarrow{\deg} \mathbb{Z} \longrightarrow 0$$

and as a consequence of Proposition 3.6

$$0 \longrightarrow \mathbb{Z} \xrightarrow{2\pi i} \Gamma(\mathbb{C}^*,O) \xrightarrow{\exp} \Gamma(\mathbb{C}^*,O^*) \xrightarrow{\deg} \mathbb{Z} \longrightarrow 0$$

The degree can be calculated by the formula

3.9
$$\boxed{\quad \deg(f) = \operatorname{Res}\!\left(\frac{f'}{f};0\right) \quad}$$

    Proof. We can write f in the form

$$f(z) = z^n\exp(g(z)) \qquad ; \qquad z \in \mathbb{C}^*$$

where g is analytic in $\mathbb{C}^*$. This gives

$$\frac{f'(z)}{f(z)} = n\,\frac{1}{z} + g'(z)$$

Note that $\operatorname{Res}(g',0) = 0$, and the result follows.

Q.E.D.

IV.4 <u>Complex curve integrals</u>. <u>The monodromy theorem</u>.

The complex plane comes equipped with the following sequence of sheaves

$$4.1 \qquad 0 \to \underset{\sim}{\mathbb{C}} \to \mathcal{O} \overset{D}{\to} \mathcal{O} \to 0$$

where $D = d/dz$ is the complex differentiation. This sequence is exact as it follows from the theory of complex power series Let there be given an open subset U of $\mathbb{C}$, a continuous curve $\gamma: [a,b] \to U$ and an analytic function $f(z)$ on U. We are going to make sense to the symbol

$$\int_\gamma f(z)\,dz$$

Consider the pull back along γ of the sequence above

$$0 \to \underset{\sim}{\mathbb{C}} \to \gamma*\mathcal{O} \overset{D}{\to} \gamma*\mathcal{O} \to 0$$

Since $H^1([a,b],\mathbb{C}) = 0$ we get an exact sequence

$$0 \to \mathbb{C} \to \Gamma([a,b],\gamma*\mathcal{O}) \to \Gamma([a,b],\gamma*\mathcal{O}) \to 0$$

The adjunction morphism $\mathcal{O} \to \gamma_*\gamma*\mathcal{O}$ induces a map

$$\Gamma(U,\mathcal{O}) \to \Gamma([a,b],\gamma*\mathcal{O})$$

Let $\gamma*f$ denote the image of f under this map, choose $g \in \Gamma([a,b],\gamma*\mathcal{O})$ with $Dg = \gamma*f$ and put

$$\int_\gamma f(z)\,dz = g(b) - g(a)$$

where $g(t)$ denotes evaluation of the stalk $g_t \in 0_{\gamma(t)}$ at $\gamma(t)$.

Let us establish the basic rules for this symbol. For an analytic function $h(z)$ on U we have

$$4.2 \qquad \int_\gamma Dh(z)\,dz = h(\gamma(b)) - h(\gamma(a))$$

as it follows by considering $\gamma^* h \in \Gamma([a,b], \gamma^* 0)$. –

Subdivide $[a,b]$ into two intervals $[a,c]$ and $[c,b]$ and let γ_1, γ_2 denote the respective restrictions. We leave it to the reader to show that

$$4.3 \qquad \int_\gamma f(z)\,dz = \int_{\gamma_1} f(z)\,dz + \int_{\gamma_2} f(z)\,dz$$

In case $\gamma: [a,b] \to U$ is continuously differentiable, our symbol can be evaluated as follows

$$4.4 \qquad \int_\gamma f(z)\,dz = \int_a^b f(\gamma(t))\gamma'(t)\,dt$$

Proof. Suppose first that $f = Dh$. In that case

$$\int_a^b f(\gamma(t))\gamma'(t)\,dt = \int_a^b \frac{d}{dt} h(\gamma(t))\,dt = h(\gamma(b)) - h(\gamma(a))$$

and the result follows from 4.2. In the general case we can use 4.3 to make a reduction to this case by a finite subdivision, using Borel-Heine.

$$Q.E.D.$$

218

Proposition 4.5. Let $\beta, \gamma: [a,b] \to U$ be continuous curves which are homotopic with fixed endpoints. Then for any analytic function $f(z)$ on U

$$\int_\gamma f(z)\,dz = \int_\beta f(z)\,dz$$

Proof. Put $K = [a,b]$ and $I = [0,1]$ and let $\Pi: K \times I \to U$ be the required homotopy i.e. a continuous map

(a,1)   γ   (b,1)
(a,0)   β   (b,0)

which is constant on the two vertical lines, and such that $\Pi(s,0) = \beta(s)$, $\Pi(s,1) = \gamma(s)$ for $s \in K$. Consider the pull back of the exact sequence 4.1 along $\Pi: K \times I \to U$

$$0 \to \underset{\sim}{\mathbb{C}} \to \Pi^* \mathcal{O} \to \Pi^* \mathcal{O} \to 0$$

Using $H^1(K \times I, \mathbb{C}) = 0$ we derive an exact sequence

$$0 \to \mathbb{C} \to \Gamma(K \times I, \Pi^*\mathcal{O}) \xrightarrow{D} \Gamma(K \times I, \Pi^*\mathcal{O}) \to 0$$

Let $\Pi^* f \in \Gamma(K \times I, \Pi^*\mathcal{O})$ denote the transform of f by the adjunction morphism $\mathcal{O} \to \Pi_* \Pi^*\mathcal{O}$, and choose $G \in \Gamma(K \times I, \Pi^*\mathcal{O})$ with $DG = \Pi^* f$. For the curve $\gamma_t: K \to U$, $\gamma_t(s) = \Pi(s,t)$, we have

$$\int_{\gamma_t} f(z)\,dz = G(b,t) - G(a,t) \qquad ; \; t \in [0,1]$$

Notice that the stalks $G_{(b,t)} \in \mathcal{O}_{\pi(b,0)}$ and $G_{(a,t)} \in \mathcal{O}_{\pi(a,0)}$ are independent of t since $[b] \times I$ and $[a] \times I$ are connected. In particular $G(b,t) - G(a,t)$ is independent of $t \in I$.

Q.E.D.

Let $\gamma\colon [a,b] \to \mathbb{C} - \{0\}$ denote a continuous curve. Choose a continuous function $\theta\colon [a,b] \to \mathbb{R}$ such that, III.4.5

$$\gamma(t) = |\gamma(t)| \, \exp i \, \theta(t) \qquad ; \quad t \in [a,b]$$

We shall establish the formula

$$4.6 \qquad \int_\gamma \frac{dz}{z} = \log|\gamma(b)\gamma(a)^{-1}| + i(\theta(b) - \theta(a))$$

__Proof.__ According to our discussion of complex logarithms IV.3 we can find $log \in \Gamma([a,b], \gamma^{*}\mathcal{O})$ with $D\,log = \gamma^{*}(z \mapsto z^{-1})$ and

$$log(t) = \log|\gamma(t)| + i\theta(t) \qquad ; \quad t \in [a,b]$$

from which the result follows.

Q.E.D.

In particular, for a closed curve $\gamma\colon [a,b] \to \mathbb{C} - \{0\}$, we can put

$$4.7 \qquad I(\gamma,0) = \frac{1}{2\pi}(\theta(b) - \theta(a))$$

220

and we find from 4.6

4.8
$$\frac{1}{2\pi i}\int_{\gamma}\frac{dz}{z} = I(\gamma;0)$$

<u>Definition 4.9</u>. Let $s \in \mathbb{C}$ and $\gamma: [a,b] \to \mathbb{C} - \{s\}$ be a continuous curve. The <u>winding number</u> of γ with respect to s is given by

$$I(\gamma;s) = I(\gamma-s;0)$$

By a simple translation argument, we find from 4.8, that

4.10
$$\frac{1}{2\pi i}\int_{\gamma}\frac{dz}{z-s} = I(\gamma;s)$$

<u>Logarithmic derivatives</u>

A zerofree analytic function g on an open set U of $\mathbb{C}$ has a logarithmic derivative $D\log g = g'/g$. Notice, that

$$D\log gh = D\log g + D\log h \qquad ; \ g,h \in \Gamma(U,\mathcal{O}*)$$

This gives a sequence of sheaves on $\mathbb{C}$

4.11
$$0 \longrightarrow \underset{\sim}{\mathbb{C}}* \longrightarrow \mathcal{O}* \xrightarrow{\ D\log\ } \mathcal{O} \longrightarrow 0$$

which is exact: Given an open subset U of $\mathbb{C}$, a point x of U and $f \in \Gamma(U,\mathcal{O})$. Choose a smaller neighbourhood V of x and $F \in \Gamma(V,\mathcal{O})$ with $F' = f$; this gives $D\log \exp F = f$ by direct calculation. - From 4.11 we deduce an exact sequence

$$\Gamma(U,\mathcal{O}*) \xrightarrow{\ D\log\ } \Gamma(U,\mathcal{O}) \longrightarrow H^{1}(U,\mathbb{C}*)$$

1. Order differential equation

Let X denote an open subset of $\mathbb{C}$ and f an analytic function on X. We shall analyse the first order differential equation

$$Dv = fv$$

According to our analysis of the logarithmic derivative any point of X has a neighbourhood U in which the equation admits a zerofree analytic solution v, say. - For a second solution w on U we get

$$D(wv^{-1}) = (Dw)v^{-1} - wv^{-2}Dv = fwv^{-1} - wv^{-2}fv = 0$$

In particular, if U is connected we find, that wv^{-1} is constant. In conclusion we have shown

<u>Proposition 4.12</u>. The solution sheaf to the 1. order equation Dv = fv is a locally constant $\mathbb{C}$-sheaf of rank 1. - The equation admits a zerofree solution on any contractible open subset U of $\mathbb{C}$.

Let us specialize to the case $X = \mathbb{C} - \{0\}$. Let U_+ be the complement to $i\,]-\infty,0]$ in $\mathbb{C}$ and U_- the complement to $i[0,+\infty[$. Choose a solution $v_+ \neq 0$ on U_+ and a solution $v_- \neq 0$ on U_- such that

$$v_+(z) = v_-(z) \quad ; \quad \mathrm{Re}\,z < 0, \quad z \in U_+ \cap U_-$$

Consider finally the number $m \in \mathbb{C}^*$ given by

$$v_-(z) = m\,v_+(z) \quad ; \quad \mathrm{Re}\,z > 0, \quad z \in U_+ \cap U_-$$

The number $m \in \mathbb{C}^*$ is called the <u>monodromy</u> of the differential equation. The monodromy depends in fact only on the solution sheaf. In particular, the solution sheaf is constant if and only if $m = 1$.

The monodromy of the equation is

4.13
$$\boxed{m = \exp(2\pi i \ \mathrm{Res}(f;0))}$$

Proof. Let us make the two choices

$$F_+ \in \Gamma(U_+, 0) \quad \text{with} \quad DF_+ = f$$

$$F_- \in \Gamma(U_-, 0) \quad \text{with} \quad DF_- = f$$

We shall assume this done in such a way that

$$F_+(z) = F_-(z) \quad ; \quad z \in U_+ \cap U_-, \ \mathrm{Re} \ z < 0$$

To calculate the monodromy we can use the solutions $v_+ = \exp F_+$ and $v_- = \exp F_-$. We have

$$m = v_-(1) v_+(1)^{-1} = \exp(F_-(1) - F_+(1))$$

$$= \exp(F_-(1) - F_-(-1) + F_+(-1) - F_+(1))$$

$$= \exp \int_{S^1} f(z) \, dz$$

It remains to quote the formula

$$\int_{S^1} f(z) \, dx = 2\pi i \ \mathrm{Res}(f;0)$$

<u>The principle of unique continuation 4.14</u>. Let E be a sheaf on the topological space X which satisfies the principle of unique continuation i.e. <u>any section of E over any open set has open support</u>. Given an open connected subset U of E and two sections $s, t \in \Gamma(U, E)$. If $s_x = t_x$ for one point x of U , then s = t.

Given a continuous map $f: W \to X$ then the sheaf f*E on W will likewise satisfy the principle of unique continuation. In particular, constant sheaves or more generally <u>locally constant sheaves</u> satisfy the principle of unique continuation.

We are especially interested in the fact that the <u>sheaf $\mathcal{O}$ of analytic functions</u> on the complex plane satisfies the principle of unique continuation.

<u>Analytic continuation 4.15</u>. Let c be a point of the complex plane and $s \in \mathcal{O}_c$ an analytic germ, i.e. a convergent power series. Given a continuous curve $\gamma: [a,b] \to \mathbb{C}$ with $\gamma(a) = c$. By an analytic continuation of s along γ we understand a section

$$g \in \Gamma([a,b], \gamma*\mathcal{O}) \qquad ; \quad g_a = s$$

An analytic continuation of s along γ is unique if it exists by the principle of unique continuation.

<u>Monodromy theorem 4.16</u>. Let X denote a connected open subset of the complex plane, c a point of X and $s \in \mathcal{O}_c$ an analytic germ, which admits analytic continuation along any continuous curve $\theta: [a,b] \longrightarrow X$ with $\theta(a) = c$. - Given continuous curves

$\beta, \gamma: [a,b] \to X$ with $\beta(a) = \gamma(a) = c$ and $\beta(b) = \gamma(b)$, which are homotopic in X with fixed end points. Analytic continuations of s along β and γ respectively have the same stalk at the other endpoint b.

$\underline{\text{Proof}}$. Let $\Pi: [a,b] \times [0,1] \to X$ denote a homotopy as in the proof of 4.5. The sheaf $A = \Pi^* 0$ has the section s over $\{a\} \times [0,1]$ which can be extended to a section over $[a,b] \times \{t\}$ for all $t \in [0,1]$. By proper base change 1.4, such a section can be extended to a section s_t of A over $[a,b] \times V_t$ where V_t is an open neighbourhood of t in $[0,1]$. By the principle of unique continuation, these sections will determine a global section of A. Conclusion by the fact that the restriction of A to $\{b\} \times [0,1]$ is constant.

Q.E.D.

A wide class of examples, where the monodromy theorem applies, is the case where s satisfies a linear differential equation

$$4.17 \qquad s^{(n)} + a_1 s^{(n-1)} + \ldots + a_{n-1} s + a_n = 0$$

where $a_1, \ldots, a_n$ are analytic functions on X.

To see this let us remark that the solution sheaf E to the equation 4.17 is a locally constant $\mathbb{C}$-sheaf of rank n as it follows from the theory of differential equations over an open disc. Given a continuous curve $\gamma: [a,b] \to X$. Then $\gamma^* E$ is a constant sheaf on $[a,b]$, as it follows from the following lemma, and consequently any germ over a can be continued over the whole interval.

__Lemma 4.18__. Any locally constant sheaf on $[0,1]$ is constant.

__Proof__. Use the procedure from the proof of III.4.5 to find a finite subdivision of $[0,1]$ such that the sheaf is constant on each subinterval and proceed by a simple induction.

$$Q.E.D.$$

__Proposition 4.20__. Let X be a simply connected and locally path-connected topological space. Any locally constant sheaf on X is constant.

__Proof__. Let E be a locally constant sheaf on X and c a point of X. The restriction map $\Gamma(X,E) \to E_c$ is injective by the principle of unique continuation. Let us prove that a given $s \in E_c$ extends to a global section of E.

For $x \in X$ choose a path γ_x from c to x and choose an open, path connected neighbourhood U_x of x such that the restriction of E to U_x is constant. Let $s_x \in \Gamma(U_x,E)$ be the continuation of s along γ_x.

We must show that given x and y in X, then s_x and s_y have the same stalk at any point z of $U_x \cap U_y$. To this end connect c and z in two ways: Follow γ_x from c to x and pass from x to z by a path in U_x. Similarly, follow γ_y from c to y and pass from y to z by a path in U_y. - The two paths from c to z are homotopic and the result follows from the monodromy principle.

$$Q.E.D.$$

__Definition 4.21.__ Let E be a locally constant sheaf on the topological space X and $\gamma: [0,1] \to X$ a loop with $\gamma(0) = \gamma(1) = c$. The linear transformation

$$E_c \overset{\sim}{\longrightarrow} \gamma^*E_0 \overset{r_0^{-1}}{\longrightarrow} \Gamma([0,1], \gamma^*E) \overset{r_1}{\longrightarrow} \gamma^*E_1 \overset{\sim}{\longrightarrow} E_c$$

where r_0 and r_1 denote restriction to the stalks, is called the __monodromy__ along γ.

__Euler's equation 4.22.__ The differential equation

$$z^2 w'' + azw' + bw = 0 \qquad ; \quad a,b \in \mathbb{C}$$

has a solution sheaf on $\mathbb{C}-\{0\}$ which is a locally constant $\mathbb{C}$-sheaf of rank two. In order to calculate the monodromy matrix we introduce the operator

$$L(w) = z^2 w'' + azw' + bw$$

and the polynomial

$$q(\nu) = \nu(\nu-1) + a\nu + b$$

By explicit calculation we find that

$$L(z^\nu) = q(\nu) z^\nu$$

and by applying $\dfrac{\partial}{\partial \nu}$ to this formula

$$L(z^\nu \log z) = q'(\nu) z^\nu + q(\nu) z^\nu \log z$$

If the indicial equation $q(\nu) = 0$ has two solutions r and s, the Euler equation has solutions z^r and z^s. If r is a double root of the indicial equation, then we have solutions z^r and $z^r \log z$. By analytic continuation and counter clockwise

$$z^r \longmapsto e^{2\pi i r} z^r$$

$$\log z \longmapsto \log z + 2\pi i$$

Consequently we find the monodromy matrix

$r \neq s$	$r = s$
$\begin{Bmatrix} e^{2\pi i r} & 0 \\ 0 & e^{2\pi i s} \end{Bmatrix}$	$e^{2\pi i r} \begin{Bmatrix} 1 & 1 \\ 0 & 1 \end{Bmatrix}$ *)

*) Notice that $\begin{Bmatrix} 1 & 2\pi i \\ 0 & 1 \end{Bmatrix}$ is conjugated to $\begin{Bmatrix} 1 & 1 \\ 0 & 1 \end{Bmatrix}$.

Bessel's equation 4.23.

$$z^2 w'' + z w' + (z^2 - n^2) w = 0$$

has monodromy matrix

$n \notin \mathbb{Z}$	$n \in \mathbb{Z}$
$\begin{Bmatrix} e^{2\pi i n} & 0 \\ 0 & e^{-2\pi i n} \end{Bmatrix}$	$\begin{Bmatrix} 1 & 1 \\ 0 & 1 \end{Bmatrix}$

In case $n = 0$ this can be seen from the explicit solutions

$$J_0(z) = \sum_{m=0}^{+\infty} \frac{(-1)^m}{(m!)^2} \left(\frac{z}{2}\right)^{2m}$$

$$K_0(z) = - \sum_{m=0}^{\infty} \frac{(-1)^m}{(m!)^2} \left(1 + \frac{1}{2} + \ldots + \frac{1}{n}\right)\left(\frac{z}{2}\right)^{2m} + \log z \, J_0(z)$$

IV.5 The inhomogeneous Cauchy-Riemann equations

A smooth $\mathbb{C}$-valued function f on an open subset of the complex plane is known to be analytic if and only if $f = u+iv$ satisfies the Cauchy-Riemann equations

$$5.1 \qquad \frac{\partial u}{\partial x} = \frac{\partial v}{\partial y} \;,\quad \frac{\partial u}{\partial y} = -\frac{\partial v}{\partial x}$$

Introducing the operator $\dfrac{\partial}{\partial \bar{z}} = \dfrac{1}{2}\left(\dfrac{\partial}{\partial x} + i\dfrac{\partial}{\partial y}\right)$ this can be expressed by $\dfrac{\partial}{\partial \bar{z}} f = 0$.

Theorem 5.2. Let g be a smooth $\mathbb{C}$-valued function on the open subset X of $\mathbb{C}$. Then the partial differential equation

$$\frac{\partial}{\partial \bar{z}} f = g$$

can be solved with a smooth function f on X.

Proof. Let us introduce the sheaf C of $\mathbb{C}$-valued smooth functions on the complex plane. We shall be concerned with the morphism $\dfrac{\partial}{\partial \bar{z}} : C \to C$.

Functions with compact support

The point of departure is the existence of an operator

$$T: \Gamma_c(\mathbb{C}, C) \longrightarrow \Gamma(\mathbb{C}, C)$$

with $(\dfrac{\partial}{\partial \bar{z}} \circ T)(f) = f$ for all $f \in \Gamma_c(\mathbb{C}, C)$. Such an operator is given by

$$Tf(w) = \frac{1}{2\pi i} \iint \frac{f(z)}{z-w} \, dz \wedge d\bar{z} \qquad ; \quad w \in \mathbb{C}$$

See this chapter 7.20 for an elementary discussion.

A compact subset K of $\mathbb{C}$

The induced operator

$$\frac{\partial}{\partial \bar{z}}: \ \Gamma(K,C) \ \longrightarrow \ \Gamma(K,C)$$

is surjective as it follows from the
accompanying commutative diagram,
noticing that the composite of the
two vertical arrows to the extreme

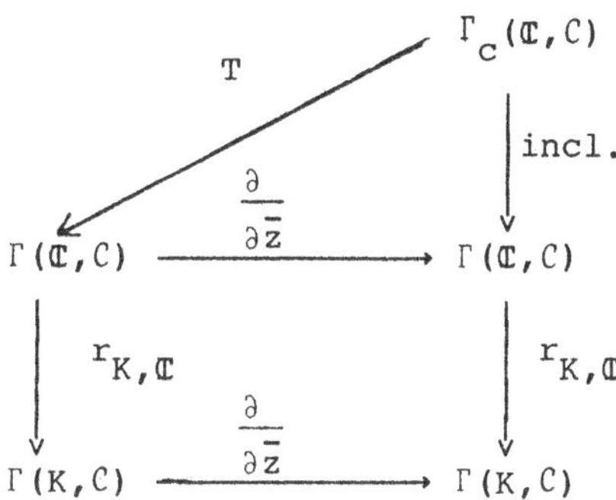

right is surjective, as it follows from the fact that C is a

soft sheaf.

The case X = $\mathbb{C}$

Given $g \in \Gamma(\mathbb{C},C)$ we are looking for $f \in \Gamma(\mathbb{C},C)$ with
$\frac{\partial}{\partial \bar{z}} f = g$. Choose a sequence of compact discs $K^0,\ldots,K^n,\ldots$ with
center in the origin and union $\mathbb{C}$. We are going to construct a
sequence $f_0,f_n,\ldots$ where $f_n \in \Gamma(K^n,C)$ with $\frac{\partial}{\partial \bar{z}} f_n = g$ and

$$\sup_{K^n} |f^{n+1}-f^n| \leq 2^{-n}$$

Suppose we have already constructed $f_0,\ldots,f_n$. Choose
$\varphi_{n+1} \in \Gamma(K^{n+1},C)$ with $\frac{\partial}{\partial \bar{z}} \varphi_{n+1} = g$. Notice that $\frac{\partial}{\partial \bar{z}} (\varphi_{n+1}-f_n) = 0$
which proves that $\varphi_{n+1}-f_n$ is analytic on K^n. Expand this
function in a power series around the origin to see that there
exists a polynomial p_{n+1} such that

$$\sup_{K^n} |\varphi_{n+1} - f - p_{n+1}| \leq 2^{-n}$$

230

and put $f_{n+1} = \varphi_{n+1} - P_{n+1}$. Define $f(z) = \lim\limits_{n\to\infty} f_n(z)$. Notice for $z \in K^n$ we have

$$f(z) = \lim\limits_{p\to\infty}(f_p(z) - f_n(z)) + f_n(z)$$

which shows that $f - f_n$ is analytic on K^n and consequently

$$\frac{\partial}{\partial \bar{z}} f(z) = \frac{\partial}{\partial \bar{z}} f_n(z) = g(z).$$

<u>The general open subset X of $\mathbb{C}$</u>

We are going to follow the proof scheme of the previous case. Thus we must show the existence of an increasing sequence $K^0,\ldots,K^n$ of compact subsets of X, such that every compact subset of X is contained in one of the K^n's and such that any $f \in \Gamma(K^n, 0)$ can be uniformly approximated by analytic functions on X.

We shall use Runge's approximation theorem 5.3 below. For $n \in \mathbb{N}$ let K^n be the complement in $\mathbb{C}$ to the set

$$\bigcup_{a \in \mathbb{C}-X} D(a;\tfrac{1}{n}) \cup \{x \in \mathbb{C} \mid |x| > n\}$$

Let us prove that any bounded connected component V of $\mathbb{C}-K^n$ meets $\mathbb{C}-X$: A point x of V must be contained in one of the discs $D(a,\tfrac{1}{n})$. This disc must be contained in V since V is a connected component, while $a \in V$. Conlusion by Runge's theorem.

Q.E.D.

<u>Runge's approximation theorem 5.3</u>. Let K be a compact subset of the complex plane, and P a subset of $\mathbb{C}-K$ consisting in one point from each bounded, connected component of $\mathbb{C}-K$. Any analytic function defined in a neighbourhood of K can be uniformly approximated on K by rational functions with poles in P.

An elementary reference is Burckel (1) VIII.§2.8.11.

IV.6 Existence theorems for analytic functions

We shall analyse a number of classical existence theorems in complex analysis by means of the following basic

Theorem 6.1. The sheaf 0 of analytic functions on the complex plane satisfies

$$H^i(X,0) = 0 \qquad ; \; i \geq 1$$

for any open subset X of $\mathbb{C}$.

Proof. Let C denote the sheaf of $\mathbb{C}$-valued smooth functions on the complex plane. The sequence

$$0 \longrightarrow 0 \longrightarrow C \xrightarrow{\frac{\partial}{\partial \bar{z}}} C \longrightarrow$$

is exact as it follows from the local part of 5.2. We are in fact dealing with a soft resolution of the sheaf 0. Thus $H^\cdot(X,0)$ can be calculated as the cohomology groups of the complex

$$0 \longrightarrow \Gamma(X,C) \xrightarrow{\frac{\partial}{\partial \bar{z}}} \Gamma(X,C) \longrightarrow 0$$

Conclusion by 5.2.

$$Q.E.D.$$

Primitives

The exact sequence 4.1

$$0 \longrightarrow \underset{\sim}{\mathbb{C}} \longrightarrow 0 \xrightarrow{D} 0 \longrightarrow 0$$

gives rise to a cohomology sequence

6.2
$$\Gamma(X,0) \xrightarrow{D} \Gamma(X,0) \to H^1(X,\mathbb{C}) \to 0$$

which shows how the problem of primitives depends on the topology of X.

Logarithms

The exponential sequence 3.6

$$0 \to \mathbb{Z} \to 0 \to 0* \to 0$$

gives rise to an exact sequence

6.3
$$\Gamma(X,0) \longrightarrow \Gamma(X,0*) \to H^1(X,\mathbb{Z}) \to 0$$

and an isomorphism

6.4
$$H^1(X,0*) \xrightarrow{\sim} H^2(X,\mathbb{Z})$$

We shall prove in VI.6.8 that $H^2(X,\mathbb{Z}) = 0$.

Example 6.5. Let us consider the case where X is a disc with its center s removed. An analytic function in X can be expanded in a Laurent series

$$f(z) = \sum_{n=-\infty}^{+\infty} a_n(z-s)^n \qquad ; \; z \in X$$

where $\displaystyle\sum_{n=0}^{+\infty} a_n(z-s)^n$ converges for all $z \in X$ while the principal part $\displaystyle\sum_{m=1}^{+\infty} a_{-m}(z-s)^{-m}$ converges for all $z \in \mathbb{C}-\{s\}$. The residue of f at s is
$$\mathrm{Res}(f;s) = a_{-1}$$

By inspection we see that f has a primitive in X if and only if $\mathrm{Res}(f;s) = 0$. Thus the residue will induce an iso-morphism

$$\mathrm{Res}(-;s) : H^1(X,\mathbb{C}) \xrightarrow{\sim} \mathbb{C}$$

Recall that f is said to be <u>meromorphic</u> at s if the principal part of the Laurent expansion of f at s is finite, i.e. the expansion has the form

$$f(z) = \sum_{n=m}^{+\infty} a_n (z-s)^n \; ; \qquad a_m \neq 0$$

The integer m is called the <u>degree</u> (or multiplicity) of f at s and is denoted $\mathrm{div}(f;s)$

<u>Mittag-Leffler theorem 6.6</u>. Given a <u>discrete</u> subset S of X which is closed relative to X, and let there for each $s \in S$ be given a <u>finite</u> series

$$\frac{a_1}{(z-s)^1} + \frac{a_2}{(z-s)^2} + \ldots$$

Then there exists an analytic function on $U-S$ which for each $s \in S$ has the above series as the principal part of the Laurent expansion at s.

<u>Proof</u>. Let us describe a general construction departing from a locally compact space X and an abelian group D. Let $\mathcal{D}$ denote the sheaf of D-valued <u>functions with discrete support</u>: for an open subset U of X we can describe $\Gamma(U,\mathcal{D})$ as the set of functions $s: U \to D$ for which the set $\{x \in U \mid s(x) \neq 0\}$ is discrete and closed relative to U. Remark that the sheaf $\mathcal{D}$

234

is <u>soft</u> (not flabby in general).

Let $\mathbb{C}^{\oplus\mathbb{N}}$ denote the direct sum of countably many copies of $\mathbb{C}$. We can think of an element of $\mathbb{C}^{\oplus\mathbb{N}}$ as a sequence $a_1, a_2, \ldots$ of complex numbers of which only finitely many are different from zero. By the <u>sheaf of Laurent expansions, *Lau*</u> we understand the sheaf of $\mathbb{C}^{\oplus\mathbb{N}}$-valued functions on $\mathbb{C}$ with discrete support. Thus our initial data is precisely a section of *Lau* over X.

The sheaf <u>*M* of meromorphic functions on $\mathbb{C}$</u> is the sheaf associated the presheaf whose sections over the open set U is

$$\{f/g \mid f, g \in \Gamma(U), \quad g_x \neq 0 \quad \text{for all} \quad x\}$$

Given $m \in \Gamma(U, M)$ and $s \in U$. In a neighbourhood of s we can represent $m = f/g$ where f and g are analytic functions with $g_x \neq 0$ for all $x \in U$. It follows that m can be represented by an analytic function in a punctured neighbourhood of s which is meromorphic at s in the sense of 6.5. The sheaf M is a sheaf of rings on $\mathbb{C}$. Here we shall only be concerned with the underlying additive sheaf. Taking the principal part of the Laurent expansion defines a morphism of sheaves pp: $M \to Lau$, which fits into an exact sequence of sheaves

$$0 \longrightarrow 0 \longrightarrow M \xrightarrow{pp} Lau \longrightarrow 0$$

Theorem 6.1 provides us with an exact sequence

$$0 \longrightarrow \Gamma(X,0) \longrightarrow \Gamma(X,M) \xrightarrow{pp} \Gamma(X,Lau) \longrightarrow 0$$

from which the result can be read off.

Q.E.D.

We shall now present a theorem of Weierstrass which ante-
dates the theorem of Mittag-Leffler.

Weierstrass' theorem 6.7. Given a discrete subset S of X
closed relative to X and let there for each $s \in S$ be given
an integer $d(s) \in \mathbb{Z}$. - There exists an analytic function f
on $X-S$ which has no zeroes on $X-S$ and which is meromorphic
at each $s \in S$ of degree $d(s)$.

Proof. The sheaf of divisors, $\mathcal{D}\iota\nu$ is the sheaf of $\mathbb{Z}$-valued
functions on the complex plane with discrete support.

Let $M*$, the **multiplicative sheaf of moromorphic functions**
denote the sheaf associated to the presheaf whose sections over
the open set U is

$$\{f/g \mid f,g \in \Gamma(U,\mathcal{O}), \ f_x \neq 0 \ \text{ and } \ g_x \neq 0 \ \text{ for all } \ x \in U\}$$

Alternatively $\Gamma(U,M*)$ can be described at the group of multiplicative units in $\Gamma(U,M)$, where M is the sheaf of meromorphic
functions. We can fit $M*$ into an exact sequence

$$0 \longrightarrow \mathcal{O}* \longrightarrow M* \xrightarrow{\text{div}} \mathcal{D}\iota\nu \longrightarrow 0$$

where div is defined at the end of 6.5. From this we deduce
the cohomology sequence

$$0 \longrightarrow \Gamma(X,\mathcal{O}*) \longrightarrow \Gamma(X,M*) \xrightarrow{\text{div}} \Gamma(X,\mathcal{D}\iota\nu) \longrightarrow H^1(X,\mathcal{O}*)$$

236

Recall from the discussion of logarithms 6.4 that

$$H^1(X, \mathcal{O}*) \xrightarrow{\sim} H^2(X, \mathbb{Z})$$

Thus the result follows from VI.6.8.

Q.E.D.

Corollary 6.8. Any meromorphic function on a connected open subset X of the complex plane is a quotient of two analytic functions on X.

Proof. Let m be a non zero meromorphic function on X. Write the divisor of m as

$$\mathrm{div}(m) = F-G$$

where F and G are positive divisors on X, i.e. assume positive values only. According to 6.7 we can find non zero meromorphic functions f and g on X, with div(f) = F and div(g) = G. Since F and G are positive, it follows that f and g are analytic functions on X, moreover

$$\mathrm{div}(mf^{-1}g) = 0$$

i.e. $mf^{-1}g$ is a zerofree analytic function on X.

Q.E.D.

IV.7 De Rham theorem

Let us describe the basic sheaves of differential geometry. To begin consider the sheaf C of smooth $\mathbb{R}$-valued functions on a finite dimensional real vector space E. Let $C(E)$ denote the <u>sheaf of vector fields</u> on E: the sections over an open subset U of E is the set of smooth maps from U to E. Given $f \in \Gamma(U,C)$ and $v \in \Gamma(U,C(E))$ define $D_v f \in \Gamma(U,C)$ to be the <u>directional derivative</u>, i.e.

$$7.1 \qquad D_v f(x) = \lim_{t \to 0} \frac{f(x+tv(x))-f(x)}{t-0} \quad ; \quad x \in U$$

Let $\mathcal{D}er_{\mathbb{R}}(C,C)$ denote the <u>sheaf of $\mathbb{R}$-derivations</u> in C: Let $j: U \to E$ denote the inclusion of an open set, a section of $\mathcal{D}er_{\mathbb{R}}(C,C)$ over U is a morphism of $\mathbb{R}$-sheaves

$$D: \; j*C \longrightarrow j*C$$

such that for any open set V of U

$$7.2 \qquad D(fg) = gD(f) + fD(g) \qquad ; \; f,g \in \Gamma(V,C).$$

<u>Proposition 7.3</u>. The directional derivative induces an isomorphism of sheaves

$$C(E) \longrightarrow \mathcal{D}er_{\mathbb{R}}(C,C)$$

<u>Proof</u>. Let us assume that $E = \mathbb{R}^n$. Let U be an open subset of $\mathbb{R}^n$ and $v = (v_1,\ldots,v_n)$ a smooth vector field on U.

238

We have

$$D_v = v_1 \frac{\partial}{\partial x_1} + \ldots + v_n \frac{\partial}{\partial x_n}$$

as it follows from the chain rule. This proves that our map is injective. Consider an arbitary section D of $\text{Der}_{\mathbb{R}}(C,C)$ over U. Put

$$P = D - \Sigma D(x_i) \frac{\partial}{\partial x_i}$$

Notice, that P is section of $\mathcal{D}er_{\mathbb{R}}(C,C)$ over U with $P(x_i) = 0$, $i = 1,\ldots,n$. Let us prove that $P = 0$. So let there be given a smooth function f on an open subset V of U. We must show that $Pf(a) = 0$ for any $a \in V$. According to Lemma 7.4 below we can write

$$f(x) = f(a) + \Sigma(x_i - a_i)h_i(x)$$

where $h_1,\ldots,h_n$ are smooth functions defined in a neighbourhood of a. If we confuse f with its restriction to that neighbourhood we get

$$Pf(x) = \Sigma(x_i - a_i)Ph_i(x)$$

which shows that $Pf(a) = 0$.

$$Q.E.D.$$

__Lemma 7.4.__ Let D be an open disc in $\mathbb{R}^n$ with center a. Any smooth function f on D can be written

$$f(x) = f(a) + \Sigma(x_i - a_i)h_i(x)$$

where $h_1,\ldots,h_n$ are smooth functions on D.

<u>Proof</u>. By means of the chain rule we get

$$f(x) - f(a) = \int_0^1 \frac{d}{dt} f(t(x-a) + a)\,dt = \Sigma (x_i - a_i) \int_0^1 \frac{\partial f}{\partial x_i}(t(x-a)+a)\,dt$$

Q.E.D.

Let X denote a <u>differentiable manifold</u> and C the sheaf of $\mathbb{R}$-valued smooth functions on X. The <u>tangent sheaf</u> of X is by definition the sheaf $T = \mathcal{D}er_{\mathbb{R}}(C,C)$. The sections of T are called <u>tangent vector fields</u>.

Given two tangent vector fields v and w over the same open subset U of X. We define the <u>bracket</u> $[v,w] \in \Gamma(U,T)$ by the formula

7.5
$$[v,w] = v \circ w - w \circ v$$

The De Rham complex

For $p \in \mathbb{N}$ we put

$$\Omega^p = Hom_C^{(p)}(T,C)$$

the sheaf of alternating p-forms on T. To describe this let us first remark that $\Omega^0 = C$. For $p \geq 1$ consider the inclusion $j: U \to X$ of an open set. The sections of Ω^p over U is the set of morphisms of sheaves

$$\omega: j^*T \times \ldots \times j^*T \to j^*C \qquad \text{(p-factors)}$$

such that for each open set V of U, $\omega(v)$ is an alternating p-form on the $C(V)$-module $T(V)$. - Recall the presence of a pairing <u>exterior product</u>

$$\Gamma(U,\Omega^p) \times \Gamma(U,\Omega^q) \to \Gamma(U,\Omega^{p+q})$$

given by the formula

7.6 $\quad \alpha \wedge \beta(v_1,\ldots,v_{p+q}) = \dfrac{1}{p!q!} \sum_{\sigma} \text{sign}(\sigma)\alpha(v_{\sigma(1)},\ldots,v_{\sigma(p)})\beta(v_{\sigma(p+1)},\ldots,v_{\sigma(p+q)})$

the sum being over all permutations σ of $[1,\ldots,p+q]$. This makes $\Omega^{\cdot}$ into a sheaf of graded algebras which is anti commutative:

7.7 $\qquad\qquad\qquad \alpha^p \wedge \beta^q = (-1)^{pq}\beta^q \wedge \alpha^p$

The representation $T = \mathcal{D}er_{\mathbb{R}}(C,C)$ and simple evaluation give rise to a morphism of sheaves

$$d:\ C \longrightarrow \Omega^1$$

in fact an $\mathbb{R}$-derivation. This can be extended to $d:\Omega^p \to \Omega^{p+1}$ by the formula

7.8 $\qquad d\omega(v_1,\ldots,v_{p+1}) = \sum_{i} (-1)^{i+1}v_i\omega(v_1,\ldots,\hat{v}_i,\ldots,v_{p+1})$

$$+ \sum_{i<j} (-1)^{i+j}\,\omega([v_i,v_j],v_1,\ldots,\hat{v}_i,\ldots,\hat{v}_j,\ldots,v_{p+1})$$

This satisfies $d \circ d = 0$ and

7.9 $\qquad\qquad d(\omega^p \wedge \omega^q) = d\omega^p \wedge \omega^q + (-1)^p\omega^p \wedge d\omega^q$

The <u>exterior derivative</u> d and $\Omega^{\cdot}$ constitutes the <u>de Rham complex</u> of X.

<u>Theorem 7.10</u>. Let X be a differentiable manifold. The de Rham complex $\Omega^{\cdot}$ is a soft resolution of the sheaf $\underset{\sim}{\mathbb{R}}$.

<u>Proof</u>. Recall from III.2.8 that soft is a local notion. Thus it follows from III.3.2 that C is a soft sheaf on X. This implies that any sheaf of C-modules is soft III.2.9.

Thus it remains to be shown that for any point $x \in X$, $\Omega^{\cdot}_x$ is a resolution of $\mathbb{R}$. By a suitable choice of coordinate system it suffices to treat the case where $X = \mathbb{R}^n$ and x is the origin. Conclusion by 7.11.

$$Q.E.D.$$

<u>Poincaré lemma 7.11</u>. Let $\Omega^{\cdot}$ denote the de Rham complex on $\mathbb{R}^n$. For an open disc D with center in 0 we have

$$H^i \Gamma(D, \Omega^{\cdot}) = 0 \qquad \text{for} \quad i \geq 1$$

<u>Proof</u>. For $p \geq 1$ let k^p denote the operator

$$k^p : \Gamma(D, \Omega^p) \longrightarrow \Gamma(D, \Omega^{p-1})$$

which maps the differential

$$\omega = f(x_1, \ldots, x_n) \, dx_{i_1} \wedge dx_{i_2} \cdots \wedge dx_{i_p} \qquad ; \ i_1 < i_2 < \ldots < i_p$$

into the differential form

$$k^p \omega = x_{i_1} \left(\int_0^1 f(0,\ldots,0,tx_{i_1},x_{i_1+1},\ldots,x_n)\, dt \right) dx_{i_2} \wedge \ldots \wedge dx_{i_p}$$

We must verify the formula $\qquad k^{p+1}\partial^p + \partial^{p-1}k^p = 1 \qquad ; \quad p \geq 1$

To this end we offer the following details:

$$\partial^{p-1}k^p \omega = \left(\int_0^1 \frac{d}{dt}(tf(..,0,tx_{i_1},x_{i_1+1},..))\, dt \right) dx_{i_1} \wedge dx_{i_2}..$$

$$+ \sum_{j>i_1} x_{i_1} \left(\int_0^1 f'_j(\ldots,0,tx_{i_1},x_{i_1+1},\ldots)\, dt \right) dx_j \wedge dx_{i_2}..$$

$$k^{p+1}\partial^p \omega = \sum_{j<i_1} \left(\int_0^1 x_j f'_j(..,0,tx_j,x_{j+1},..)\, dt \right) dx_{i_1} \wedge dx_{i_2}..$$

$$- \sum_{j>i_1} x_{i_1} \left(\int_0^1 f'_j(..,0,tx_{i_1},x_{i_1+1},..)\, dt \right) dx_j \wedge dx_{i_2}..$$

leaving it to the reader to evaluate the integrals in the first and the third line.

Q.E.D.

De Rham theorem 7.12. Let X denote a differentiable manifold which is countable at infinity. Then

$$H^{\bullet}(X,\mathbb{R}) = H^{\bullet}\Gamma(X,\Omega^{\bullet})$$

Moreover, the cup product is induced by exterior product.

Proof. Combine 7.10 and 2.2 with I.7.5. The statement about the cup product follows from II.10.14.

Q.E.D.

<u>Integration 7.13</u>. Let Ω^n denote the sheaf of smooth n-forms on $\mathbb{R}^n$. We can define

$$\int : \Gamma_c(\mathbb{R}^n, \Omega^n) \longrightarrow \mathbb{R}$$

by means of the usual integral on $\mathbb{R}^n$

$$\int f(x_1, \ldots, x_n)\, dx_1 \wedge \cdots \wedge dx_n = \int f(x_1, \ldots, x_n)\, dx_1 \cdots dx_n$$

Let us prove the fundamental formula

$$\int d\omega = 0 \qquad ; \quad \omega \in \Gamma_c(\mathbb{R}^n, \Omega^{n-1})$$

It suffices to treat the case $\omega = g(x_1, \ldots, x_n)\, dx_1 \cdots \wedge \widehat{dx_i} \cdots \wedge dx_n$ where g is a smooth function on $\mathbb{R}^n$ with compact support

$$\int d\omega = (-1)^{i-1} \int \frac{\partial g}{\partial x_i}(x_1, \ldots, x_n)\, dx_1 \wedge \cdots \wedge dx_n$$

$$= (-1)^{i-1} \int_{-\infty}^{+\infty} dx_1 \cdots \int_{-\infty}^{+\infty} \frac{\partial g}{\partial x_i}(x_1, \ldots, x_n)\, dx_i = 0$$

In conclusion the integral induces an isomorphism

$$\int : H_c^n(\mathbb{R}^n, \mathbb{R}) \longrightarrow \mathbb{R}$$

More generally let X be an oriented smooth n-manifold, i.e. X admits an atlas (U_α, ϕ_α) whose transition functions $\phi_\alpha \circ \phi_\beta^{-1}$ have positive Jacobi determinants. Then we can construct integrals

$$\int : \Gamma_c(U_\alpha, \Omega^n) \rightarrow \mathbb{R}$$

which agrees on the overlaps. From the fact that Ω^n is soft

follows V.1.2 that the presheaf

$$U \;\mapsto\; \Gamma_c(U,\Omega^n)^{\vee}$$

is a sheaf. Thus we obtain an integral

$$\int : \; \Gamma_c(X,\Omega^n) \;\to\; \mathbb{R}$$

which satisfies the relation

$$\int d\omega \;=\; 0 \qquad ; \;\; \omega \in \Gamma_c(X,\Omega^{n-1})$$

using the fact that the presheaf $\quad U \;\mapsto\; \Gamma_c(U,\Omega^{n-1})^{\vee}$

is a sheaf, we can reduce the problem to $\mathbb{R}^n$, where it allready

has been delt with.

Example 7.14. Let V denote a finite dimensional real

vector space. The exact sequence in cohomology with compact

support gives rise to a boundary map

$$H_c^{\;0}(\{0\},\mathbb{R}) \;\longrightarrow\; H_c^{\;1}(V,\mathbb{R})$$

Let us seek a de Rham representation of the image of the constant

function 1. To this end consider the commutative diagram

$$
\begin{array}{ccccccccc}
0 & \longrightarrow & \Gamma_c(V-\{0\},\Omega^0) & \longrightarrow & \Gamma_c(V,\Omega^0) & \longrightarrow & \Gamma_c(\{0\},\Omega^0) & \longrightarrow & 0 \\
 & & \downarrow{\scriptstyle d} & & \downarrow{\scriptstyle d} \quad {\scriptstyle \rho \;\mapsto\; 1} & & \downarrow{\scriptstyle d} & & \\
0 & \longrightarrow & \Gamma_c(V-\{0\},\Omega^1) & \longrightarrow & \Gamma_c(V,\Omega^1) & \longrightarrow & \Gamma_c(\{0\},\Omega^1) & \longrightarrow & 0
\end{array}
$$

Take   any smooth function ρ on V with compact support, which
is constant 1 in a neighbourhood of 0 in V. The restriction of
$d\rho$ to V-{0} represents the canonical cohomology class.

Example 7.15. Let X denote an open subset of $\mathbb{C}$ and f(z)
an analytic function on X. The exact sequence 4.1 gives rise to
a boundary map

$$H^0(X,0) \longrightarrow H^1(X,\mathbb{C})$$

The image of f(z) by this map can be realized as a de Rham cohomology class by the differential form $-f(z)dz$. To see this
consider the commutative diagram

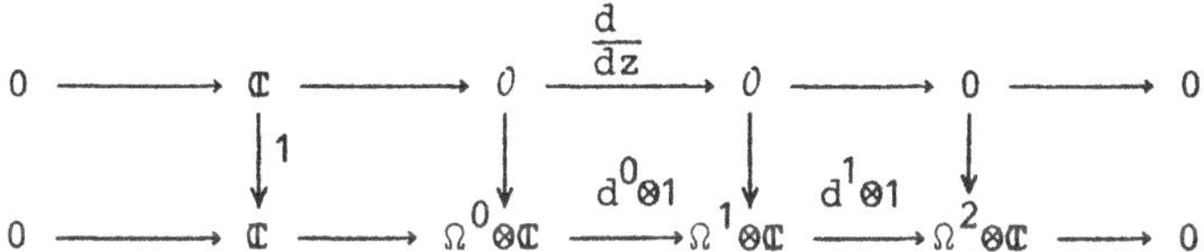

where the second vertical arrow is separation in real and
imaginary part while the third vertical arrow can be described
by decomposing the analytic function f into

$$f(x+iy) = u(x,y) + iv(x,y) \qquad ; \ x+iy \in X$$

where u and v are smooth real functions on X. To f(z) we
associate the differential form

$$f(z)dz = (u(x,y)dx - v(x,y)dy) + i(u(x,y)dy + v(x,y)dx)$$

The commutativity of the diagram follows from the Cauchy-Riemann
equations 5.1. Conclusion by XI.4.9.

246

$\underline{\text{Example 7.16}}$. Let ρ be a smooth real function on $\mathbb{C}$ with compact support which is constant in a neighbourhood of 0. Let us show that

$$\int_{\mathbb{C}*} \frac{1}{2\pi i} \frac{dz}{z} \wedge d\rho = \rho(0)$$

In case $\rho(0) = 0$ the function ρ has compact support on $\mathbb{C}^*$ and we find that

$$\int_{\mathbb{C}*} \frac{1}{2\pi i} \frac{dz}{z} \wedge d\rho = -\frac{1}{2\pi i} \int_{\mathbb{C}*} d(\rho\frac{dz}{z}) = 0$$

As a consequence the value of the integral depends only on $\rho(0)$. Consider a smooth function $\rho: [0,+\infty] \to \mathbb{R}$ with compact support which is constant in a neighbourhood of 0. The composite of ρ and $r = \sqrt{x^2+y^2}$ defines a function which we also denote by ρ. We have

$$d\rho = \frac{d\rho}{dr} \left(\frac{\partial r}{\partial x} dx + \frac{\partial r}{\partial y} dy \right) = \frac{1}{r} \frac{d\rho}{dr}(xdx + ydy)$$

$$\frac{1}{z} dz = \tfrac{1}{2} d\log (x^2+y^2) + \frac{1}{r^2} i(-ydx + xdy)$$

Using polar coordinates we get

$$\int \frac{1}{2\pi i} \frac{dz}{z} \wedge d\rho = -\frac{1}{2\pi} \frac{1}{r^3} \int \frac{d\rho}{dr} r^2 dx \wedge dy =$$

$$-\frac{1}{2\pi} \iint \frac{1}{r} \frac{d\rho}{dr} dxdy = -\frac{1}{2\pi} \int_0^{2\pi} d\theta \int_0^{+\infty} \frac{1}{r} \frac{d\rho}{dr} rdr = \rho(0)$$

Example 7.17. The previous example can be generalized to $\mathbb{R}^n$ as follows. Put

$$\theta(x) = r(x)^{-n} \sum_{i=1}^{n} (-1)^{i-1} x_i \, dx_1 \wedge \cdots \wedge \widehat{dx_i} \wedge \cdots \wedge dx_n$$

where $r(x) = (x_1^2 + \ldots + x_n^2)^{\frac{1}{2}}$. This is a closed differential form on the space $\mathbb{R}^n - \{0\}$:

$$d\theta = dr^{-n} \wedge r^n \theta + r^{-n} dr^n \theta = -\frac{n}{2} r^{-n-2} dr^2 \wedge r^n \theta + nr^{-n} dx_1 \ldots dx_n$$

which is zero by direct calculation of the first term. For a smooth function $\rho: \mathbb{R}^n \to \mathbb{R}$ with compact support and constant in a neighbourhood of the origin we have

$$\sigma_{n-1} \rho(0) = (-1)^n \int_{\mathbb{R}^n - \{0\}} \theta \wedge d\rho$$

where $\sigma_{n-1} = n\mu_n$, i.e. n times the volume of the unit ball in $\mathbb{R}^n$. The formula can be proved in a way similar to the proof of the formula 7.16.

Example 7.18. The formula

$$\int_{\mathbb{C}^*} \frac{1}{2\pi i} \frac{dz}{z} \wedge d\rho = \rho(0)$$

is true for any smooth function ρ with compact support: According to 7.16 it suffices to treat the case where $\rho(0) = 0$. Using Stoke's theorem we get

$$\int_{|z| \geq \epsilon} \frac{1}{2\pi i} \frac{dz}{z} \wedge d\rho = \int_{|z| \geq \epsilon} \frac{1}{2\pi i} d(\rho \wedge \frac{dz}{z}) = \int_{|z| = \epsilon} \frac{1}{2\pi i} \rho \frac{dz}{z}$$

248

write out the last integral to get the   estimate

$$\left| \int_{|z| \geq \varepsilon} \frac{1}{2\pi i} \frac{dz}{z} \wedge d\rho \right| \leq \sup_{|z|=\varepsilon} |\rho(z)| .$$

from which the result follows

Example 7.19. For a smooth function $f: \mathbb{C} \to \mathbb{C}$ with compact support er get from 7.18

$$f(w) = \frac{1}{2\pi i} \int \frac{dz}{z-w} \wedge df(z) \quad .$$

or with $df(z) = \frac{\partial f}{\partial z}(z)\,dz + \frac{\partial f}{\partial \bar{z}}(z)\,d\bar{z}$

$$f(w) = \frac{1}{2\pi i} \int \frac{\partial f}{\partial \bar{z}}(z) \frac{1}{z-w} dz \wedge d\bar{z} .$$

Example 7.20. For a smooth function $f: \mathbb{C} \to \mathbb{C}$ with compact support we have

$$\frac{\partial}{\partial \bar{w}} \int f(z) \frac{1}{z-w} \, dz \wedge d\bar{z} = f(w) .$$

According to 7.19 it suffices to prove that

$$\frac{\partial}{\partial \bar{w}} \int f(z) \frac{1}{z-w} \, dz \wedge d\bar{z} = \int \frac{\partial f}{\partial \bar{z}}(z) \frac{1}{z-w} \, dz \wedge d\bar{z}$$

which is easily checked by changing into polar coordinates

$$\int f(z) \frac{1}{z-w} \, dz \wedge d\bar{z} = -\frac{1}{\pi} \int_0^{2\pi} d\theta \int_0^{+\infty} f(w+re^{i\theta}) e^{i\theta} dr$$

IV.8   Relative cohomology

Let Z denote a closed subset of the topological space X and $U = X-Z$ the complement. The inclusions are denoted $i: Z \to X$ and $j: U \to X$ respectively.

Definition 8.1. For a sheaf F on X we put

$$H^p(X,Z;F) = H^p(X,j_!j^*F) \qquad ; \quad p \in \mathbb{Z}$$

From the short exact sequence II.6.11 we deduce by means of II.5.4 a long exact sequence

$$8.2 \qquad \to H^{p-1}(Z,F) \to H^p(X \; Z,F) \to H^p(X,F) \to H^p(Z,F)$$

For closed subsets A and B of X we derive from the first row of II.6.17 an exact sequence

$$8.3. \qquad \to H^p(X,A\cup B;F) \to H^p(X,A;F) \to H^p(B,B\cap A;F) \to$$

Proposition 8.4. Suppose X is a locally compact space, countable at infinity and Z a closed subspace of X. Then for any soft sheaf S on X, $\quad H^p(X,Z;S) = 0 \qquad ; \; p \geq 1$

Proof. The restriction i^*F of F to Z is a soft sheaf III.2.5. It follows from 2.2 that $H^p(Z,S) = 0 \; ; \; p \geq 1$. The result follows from the long exact sequence 8.2 using that

$$H^0(X,S) \longrightarrow H^0(Z,S)$$

is surjective, 2.3.

Q.E.D.

IV.9 Classification of locally constant sheaves

Let us introduce the notion of equivariant k-sheaves. Given a commutative ring k and a topological space X on which the group G acts from the right. By a __G-action__ on the sheaf E we understand data $(a_\sigma)_{\sigma \in G}$ where $a_\sigma \colon E \to \sigma_* E$ is a morphism of k-sheaves subject to the conditions

9.1
$$a_{\sigma\tau} = \tau_* a_\sigma \circ a_\tau \qquad ; \quad \sigma, \tau \in G$$

By a __G-equivariant__ sheaf E we understand a sheaf equipped with a G-action. By a morphism $h \colon E \to F$ of G-equivariant sheaves we understand a morphism of k-sheaves which makes the following diagram commutative

$$
\begin{array}{ccc}
E & \longrightarrow & \sigma_* E \\
\downarrow h & & \downarrow \sigma_* h \\
F & \longrightarrow & \sigma_* F
\end{array}
\qquad ; \quad \sigma \in G
$$

The category of G-equivariant sheaves on X will be denoted $Sh^G(X,k)$. In the following we put $B = X/G$ and let $f \colon X \to B$ denote the projection. We shall introduce two functors

9.2
$$Sh^G(X,k) \underset{Inv^G}{\overset{f^*}{\longleftrightarrow}} Sh(B,K)$$

Let us first remark that a G-action on a k-sheaf E on X can be described in terms of isomorphisms $a^\sigma \colon \sigma^* E \to E$ as it follows by adjunction. Thus for a sheaf F on B the canonical isomorphism

$$\sigma^* f^* F \longrightarrow f^* F \qquad ; \quad \sigma \in G$$

defines a G-action on p*F. - Given a G-sheaf E on X. In
order to describe the sections of $\text{Inv}^G E$ over an open set V
of B we can remark that the formula 9.1 defines a left action
of G on $\Gamma(f^{-1}(V),E)$. We put

$$\Gamma(V,\text{Inv}^G E) = \Gamma(f^{-1}(V),E)^G$$

i.e. the G-invariant sections of E over $f^{-1}(V)$. We leave it
to the reader to prove that f* is a left adjoint to Inv, i.e.

9.3 $$\text{Hom}_G(f*F,E) = \text{Hom}(F,\text{Inv}^G E)$$

To a k-module N with a left G-action we can associate a G-action
on N: To $\sigma \in G$ and the open set U of X we associate
the map

9.4. $$\text{Hom}(U,N) \to \text{Hom}(\sigma^{-1}(U),N) \quad ; \quad \varphi \longmapsto \sigma_N \circ \varphi \circ \sigma$$

Lemma 9.5. In case X is connected, the functor $E \to \Gamma(X,E)$
induces an equivalence between the category of G-sheaves on X,
whose underlying k-sheaf is constant, and the category of k-modules
with a left G-action.

Proof. The inverse functor is furnished by 9.4.

Q.E.D.

Lemma 9.6. Suppose every point $b \in B$ admits an open neighbourhood V such that $f^{-1}(V) \to V$ is G-equivariantly isomorphic
to pr: $V \times G \to V$. Then f* and Inv^G induces an equivalence
between the category of locally constant k-sheaves on B and G-
sheaves whose underlying k-sheaf is locally constant.

<u>Proof</u>. It suffices to prove a similar statement for
$B \times G \to B$ with locally constant replaced by constant. This is
left to the reader.

Q.E.D.

Let us turn to the main theme of this section. Given a
topological space B and a basepoint $b \in B$. Regarding the fundamental group $\pi_1(B,b)$ we use the convention proposed by Deligne
(1) that $\alpha * \beta$ denotes the loop obtained by first running through
the loop β. Given a locally constant k-sheaf E on B, then
monodromy 4.21 defines a <u>left</u> action of $\pi_1(B,b)$ on E_b, the
stalk of E at b.

<u>Theorem 9.7</u>. Let B be a connected, locally path-connected
space for which every point $y \in B$ admits an open neighbourhood V
such that $\pi_1(V,y) \to \pi_1(B,y)$ is trivial. Then, monodromy is an
equivalence between the category of locally constant k-sheaves on
B and the category of left $\pi_1(B,b)$-equivariant k-modules.

<u>Proof</u>. Put $G = \pi_1(B,b)$ and let $f: X \to B$ denote the uni-
versal covering space of X. The space X is simply connected,
Spanier (1) 2.5.15. Thus it follows from 4.20 that any locally
constant sheaf on X is constant. A combination of the two previous lemmas gives an equivalence between the locally constant
k-sheaves and the G-equivariant k-modules.

In order to identify this correspondence with the monodromy
correspondence, consider the base point x of X, $f(x) = b$.
For a locally constant sheaf E on B, we want to transport the

G-action on $\Gamma(X,f^*E)$ to the stalk E_b of E at b. To this end introduce names to the canonical map

$$E_b \xrightarrow{\ f_x\ } f^*E_x \xleftarrow{\ r_x\ } \Gamma(X,f^*E)$$

The result is quite simply

$$9.8 \qquad \sigma \longmapsto f_x^{-1}\, r_x\, \sigma\, r_x^{-1}\, f_x$$

For a fixed $\sigma \in G$ the action on E_b may be calculated as the composite of the maps

$$E_b \xrightarrow{\ f_x\ } f^*E_x \xrightarrow{\ r_x^{-1}\ } \Gamma(X,f^*E) \xrightarrow{\ r_{\sigma(x)}\ } f^*E_{\sigma(x)} \xrightarrow{\ f_{\sigma(x)}^{-1}\ } E_b$$

i.e. by the formula

$$9.9 \qquad \sigma \longmapsto f_{\sigma(x)}^{-1}\, r_{\sigma(x)}\, r_x^{-1}\, f_x \qquad ; \qquad \sigma \in G$$

The formula 9.9 follows from 9.8 using the following commutative diagram

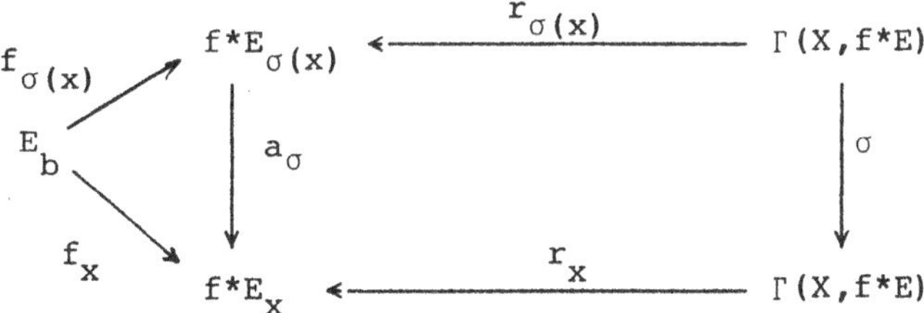

It remains to identify 9.8 with the recipe 4.21.

$$\text{Q.E.D.}$$

V. Duality with Coefficient in a Field

In this chapter we shall work with a fixed field k. For
a k-vector space N we let $N^{\vee}$ denote the dual vector space, i.e.
$N^{\vee} = \mathrm{Hom}_k(N,k)$.

Let X be a locally compact space and S a k-sheaf on X.
The inclusion $j: V \to U$ of two open subsets of X will induce
a map

$$j_!: \Gamma_c(V,S) \to \Gamma_c(U,S)$$

"extension by zero". The k-linear dual of this

$$j_!^{\vee}: \Gamma_c(V,S)^{\vee} \to \Gamma_c(U,S)^{\vee}$$

gives rise to restriction maps in a <u>presheaf $S^{\vee}$</u> defined by

1.1
$$\Gamma(U,S^{\vee}) = \Gamma_c(U,S)^{\vee}$$

<u>Proposition 1.2</u>. For a soft k-sheaf S on the locally
compact space X, the presheaf $S^{\vee}$ is a sheaf.

<u>Proof</u>. Let us first prove that for any two open sets W
and V the sequence

$$0 \to \Gamma_c(V \cup W, S^{\vee}) \to \Gamma_c(V,S^{\vee}) \oplus \Gamma_c(W,S^{\vee}) \to \Gamma_c(V \cap W, S^{\vee})$$

formed by the sum and difference between two restriction maps is exact. - Consider the Mayer-Vietoris sequence III.7.5

$$0 \to \Gamma_c(V \cap W, S) \to \Gamma_c(V, S) \oplus \Gamma_c(W, S) \to \Gamma_c(V \cup W, S) \to H_c^1(V \cap W, S)$$

and notice that $H_c^1(V \cap W, S) = 0$, since the restriction of S to $V \cap W$ is soft. The result now follows by taking the k-linear dual of the Mayer-Vietoris sequence.

Let us next consider a set $\mathcal{U}$ of open subsets of X with the property, that for any pair $V, W \in \mathcal{U}$ there exists $U \in \mathcal{U}$ containing V and W. We shall prove that S^V has the sheaf property relative to such a directed covering: Let $\mathcal{O}$ denote the union of all the sets from $\mathcal{U}$. It follows rather directly from Borel-Heine that

$$\varinjlim_{U \in \mathcal{U}} \Gamma_c(U, S) = \Gamma_c(\mathcal{O}, S)$$

Using the universal mapping property of $\varinjlim$ it follows directly from this formula that S^V has the sheaf property relative to $\mathcal{U}$.

In order to show that S^V has the sheaf property relative to an arbitrary family V of open sets, we can introduce the family $\mathcal{U}$ of open sets which are unions of finitely many open sets from V, and then we use the previous result.

Q.E.D.

<u>Proposition 1.3</u>. Let S be a soft k-sheaf on the finite dimensional locally compact space X. For any k-sheaf F on X, the sheaf $S \otimes_k F$ is soft.

256

$\underline{\text{Proof}}$. Let $j: U \to X$ be the inclusion of an open subset.
Let us establish a canonical isomorphism for any k-sheaf S

$$1.4 \qquad S \otimes_k j_! \underset{\sim}{k} = j_! j^* S$$

Consider an open subset V of X. Scalar multiplication

$$\Gamma(V,S) \otimes_k \Gamma(V \cap U, \underset{\sim}{k}) \to \Gamma(V \cap U, S)$$

will induce a linear map II.6.1

$$\Gamma(V,S) \otimes_k \Gamma(V, j_! \underset{\sim}{k}) \to \Gamma(V, j_! j^* S)$$

and by variation of V a morphism of sheaves

$$S \otimes_k j_! \underset{\sim}{k} \to j_! j^* S$$

By localization this is seen to be an isomorphism, II.6.3.

Let the dimension of X be n and consider a resolution
of F by k-sheaves II.7.4

$$P_{n-1} \xrightarrow{\partial_{n-1}} P_{n-2} \cdots P_1 \xrightarrow{\partial_1} P_0 \longrightarrow F \dashrightarrow 0$$

where the P's are direct sums of k-sheaves of the form $j_! \underset{\sim}{k}$
where $j: U \to X$ is the inclusion of an open subset. From the
formula 1.4 and III.7.2 follows that $S \otimes_k j_! k$ is soft. Let
us remark that a direct sum of soft sheaves is soft as it follows
from III.5. Using this we conclude that $S \otimes_k P_i$ is soft.

Let K denote the kernel of ∂_{n-1}. From the resolution
above we obtain an exact sequence of sheaves

$$0 \to S \otimes K \to S \otimes P_{n-1} \to S \otimes P_{n-2} \to \ldots \to S \otimes P_0 \to S \otimes F \to 0$$

Using III.9.9. we conclude that $S \otimes F$ is soft.

$$\text{Q.E.D.}$$

$\underline{\text{Proposition 1.5}}$. Let X be a finite dimensional locally compact space and S a soft k-sheaf on X. There is a natural isomorphism

$$\Gamma_c(X, F \otimes S)^\vee \overset{\sim}{\to} \operatorname{Hom}(F, S^\vee)$$

as F varies through $Sh(X,k)$.

$\underline{\text{Proof}}$. Let $j: U \to X$ denote the inclusion of an open subspace. Consider the natural maps

$$\Gamma(U,F) \otimes \Gamma_c(U,S) \to \Gamma_c(U, F \otimes S) \xrightarrow{\ j_!\ } \Gamma_c(X, F \otimes S)$$

The dual of the composite can be written

$$\Gamma_c(X,\ F \otimes S)^\vee \longrightarrow \operatorname{Hom}(\Gamma(U,F), \Gamma_c(U,S)^\vee)$$

By variation of U this defines a map

$$1.6 \qquad\qquad \Gamma_c(X, F \otimes S)^\vee \longrightarrow \operatorname{Hom}(F, S^\vee)$$

We shall prove that this is an isomorphism.

Let us first check the case $F = j_! \underset{\sim}{k}$ where j is the inclusion of an open subspace. In order to evaluate the left hand side of 1.6 we use formula 1.4 and III.7.1 to get

$$\Gamma_c(X, j_! \underset{\sim}{k} \otimes S)^\vee = \Gamma_c(X, j_! j^*S)^\vee = \Gamma_c(U,S)^\vee = \Gamma(U, S^\vee)$$

We can evaluate the right hand side of 1.6 by II.6. and II.7.2

$$\operatorname{Hom}(j_! \underset{\sim}{k}, S^\vee) = \operatorname{Hom}(\underset{\sim}{k}, j^*S^\vee) = \Gamma(U, j^*S^\vee) = \Gamma(U, S^\vee)$$

258

It is left to the reader to check that these identifications
transform the map 1.6 into the identity.

The general case can be handled by choosing a presentation
of F of the form

$$P \longrightarrow Q \longrightarrow F \longrightarrow 0$$

where P and Q are direct sums of sheaves of the form just
described. The two functors involved in 1.6 transform direct
sums into direct products. It follows that the two vertical
maps to the right in the commutative diagram below are isomorphisms

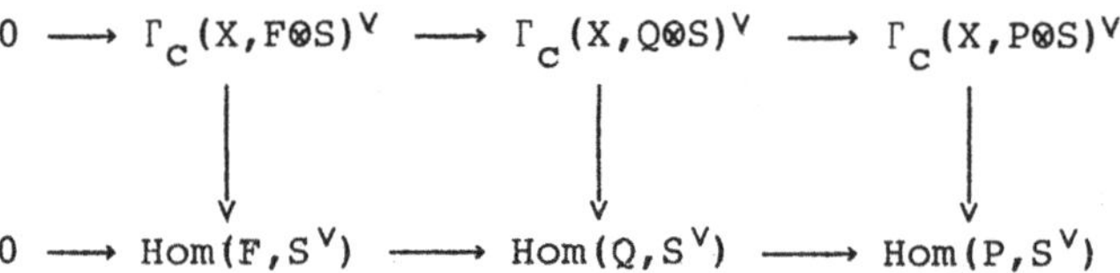

It is easily seen that the diagram is exact. From this follows
that the vertical arrow to the left is an isomorphism.

Q.E.D.

Corollary 1.7. The sheaf $S^\vee$ is injective in Sh(X,k).

Proof. According to 1.5 we must show that

$$F \longmapsto \Gamma_c(X, F \otimes S)^\vee$$

is an exact functor. This follows from 1.3 and III.2.7.

Q.E.D.

V.2. Verdier duality

Let X denote a locally compact space of finite dimension and let k denote a fixed field. The category $Sh(X,k)$ of k-sheaves on X is an abelian category with enough injectives. We let $D^+(X,k)$ denote the homotopy category of bounded below complexes of injective k-sheaves.

To an object $I^\cdot$ of $D^+(X,k)$ we can associate $[\Gamma_c(X,I^\cdot),k]$ the set of homotopy classes of morphisms of complexes of k-vector-spaces from $\Gamma_c(X,I^\cdot)$ to k. More naively

$$[\Gamma_c(X,I^\cdot),k] = H^0\Gamma_c(X,I^\cdot)^\vee$$

The assignment $I^\cdot \mapsto [\Gamma_c(X,I^\cdot),k]$ defines a contravariant functor from $D^+(X,k)$ to the category of k-vectorspaces which is representable:

Verdier duality 2.1. There exists $\mathcal{D}^\cdot$ in $D^+(X,k)$ and a natural isomorphism

$$[I^\cdot,\mathcal{D}^\cdot] \xrightarrow{\sim} [\Gamma_c(X,I^\cdot),k]$$

as $I^\cdot$ varies through $D^+(X,k)$.

Proof. Let us first establish the proper sign conventions. For a complex $L^\cdot$ of k-vectorspaces we put $L^{\cdot\vee} = \text{Hom}^\cdot(L^\cdot,k)$ with the notation of I.4.3. Notice that $L^{\cdot\vee}$ is a complex of k-vectorspaces whose p'th differential is given by

$$(-1)^{p+1}(\partial^{-p-1})^\vee : (L^{-p})^\vee \to (L^{-p-1})^\vee$$

This formula will also be used to extend the functor $S \mapsto S^{\vee}$ on $Sh(X,k)$ given by 1.1 to complexes of k-sheaves, see I.11.9.

Let $S^{\boldsymbol{\cdot}}$ denote a bounded soft resolution of the constant sheaf $\underset{\sim}{k}$ on X. Recall from 1.7 that $S^{\boldsymbol{\cdot}\,\vee}$ is a bounded sheaf of injective k-sheaves. For $I^{\boldsymbol{\cdot}}$ in $D^{+}(X,k)$ and integers p and q we have a canonical isomorphism, 1.5

$$\Gamma_{c}(X,I^{p} \otimes S^{q})^{\vee} = \mathrm{Hom}(I^{p},S^{q\vee})$$

Taking the direct sum over all p,q with p+q = -n we deduce a canonical isomorphism of k-vector spaces

$$[\Gamma_{c}(X,I^{\boldsymbol{\cdot}} \otimes S^{\boldsymbol{\cdot}})^{\vee}]^{n} \xrightarrow{\sim} \mathrm{Hom}^{n}(I^{\boldsymbol{\cdot}},S^{\boldsymbol{\cdot}\vee})$$

It is left to the reader to check the signs to see that this in fact gives an isomorphism of complexes

2.2 $$\Gamma_{c}(X,I^{\boldsymbol{\cdot}} \otimes S^{\boldsymbol{\cdot}})^{\vee} = \mathrm{Hom}^{\boldsymbol{\cdot}}(I^{\boldsymbol{\cdot}},S^{\boldsymbol{\cdot}\vee})$$

From the quasi-isomorphism $\underset{\sim}{k} \to S^{\boldsymbol{\cdot}}$ we deduce a quasi-isomorphism

$$I^{\boldsymbol{\cdot}} \longrightarrow I^{\boldsymbol{\cdot}} \otimes S^{\boldsymbol{\cdot}}$$

Noticing that both complexes are soft 1.3, we deduce still another quasi-isomorphism I.7.5

$$\Gamma_{c}(X,I^{\boldsymbol{\cdot}}) \longrightarrow \Gamma_{c}(X,I^{\boldsymbol{\cdot}} \otimes S^{\boldsymbol{\cdot}})$$

The k-linear dual of this

$$\Gamma_C(X,I^\bullet \otimes S^\bullet)^V \longrightarrow \Gamma_C(X,I^\bullet)^V$$

combined with the isomorphism 2.2 yields a final quasi-iso-morphism

$$2.3 \qquad\qquad \operatorname{Hom}^\bullet(I^\bullet,S^{\bullet V}) \to \Gamma_C(X,I^\bullet)^V$$

Passing to H^0 this yields an isomorphism

$$[I^\bullet,S^{\bullet V}] \xrightarrow{\sim} H^0\Gamma_C(X,I^\bullet)^V$$

Finally put $\mathcal{D}^\bullet = S^{\bullet V}$.

$$\text{Q.E.D.}$$

The complex $\mathcal{D}^\bullet$ from 2.1 is called <u>the dualizing complex</u>. This is a bounded below complex of injective k-sheaves uniquely determined up to homotopy, in particular the cohomology sheaves $H^p\mathcal{D}^\bullet$, $p \in \mathbb{Z}$, are uniquely determined up to isomorphism. We shall now calculate these: Recall that an inclusion of open subspaces $j: V \to U$ give rise to a map

$$j_! : H_C^p(V,\underset{\sim}{k}) \to H_C^p(U,\underset{\sim}{k})$$

"extension by zero". The k-linear dual of this

$$j_!^V : H_C^p(U,\underset{\sim}{k})^V \to H_C^p(V,\underset{\sim}{k})^V \qquad ; \quad p \in \mathbb{Z}$$

gives rise to a presheaf $U \longmapsto H_C^p(U,\underset{\sim}{k})^V$.

Proposition 2.4. Let $\mathcal{D}^\bullet$ denote the dualizing complex for the locally compact space X. For any integer p, the cohomology sheaf $H^{-p}\mathcal{D}^\bullet$ is the sheaf associated to the presheaf

$$U \longmapsto H_c^p(U,\underset{\sim}{k})^\vee$$

Proof. Let us first establish a canonical isomorphism for all integers $p \in \mathbb{Z}$

$$2.5 \qquad H_c^p(X,F)^\vee = H^{-p}\mathrm{Hom}(F,\mathcal{D}^\bullet)$$

as F runs through $\mathrm{Sh}(X,k)$. To do this let us choose an injective resolution $I^\bullet$ of F. Notice that

$$H_c^p(X,F) = H^p\Gamma_c(X,I^\bullet) = H^0\Gamma_c(X,I^\bullet[p])$$

From the duality formula 2.1 we get

$$H_c^p(X,F)^\vee = [\Gamma_c(X,I^\bullet[p]),k] = [I^\bullet[p],\mathcal{D}^\bullet] = [I^\bullet,\mathcal{D}^\bullet[-p]]$$

The quasi-isomorphism $F \to I^\bullet$ allows us to make the identification

$$[I^\bullet,\mathcal{D}^\bullet[-p]] = H^{-p}\mathrm{Hom}^\bullet(I^\bullet,\mathcal{D}^\bullet) = H^{-p}\mathrm{Hom}(F,\mathcal{D}^\bullet)$$

which proves 2.5. — Consider in particular $F = j_!\underset{\sim}{k}$ where $j: U \to X$ is the inclusion of an open subset U of X. This gives

$$H_c^p(U,\underset{\sim}{k})^\vee = H_c^p(X,j_!\underset{\sim}{k})^\vee = H^{-p}\mathrm{Hom}(j_!\underset{\sim}{k},\mathcal{D}^\bullet)$$

Using the adjunction formula II.6.6 and II.7.2

$$\mathrm{Hom}(j_!\underset{\sim}{k}, \mathcal{D}^\cdot) = \mathrm{Hom}(\underset{\sim}{k}, j*\mathcal{D}^\cdot) = \Gamma(U, \mathcal{D}^\cdot)$$

we get the formula

2.6 $$H^{-p}\Gamma(U, \mathcal{D}^\cdot) \overset{\sim}{\to} H_c^{\,p}(U, \underset{\sim}{k})^\vee$$

The formula 2.5 is functorial in F, which implies that 2.6 is compatible with restriction to an open subset of V.

$$\mathrm{Q.E.D.}$$

The next result is already contained in the proof of 2.2 but I find it enlightening to deduce it as a corollary to 2.4.

Corollary 2.7. The dualizing complex for an n-dimensional locally compact space X may be represented by a complex $\mathcal{D}^\cdot$ of injective k-sheaves where

$$\mathcal{D}^i = 0 \quad \text{for} \quad i \notin [-n, 0]$$

Proof. Let $\mathcal{D}^\cdot$ be a dualizing complex. It follows from 2.5 that

$$H^p\mathcal{D}^\cdot = 0 \quad \text{for} \quad p \notin [-n, 0]$$

From this we conclude that the canonical morphism I.5.8

$$\mathcal{D}^\cdot \longrightarrow \tau_{\geq -n} \, \mathcal{D}^\cdot$$

is a quasi-isomorphism, and that the second of these complexes consists of injectives. Thus we may assume that $\mathcal{D}^p = 0$ for $p < -n$. – Notice that the following sequence is exact

$$0 \longrightarrow \mathrm{Ker}\ \partial^0 \longrightarrow \mathcal{D}^0 \xrightarrow{\ \partial^0\ } \mathcal{D}^1 \xrightarrow{\ \partial^1\ } \mathcal{D}^2 \longrightarrow \ldots$$

Interpreting this as an injective resolution of $\mathrm{Ker}\ \partial^0$ we get for any k-sheaf F on X

$$\mathrm{Ext}^1(F, \mathrm{Ker}\ \partial^0) = H^1\ \mathrm{Hom}(F, \mathcal{D}^{\cdot})$$

Using the formula 2.6 we get

$$H^1\ \mathrm{Hom}(F, \mathcal{D}^{\cdot}) = H_c^{-1}(X, F)^{\vee} = 0$$

from which we conclude that $\mathrm{Ker}\ \partial^0$ is injective. It follows that the canonical morphism

$$\tau_{\leq 0}\ \mathcal{D}^{\cdot} \longrightarrow \mathcal{D}^{\cdot}$$

is a homotopy equivalence.

Q.E.D.

<u>Corollary 2.8</u>. On a locally compact space X of dimension n, the presheaf

$$U \longmapsto H_c^n(U, \underset{\sim}{k})^{\vee}$$

is a sheaf.

Proof. Let the dualizing complex $\mathcal{D}^{\cdot}$ be represented as in 2.7. For an open subset U of X we have an exact sequence

$$0 \to \Gamma(U, H^{-n}\mathcal{D}^{\cdot}) \to \Gamma(U, \mathcal{D}^{-n}) \to \Gamma(U, \mathcal{D}^{-n+1})$$

By the formula 2.6 we have

$$H^{-n}\Gamma(U, \mathcal{D}^{\cdot}) = H_c^{\,n}(U, \underset{\sim}{k})^{\vee}$$

A combination of the two results realizes $H_c^{\,n}(U, \underset{\sim}{k})^{\vee}$ as sections in the sheaf $H^{-n}\mathcal{D}^{\cdot}$.

$$\text{Q.E.D.}$$

Example 2.9. Let X denote a locally closed subspace of $\mathbb{R}^n$ and $\textit{or}$ the sheaf from 2.8. For a point $x \in X$ we have

$$\textit{or}_x = \begin{cases} k & \text{for} \quad x \in \overset{\bullet}{X} \\ 0 & \text{for} \quad x \in \partial X \end{cases}$$

The first statement has already been noted in III.8.14. To prove the second statement choose an open disc D with center in x such that $X \cap D$ is closed relative to D. It follows that $X \cap D$ is homeomorphic to a proper closed subspace of $\mathbb{R}^n$ and as noted in III.9.8 $H_c^{\,n}(X \cap D, k) = 0$.

V.3 Orientation of topological manifolds

Let X be a topological manifold of dimension n, i.e. a Hausdorff space locally homeomorphic to $\mathbb{R}^n$. Consider a fixed field k and recall from III.8.13 that the corresponding <u>orientation sheaf</u> or has sections over the open subset U of X given by

$$3.1 \qquad\qquad \Gamma(U, or) = H_c^n(X, \underset{\sim}{k})^\vee$$

The k-sheaf or is locally isomorphic to $\underset{\sim}{k}$ loc.cit. - Let us recall the formula II.7.5

$$H^p(X, or) = \text{Ext}^p(\underset{\sim}{k}, or) \qquad ; \quad p \in \mathbb{Z}$$

where Ext is calculated in Sh(X,k). From general principles I.8.6 we deduce for $p, q \in \mathbb{Z}$ a cupproduct

$$\alpha \cup \beta \in H_c^{p+q}(X, or) \qquad ; \quad \alpha \in H^p(X, or), \ \beta \in H_c^q(X, \underset{\sim}{k})$$

subjected to Poincaré duality:

<u>Poincaré duality 3.2</u>. There exists a linear form

$$\int : H_c^n(X, or) \to k$$

such that for each $p \in \mathbb{Z}$ the bilinear form

$$\int \alpha \cup \beta \qquad ; \quad (\alpha, \beta) \in H^p(X, or) \times H_c^{n-p}(X, k)$$

induces an isomorphism

$$H^p(X, or) \to H_c^{n-p}(X, k)^\vee$$

$\underline{\text{Proof}}$. Let $\mathcal{D}^{\cdot}$ denote a dualizing complex represented as in 2.7. For $x \in X$ we have according to 2.4

$$H^{-p}\mathcal{D}^{\cdot}{}_x = \varinjlim H_c^p(U,k) \qquad ; \quad p \in \mathbb{Z}$$

where the direct limit is taken over all open neighbourhoods U of x in X. The point x has a fundamental system of open neighbourhoods each of which is homeomorphic to $\mathbb{R}^n$. This implies, III.8.14.

$$H^{-p}\mathcal{D}^{\cdot} = 0 \qquad ; \quad p \neq n$$

According to 2.8 we have $H^{-n}\mathcal{D}^{\cdot} = o\hbar$. We can record this as a quasi-isomorphism

$$3.3 \qquad\qquad o\hbar[n] \xrightarrow{\;\sim\;} \mathcal{D}^{\cdot}$$

Otherwise expressed $o\hbar \to \mathcal{D}^{\cdot}[-n]$ is an injective resolution of the sheaf $o\hbar$. For $p \in \mathbb{Z}$ we get accordingly

$$H^p(X,o\hbar) = H^p\Gamma(X,\mathcal{D}^{\cdot}[-n]) = H^0\Gamma(X,\mathcal{D}^{\cdot}[p-n]) = [\underset{\sim}{k},\mathcal{D}^{\cdot}[p-n]]$$

Choose an injective resolution $\underset{\sim}{k} \to K^{\cdot}$ to get

$$H^p(X,o\hbar) = [K^{\cdot},\mathcal{D}^{\cdot}[p-n]] = [K^{\cdot}[n-p],\mathcal{D}^{\cdot}]$$

Let us now reorganize the duality isomorphism 2.1

$$[I^{\cdot},\mathcal{D}^{\cdot}] \xrightarrow{\sim} [\Gamma_c(X,I^{\cdot}),k]$$

Put $I^\cdot = \mathcal{D}^\cdot$ and let the image of $1 \in [\mathcal{D}^\cdot, \mathcal{D}^\cdot]$ be denoted

3.4
$$\int : \Gamma_c(X, \mathcal{D}^\cdot) \to k$$

According to the Yoneda principle for representable functors the duality isomorphism is given by

3.5
$$\varphi \longmapsto \int \circ \Gamma_c(X, \varphi) \qquad ; \; \varphi \in [I^\cdot, \mathcal{D}^\cdot]$$

Let us apply this to $I^\cdot = K^\cdot[n-p]$ to get

$$[K^\cdot[n-p], \mathcal{D}^\cdot] = [\Gamma_c(X, K^\cdot[n-p]), k] = H_c^{n-p}(X, k)$$

Finally use 3.5 to identify this isomorphism with the one described by the cup product formula.

Q.E.D.

By a <u>k-orientation</u> of a topological manifold X we understand an isomorphism

$$\underset{\sim}{k} \overset{\sim}{\longrightarrow} o\hbar$$

of k-sheaves, where $o\hbar$ is the orientation sheaf. We shall say that X is <u>orientable</u> relative to k if a k-orientation exists and <u>k-unorientable</u> in the opposite case.

We shall now give a cohomological criterion for orientability of a manifold.

<u>Proposition 3.6</u>. Let X be a <u>connected</u> topological manifold of dimension n. Then

$$H_c^n(X,\underset{\sim}{k}) \overset{\sim}{\to} k \quad \text{if} \quad X \text{ is orientable relative to } k.$$

$$H_c^n(X,\underset{\sim}{k}) \overset{\sim}{\to} 0 \quad \text{if} \quad X \text{ is unorientable relative to } k.$$

<u>Proof</u>. Suppose X is orientable relative to k, and let $or \overset{\sim}{\to} k$ be an orientation of X. According to Poincaré duality 3.2 with $p = 0$ we get

$$H^0(X,\underset{\sim}{k}) \simeq H_c^n(X,\underset{\sim}{k})$$

Since X is connected we have $H^0(X,k) = k$ and it follows that $H_c^n(X,\underset{\sim}{k}) \overset{\sim}{\to} k$.

Let us next assume that $H_c^n(X,\underset{\sim}{k}) \neq 0$. Then we get from Poincaré duality 3.2 that

$$\Gamma(X,or) \neq 0$$

Let s be a non trivial section of or over X. Recall that $\text{Supp}(s)$ is a closed subset of X for general reasons. The fact that or is locally constant of rank 1 implies that $\text{Supp}(s)$ is open. Thus $\text{Supp}(s) = X$ since X is connected, and $\text{Supp}(S) \neq \emptyset$. In resumé: $or \overset{\sim}{\to} \underset{\sim}{k}$. This in turn implies $H_c^n(X,\underset{\sim}{k}) \overset{\sim}{\to} k$.

$$\text{Q.E.D.}$$

<u>Case $k = \mathbb{F}_2$</u>. Let us prove that in case the coefficient field $k = \mathbb{F}_2$, the field on two elements, then

$$3.7 \qquad\qquad or = \underset{\sim}{\mathbb{F}}_2$$

This is a consequence of

    <u>Lemma 3.8</u>. Let X be an arbitrary topological space. Any $\mathbb{F}_2$-sheaf ω locally isomorphic to $\underset{\sim}{\mathbb{F}}_2$ is isomorphic to $\underset{\sim}{\mathbb{F}}_2$.

    <u>Proof</u>. Recall the formula II.7.5

$$\mathrm{Hom}(\underset{\sim}{\mathbb{F}}_2, \omega) = \Gamma(X, \omega)$$

This allows us to identify an isomorphism $s: \underset{\sim}{\mathbb{F}}_2 \xrightarrow{\sim} \omega$ with a global section

$$s \in \Gamma(X, \omega) \quad \text{with} \quad s_x \neq 0 \quad \text{for all} \quad x \in X.$$

Let us prove that there exists at most one such s: If t is another such section, we can find a section u of $\Gamma(X, \mathbb{F}_2^{*})$ such that t = us, but $\mathbb{F}_2^{*} = \{1\}$. For each $x \in X$, choose an open neighbourhood U_x of x and an isomorphism s_x between $\underset{\sim}{\mathbb{F}}_2$ and the restriction of ω to U_x. By the previous result s_x and s_y have the same restriction to $U_x \cap U_y$. Thus we can glue the s_x's together to a global isomorphism.

                                                       Q.E.D.

V.4 Submanifolds of $\mathbb{R}^n$ of codimension 1

The origin of this topic is the celebrated theorem of Jordan: The complement of a simple closed plane curve has two connected components.

The basis for counting connected components of a topological space is the following

Proposition 4.1[*)]. Let X be a topological space and k a field. The set $c.c\,X$ of connected components of X can be counted as follows

$$\#c.c\,X = \dim_k H^0(X,k)$$

Proof. Suppose $\#\,c.c\,X$ is finite. In that case each connected component is both open and closed. It follows that the characteristic functions for the connected components form a basis for $H^0(X,k)$.

Suppose conversely that $H^0(X,k)$ is finite dimensional. Assume $X \neq \emptyset$ and let C denote the set of non empty subsets of X which are both open and closed. Let $C \in \mathcal{C}$ be chosen such that $\dim H^0(C,k)$ is as small as possible. This implies that C is connected: otherwise we would have a partition $C = E \cup D$ of C into the non empty open subsets with $E \cap D = \emptyset$. From the formula

$$\dim H^0(C,k) = \dim H^0(E,k) + \dim H^0(D,k)$$

we derive at a contradiction. Thus we have proved that X has a connected component which is open (and closed). It is now

---

*) For a set S, we let $\#S \in \overline{\mathbb{N}} = \mathbb{N} \cup \{+\infty\}$ denote the number of elements in S. For a k-vector space V we put $\dim_k V = \#B$ where B is a basis for V.

272

easy to proceed by induction on $\dim_k H^0(X,k)$.

$$Q.E.D.$$

Corollary 4.2. Let X be a locally compact space and k a field. The set $c.c.c\,X$ of compact connected components of X can be counted as follows

$$\# \, c.c.c\,X = \dim_k H_c^0(X,k)$$

Proof. In case X is compact the result follows from 4.1. In case X is non compact we let $\hat{X}$ denote the 1-point compactification of X. The space $\hat{X}$ is compact and contains a point ∞ such that $\hat{X}-\{\infty\}$ is homeomorphic to X. The exact sequence

$$0 \to H_c^0(X,k) \to H_c^0(\hat{X},k) \to H_c^0(\{\infty\},k) \to 0$$

shows that

$$\dim_k H^0(\hat{X},k) = 1+\dim_k H_c^0(X,k)$$

Let $C(\infty)$ denote the connected component of $\hat{X}$ which contains ∞. We leave it to the reader to identify the remaining connected components of $\hat{X}$ and $c.c.c\,X$. This gives

$$\# \, c.c\,\hat{X} = 1 + \# c.c.c\,X$$

The result follows from 4.1 by combining the two formulas.

$$Q.E.D.$$

Let us now return to our main theme <u>a proper closed subset</u> <u>X of $\mathbb{R}^n$.</u> For an arbitrary field we have

4.3
$$\# \, c.c(\mathbb{R}^n - X) = 1 + \dim_k H_c^{n-1}(X,k)$$

<u>Proof</u>. Consider the exact sequence $(X \neq \mathbb{R}^n)$

$$0 \to H_c^{n-1}(X,k) \to H_c^n(\mathbb{R}^n-X,k) \to H_c^n(\mathbb{R}^n,k) \to 0$$

deduced from the III.8.3 and III.8.5 see the proof of III.8.7, and compare this with Poincaré duality 3.2 for $\mathbb{R}^n-X$

$$H^0(\mathbb{R}^n-X,k) \overset{\sim}{\to} H_c^{\,n}(\mathbb{R}^n-X,k)^{\vee}$$

to deduce the result.

Q.E.D.

<u>Jordan-Brouwer separation theorem 4.4</u>. Let $\gamma: S^{n-1} \to \mathbb{R}^n$ be an injective continuous map. The complement of $\gamma(S^{n-1})$ in $\mathbb{R}^n$ has two connected components.

<u>Proof</u>. We have $H^{n-1}(S^{n-1},\underset{\sim}{k}) \overset{\sim}{\to} k$ according to III.8.8, and the result follows from 4.3.

Q.E.D.

Let us turn to another important classical theorem

<u>Invariance of domain 4.5</u>. Any injective continuous map $f: \mathbb{R}^n \to \mathbb{R}^n$ is open, i.e. transforms open sets into open sets.

$\underline{\text{Proof}}$. It remains the same to prove that any injective continuous map $f: B^n \to \mathbb{R}$ will transform the interior D^n of B^n into an open subset of $\mathbb{R}^n$.

Let us first remark that it follows from 4.3 that $\mathbb{R}^n - f(B^n)$ is connected, whereas $\mathbb{R}^n - f(S^{n-1})$ has precisely two connected components. Notice that this set admits the partition

$$\mathbb{R}^n - f(S^{n-1}) = \mathbb{R}^n - f(B^n) \cup f(D^n)$$

Thus the connected components of $\mathbb{R}^n - f(S^{n-1})$ are $\mathbb{R}^n - f(B^n)$ and $f(D^n)$. Quite generally the connected components of an open subset of $\mathbb{R}^n$ are open. Thus $f(D^n)$ is open.

$$Q.E.D.$$

$\underline{\text{Theorem 4.6}}$. Let X be a closed, connected submanifold of $\mathbb{R}^n$ of codimension 1. The complement $\mathbb{R}^n - X$ has two connected components U and V and $X = \bar{U} \cap \bar{V}$.

$\underline{\text{Proof}}$. Let us first remark that X is oriented relative to $k = \mathbb{F}_2$ according to 3.7. Thus we conclude from 3.6 and 4.3 that $\mathbb{R}^n - X$ has two connected components, U and V say. - We have $\bar{U} \cup \bar{V} = \mathbb{R}^n$ as it follows by remarking that the complement is an open subset of $\mathbb{R}^n$ contained in X, but X has no interior points as a subset of $\mathbb{R}^n$. - Consider the Mayer-Vietoris sequence

$$H_c^{n-1}(\bar{U} \cap \bar{V}, k) \to H_c^n(\bar{U} \cup \bar{V}, k) \to H_c^n(\bar{U}, k) \oplus H_c^n(\bar{V}, k)$$

Since $\bar{U}$ and $\bar{V}$ are proper closed subsets of $\mathbb{R}^n$ we find from III.9.8 that the two groups in the direct sum are zero. We conclude that $H_c^{n-1}(\bar{U} \cap \bar{V}, k) = 0$. We can now apply lemma 4.7

below to the closed subset $\bar{U} \cap \bar{V}$ of X to get the result.

Q.E.D.

$\underline{\text{Lemma 4.7}}$. Let X be a connected n-dimensional manifold,

oriented relative to k. For a proper closed subset Z of

X we have

$$H_c^n(Z,k) = 0$$

$\underline{\text{Proof}}$. Consider the long exact sequence

$$\to H_c^n(X-Z,k) \overset{e}{\to} H_c^n(X,k) \to H_c^n(Z,k) \to 0$$

The restriction map in the orientation sheaf $\Gamma(X, o\iota) \to \Gamma(U, o\iota)$

may be identified with $e^\vee$, the linear dual of e, by 3.1.

Since $o\iota \overset{\sim}{\to} k$ and X is connected we conclude that $e^\vee$ is

injective. From this follows that e is surjective.

Q.E.D.

$\underline{\text{Theorem 4.8}}$. A closed submanifold X of $\mathbb{R}^n$ of codi-

mension $n-1$ is orientable relative to any field. Moreover

$$\# c.c(\mathbb{R}^n - X) = 1 + \# c.c\, X$$

$\underline{\text{Proof}}$. We must prove that any connected component Y of

X is orientable relative to k. According to 4.6 we find that

$\mathbb{R}^n - Y$ has two connected components. Using 4.3 we conclude that

$H_c^n(Y,k)$ is 1-dimensional, which implies that Y is orientable,

3.6. From Poincaré duality applied to the oriented manifold X

we get that $$H^0(X,k) \overset{\sim}{\longrightarrow} H_c^n(X,k)^\vee$$

Conclusion by 4.1 and 4.3.

Q.E.D.

V.5 Duality for a subspace

Let X denote a locally compact space of finite dimension and k a fixed field. The corresponding dualizing complex will be denoted $\mathcal{D}^{\cdot}$. Verdier duality is a natural isomorphism

5.1 $$[I^{\cdot},\mathcal{D}^{\cdot}] \xrightarrow{\sim} [\Gamma_c(X,I^{\cdot}),k]$$

as I varies through $D^+(X,k)$. Put $I^{\cdot} = \mathcal{D}^{\cdot}$ and let the image of $1 \in [\mathcal{D}^{\cdot},\mathcal{D}^{\cdot}]$ be denoted

5.2 $$\int_X : \Gamma_c(X,\mathcal{D}^{\cdot}) \longrightarrow k$$

which we call the __trace map__. According to the Yoneda principle for representable functors the duality isomorphism can be re-covered from the trace map by the formula

5.3 $$\varphi \longmapsto \int_X \circ\, \Gamma_c(X,\varphi) \qquad ; \varphi \in [I^{\cdot},\mathcal{D}^{\cdot}]$$

__Lemma 5.4.__ Let $S^{\cdot}$ denote a bounded below complex of soft k-sheaves. The map 5.3 induces an isomorphism

$$[S^{\cdot},\mathcal{D}^{\cdot}] \xrightarrow{\sim} [\Gamma_c(X,S^{\cdot}),k]$$

__Proof.__ Let $S^{\cdot} \to I^{\cdot}$ be an injective resolution in $Sh(X,k)$. This gives rise to a commutative diagram

$$
\begin{array}{ccc}
[I^{\cdot},\mathcal{D}^{\cdot}] & \longrightarrow & [\Gamma_c(X,I^{\cdot}),k] \\
\downarrow & & \downarrow \\
[S^{\cdot},\mathcal{D}^{\cdot}] & \longrightarrow & [\Gamma_c(X,S^{\cdot}),k]
\end{array}
$$

The vertical maps are isomorphisms according to I.6.2 and I.7.5. Conclusion by 5.1.

$$\text{Q.E.D.}$$

Let us now consider the inclusion $h: W \to X$ of a locally closed subspace W of X. The functors $h_!$ and $h^!$ of II.6 are adjoint:

$$\text{Hom}(h_! E, F) \;=\; \text{Hom}(E, h^! F)$$

for sheaves E on W and F on X. In particular we have an adjunction morphism

$$h_! h^! \mathcal{D}^\cdot \;\longrightarrow\; \mathcal{D}^\cdot$$

Applying $\Gamma_c(X,-)$ to this we get a map, III.7.1

$$\Gamma_c(W, h^! \mathcal{D}^\cdot) \;\to\; \Gamma_c(X, \mathcal{D}^\cdot)$$

The composite of this with $\displaystyle\int_X$ will be denoted

$$5.5 \qquad\qquad \int_W : \Gamma_c(W, h^! \mathcal{D}^\cdot) \;\longrightarrow\; k$$

Theorem 5.6. The complex $h^! \mathcal{D}^\cdot$ is a dualizing complex for W. The corresponding trace map $\displaystyle\int_W$ is given by 5.5 above.

278

<u>Proof</u>. Consider the commutative diagram below. The morphisms involved are those discussed above.

$$
\begin{array}{ccc}
[J^{\cdot},h^{!}\mathcal{D}^{\cdot}] & \xrightarrow{\ \sim\ } & [h_{!}J^{\cdot},\mathcal{D}^{\cdot}] \\
\downarrow {\scriptstyle \Gamma_{c}(W,-)} & & \downarrow {\scriptstyle \Gamma_{c}(X,-)} \\
[\Gamma_{c}(W,J^{\cdot}),\Gamma_{c}(W,h^{!}\mathcal{D}^{\cdot})] & \xrightarrow{\ \sim\ } & [\Gamma_{c}(X,h_{!}J^{\cdot}),\Gamma_{c}(X,\mathcal{D}^{\cdot})] \\
\downarrow {\scriptstyle [-,\int_{W}]} & & \downarrow {\scriptstyle [-,\int_{X}]} \\
[\Gamma_{c}(W,J^{\cdot}),k] & \xrightarrow{\ \sim\ } & [\Gamma_{c}(X,h_{!}J^{\cdot}),k]
\end{array}
$$

The composite of the two vertical arrows to the right is an isomorphism by Lemma 5.4. It follows that the composite of the two vertical arrows to the left is an isomorphism.

Q.E.D.

<u>Corollary 5.7</u>. Let X be a topological manifold with orientation sheaf $o\hbar_{X}$ and trace map $\int_{X}: H_{c}^{n}(X,o\hbar_{X}) \to k$. Let $j: U \to X$ be the inclusion of an open subspace. The orientation sheaf on U is $j^{*}o\hbar_{X}$ and the corresponding trace map $\int_{W}: H_{c}^{n}(U,j^{*}o\hbar_{X}) \to k$ is given by

$$
\int_{W} \alpha = \int_{X} j_{!}\, \alpha \qquad ; \ \alpha \in H_{c}^{n}(U,j^{*}o\hbar_{X}) .
$$

<u>Proof</u>. Follows from the fact that $o\hbar[n] \xrightarrow{\sim} \mathcal{D}^{\cdot}$, 3.3.

Q.E.D.

V.6 Alexander duality

Let us start with a discussion of cup products in cohomology with compact support. We consider a fixed commutative ring k.

Consider a closed subspace of a locally compact space X and a k-sheaf F on X. We shall construct a <u>cup product</u>

6.1
$$\alpha \cup \beta \in H_c^{p+q}(X,F) \quad ; \quad \alpha \in H_Z^p(X,F), \ \beta \in H_c^q(Z,k)$$

To this end we let $i: Z \to X$ denote the inclusion and use the representation II.9.8

$$H_Z^p(X,k) = \operatorname{Ext}_k^p(i_*k,F)$$

An element $\alpha \in \operatorname{Ext}_k^p(i_*k,F)$ will induce a linear map

$$\beta \longmapsto \alpha \cup \beta \quad ; \quad H_c^q(X,i_*k) \longrightarrow H_c^{p+q}(X,F)$$

The result follows from the identification III.1.7

$$H_c^{\cdot}(X,i_*k) = H_c^{\cdot}(Z,k)$$

Put $U = X-Z$ and let $i: Z \to X$ and $j: U \to X$ denote the inclusions. The cup product and the two basic long exact sequences

$$H^p(U,k) \xleftarrow{\ j^*\ } H^p(X,k) \xleftarrow{\ r\ } H_Z^p(X,k) \xleftarrow{\ \partial\ } H^{p-1}(U,k)$$
$$\qquad\qquad\qquad \gamma \qquad\qquad\quad \beta \qquad\qquad\quad \alpha$$
$$\zeta \qquad\qquad\qquad \eta \qquad\qquad\quad \xi$$
$$H_c^{n-p}(U,k) \xrightarrow{\ j_!\ } H_c^{n-p}(X,k) \xrightarrow{\ i^*\ } H_c^{n-p}(Z,k) \xrightarrow{\ \partial\ } H_c^{n-p+1}(U,k)$$

are interrelated through the three formulas

6.2 $$(-1)^p \partial\alpha \cup \xi = j_!(\alpha \cup \partial\xi)$$

6.3 $$r\,\beta \cup \eta = \beta \cup i^*\eta$$

6.4 $$j_!(j^*\gamma \cup \zeta) = \gamma \cup j_!\zeta$$

<u>Verification</u>. The canonical exact sequence

$$0 \to j_!\underset{\sim}{k} \to \underset{\sim}{k} \to i_*\underset{\sim}{k} \to 0$$

gives rise to a canonical class, I.8.4

$$\theta \in \mathrm{Ext}^1(i_*\underset{\sim}{k}, j_!\underset{\sim}{k})$$

The long exact sequence in local cohomology

$$\leftarrow H^p(U,k) \xleftarrow{j^*} H^p(X,k) \xleftarrow{r} H_Z^p(X,k) \xleftarrow{\partial} H^{p-1}(U,k) \leftarrow$$

may be identified with the Ext-sequence, I.8.5 and II.9.8

$$\longleftarrow \mathrm{Ext}^p(j_!\underset{\sim}{k},\underset{\sim}{k}) \longleftarrow \mathrm{Ext}^p(\underset{\sim}{k},\underset{\sim}{k}) \longleftarrow \mathrm{Ext}^p(i_*\underset{\sim}{k},\underset{\sim}{k}) \xleftarrow{\cup\theta(-1)^p} \mathrm{Ext}^{p-1}(j_!\underset{\sim}{k},\underset{\sim}{k})$$

while the long exact sequence in cohomology with compact support

$$H_c^{n-p}(U,\underset{\sim}{k}) \longrightarrow H_c^{n-p}(X,\underset{\sim}{k}) \longrightarrow H_c^{n-p}(Z,\underset{\sim}{k}) \longrightarrow H_c^{n-p+1}(U,\underset{\sim}{k})$$

may be identified with

$$H_c^{n-p}(X,j_!\underset{\sim}{k}) \to H_c^{n-p}(X,\underset{\sim}{k}) \longrightarrow H_c^{n-p}(X,i_*\underset{\sim}{k}) \xrightarrow{\theta\cup} H_c^{n-p+1}(X,j_!\underset{\sim}{k})$$

With these identifications the formulas become more ore less obvious.

Excision formulas for cup product

Let Z denote a closed subset of
the locally compact space X and O
an open subspace. For any k-sheaf
F we have

$$
\begin{array}{ccc}
Z \cap O & \xrightarrow{\ 1\ } & O \\
\downarrow{\scriptstyle h} & & \downarrow{\scriptstyle j} \\
Z & \xrightarrow{\ i\ } & X
\end{array}
$$

6.5
$$
\omega \cup h_! \alpha = j_!(j^*\omega \cup \alpha) \quad ; \ \omega \in H_Z^p(X,F), \alpha \in H_c^q(Z\cap O)
$$

where $j^*\omega \in H_{Z\cap O}^p(O,F)$ is the restriction of ω.

Proof. Let us recall II.6.13, that for a sheaf G on Z
we have $j^*i_*G = 1_*h^*G$. In particular if we take an injective resolution $k \to K^\cdot$ in $Sh(Z,k)$ we find that $j^*i_*K^\cdot$ is an injective
resolution of $1_*\underset{\sim}{k}$, since

$$
1_*\underset{\sim}{k} = 1_*h^*\underset{\sim}{k} = j^*i_*\underset{\sim}{k}
$$

Let $F \to J^\cdot$ be an injective resolution $Sh(X,k)$. Since

$$
H_Z^p(X,F) = \text{Ext}^p(i_*k,F) = [i_*K^\cdot,I^\cdot[p]]
$$

we can represent ω as a morphism of complexes $\omega: i_*K^\cdot \to I^\cdot[p]$.
The commutative diagram

$$
\begin{array}{ccc}
j_!j^*i_*K^\cdot & \xrightarrow{\ j_!j^*\omega\ } & j_!j^*I^\cdot[p] \\
\downarrow & & \downarrow \\
i_*K^\cdot & \xrightarrow{\ \omega\ } & I^\cdot[p]
\end{array}
$$

transforms by $\Gamma_c(X,-)$ into the diagram, compare III.7.1.

282

$$\Gamma_C(0,j*i_*K^{\cdot}) \xrightarrow{\quad \Gamma_C(0,j*\omega) \quad} \Gamma_C(0,j*I^{\cdot}[p])$$

$$\downarrow \qquad\qquad\qquad\qquad \downarrow$$

$$\Gamma_C(X,i_*K^{\cdot}) \xrightarrow{\quad \Gamma_C(X,\omega) \quad} \Gamma_C(X,I^{\cdot}[p])$$

Applying H^q to this we get the commutative diagram

$$H_C^{\,q}(0 \cap Z,k) \xrightarrow{\quad j*\omega \cup \quad} H_C^{\,p-q}(0,j*F)$$

$$\downarrow h_! \qquad\qquad\qquad\qquad \downarrow j_!$$

$$H_C^{\,q}(Z,k) \xrightarrow{\quad \omega \cup \quad} H_C^{\,p+q}(X,F)$$

from which the result follows.

Q.E.D.

Alexander Duality 6.6. Let X denote an n-dimensional topological manifold oriented relative to the field k, and let

$$\int_X : H_C^{\,n}(X,k) \longrightarrow k$$

denote the trace map. For a closed subset Z of X the bilinear form

$$\int_X \alpha \cup \beta \qquad ; \qquad \alpha \in H_Z^{\,p}(X,k), \beta \in H_C^{\,n-p}(Z,k)$$

induces an isomorphism $H_Z^{\,p}(X,k) \xrightarrow{\sim} H_C^{\,n-p}(Z,k)^{\vee}$.

Proof. The duality isomorphism

$$\varphi \longmapsto \int_X \circ \, \Gamma(X,\varphi) \, ; \quad [I^{\cdot},\mathcal{D}^{\cdot}] \xrightarrow{\sim} [\Gamma_C(X,I^{\cdot}),k]$$

can be rewritten

$$[I^{\cdot},\mathcal{D}^{\cdot}[p-n]] \xrightarrow{\sim} (H^{n-p}\Gamma_C(X,I^{\cdot}))^{\vee}$$

Applied to an injective resolution $i_*k \to I^{\cdot}$ this gives

$$[i_*k, \mathcal{D}^{\cdot}[-n][p]] \xrightarrow{\sim} H_c^{n-p}(Z,k)^{\vee}.$$

The result follows by the fact that $\mathcal{D}^{\cdot}[-n]$ is an injective resolution of $\underset{\sim}{k}$, 3.3.

Q.E.D.

Alexander duality for $\mathbb{R}^n$

Let Z denote a proper closed subspace of $\mathbb{R}^n$. We have an isomorphism

$$H_Z^p(X,k) \xrightarrow{\sim} H_c^{n-p}(Z,k)^{\vee} \quad ; \quad p \in \mathbb{N}$$

by Alexander duality. Using the long exact sequence with $U = \mathbb{R}^n - Z$

$$H^p(\mathbb{R}^n,k) \to H^p(U,k) \to H_Z^{p+1}(\mathbb{R}^n,k) \to H^{p+1}(\mathbb{R}^n,k) \to$$

we deduce the following formulas

$$\dim_k H^0(U,k) = 1 + \dim_k H_c^{n-1}(Z,k)$$

6.7

$$\dim_k H^p(U,k) = \dim_k H_c^{n-p-1}(Z,k) \quad ; \quad p \geq 1$$

The first of these has been exploited in 4.3. Let us work out the second relation in case $p = n-1$. We get according to 4.2

6.8
$$\dim_k H^{n-1}(U,k) = \# \, c.c.c(Z)$$

In case $p = n$ we get

6.9
$$H^n(U,k) = 0 \quad ; \quad U \subseteq \mathbb{R}^n$$

V.7 Residue theorem for $n-1$ forms on $\mathbb{R}^n$

In this section we shall generalize Cauchy's residue theorem to $\mathbb{R}^n$.

Local symbols

We consider a fixed orientation of $\mathbb{R}^n$ with trace map

$$\int_{\mathbb{R}^n} : H_c^n(\mathbb{R}^n, \mathbb{C}) \to \mathbb{C}$$

For a point $s \in \mathbb{R}^n$, the cup product gives a non degenerate pairing by Alexander duality 6.6

$$H_{\{s\}}^n(\mathbb{R}^n, \mathbb{C}) \times H_c^0(\{s\}, \mathbb{C}) \longrightarrow H_c^n(\mathbb{R}^n, \mathbb{C})$$

We let $1_s \in H_c^0([s], \mathbb{C})$ denote the constant 1 and denote by $\omega_s \in H^{n-1}(\mathbb{R}^n - \{s\}, \mathbb{C})$ the class such that

$$7.1 \qquad \int_{\mathbb{R}^n} \partial \omega_s \cup 1_s = 1$$

where $\partial : H^{n-1}(\mathbb{R}^n - \{s\}, \mathbb{C}) \longrightarrow H_{\{s\}}^n(\mathbb{R}^n, \mathbb{C})$ denotes the boundary in the long exact sequence of local cohomology.

For an open neighbourhood U of s and $\omega \in H^{n-1}(U - \{s\}, \mathbb{C})$ we let $\partial \omega \in H_{\{s\}}^n(U, \mathbb{C})$ denote image under the boundary map from the long exact sequence in local cohomology. By excision II.9.6 we can identify $\partial \omega$ with an element of $H_{\{s\}}^n(\mathbb{R}^n, \mathbb{C})$ which we denote

$$\partial_s \omega \in H_{\{s\}}^n(\mathbb{R}^n, \mathbb{C})$$

It follows from II.9.7 that the symbol $\partial_s \omega$ is unchanged if we replace ω by the restriction of ω to $V - \{s\}$ where V is an open neighbourhood of s in U . Finally put

7.2
$$\mathrm{Tr}(\omega; s) = \int_{\mathbb{R}^n} \partial_s \omega \cup 1_s$$

Let there be given $\gamma: S^{n-1} \to \mathbb{R}^n$ a continuous map and an orientation of S^{n-1} with trace map

$$\int_{S^{n-1}} H^{n-1}(S^{n-1}, \mathbb{C}) \to \mathbb{C}$$

For $s \in \mathbb{R}^n - \gamma(S^{n-1})$ we define the <u>index</u> by

7.3
$$I(\gamma; s) = \int_{S^{n-1}} \gamma^* \omega_s$$

<u>Theorem 7.4.</u> Let U denote an open subset of $\mathbb{R}^n$ and $\gamma: S^{n-1} \to U$ a continuous map. The function $s \mapsto I(\gamma; s)$ is locally constant on the complement Z of U in $\mathbb{R}^n$. Let $\Gamma \in H_c^0(Z, \mathbb{C})$ be the corresponding cohomology class. We have

$$\int_{S^{n-1}} \gamma^* \omega = \int_{\mathbb{R}^n} \partial \omega \cup \Gamma \qquad ; \ \omega \in H^{n-1}(U, \mathbb{C})$$

where $\partial \omega \in H_Z^n(\mathbb{R}^n, \mathbb{C})$ denotes the image of ω by the boundary map of the long exact sequence of local cohomology.

<u>Proof.</u> Choose an open subset V of U containing $\gamma(S^{n-1})$ and such that $H_c^1(V, \mathbb{C})$ is a finite dimensional vector space. Take for example V to be a finite union of open cubes. — The linear form on $H^{n-1}(V, \mathbb{C})$

$$\omega \mapsto \int_{S^{n-1}} \gamma^* \omega$$

can according to Poincaré duality 3.2 be represented

$$\int_{S^{n-1}} \gamma^* \omega = \int_V \omega \cup \varphi$$

where $\varphi \in H_c^1(V,\mathbb{C})$. Let $j: V \to U$ denote the inclusion and put $\theta = j_! \varphi$. We get for $\omega \in H^{n-1}(U,\mathbb{C})$ using 6.4 and 5.7

$$\int_U \omega \cup \theta = \int_U \omega \cup j_! \varphi = \int_U j_!(j^*\omega \cup \varphi) = \int_V j^*\omega \cup \varphi$$

which proves the formula

$$7.5 \qquad \int_{S^{n-1}} \gamma^*\omega = \int_U \omega \cup \theta \qquad ; \qquad \omega \in H^{n-1}(U,\mathbb{R})$$

Consider a point $s \notin U$ and let $j_s: U \to \mathbb{R}^n-[s]$ denote the inclusion. Proceeding as above we find

$$\int_{S^{n-1}} \gamma^*\omega = \int_{\mathbb{R}^n-\{s\}} \omega \cup j_{s!}\theta \qquad ; \qquad \omega \in H^{n-1}(\mathbb{R}^n-\{s\},\mathbb{C})$$

At this point let us introduce the commutative diagram III.7.7

$$H_c^0(\mathbb{R}^n,\mathbb{C}) \to H_c^0(\mathbb{R}^n-U,\mathbb{C}) \xrightarrow{\ \partial\ }_{\sim} H_c^1(U,\mathbb{C}) \to H_c^1(\mathbb{R}^n,\mathbb{C})$$

$$\downarrow \qquad\qquad\qquad \downarrow j_{s!}$$

$$H_c^0(\{s\},\mathbb{C}) \xrightarrow{\ \partial\ }_{\sim} H_c^1(\mathbb{R}^n-\{s\},\mathbb{C})$$

Put $\Gamma = (-1)^n \partial^{-1}\theta$ to get $j_{s!}\theta = (-1)^n \Gamma(s)\partial 1_s$ and

$$\int_{S^{n-1}} \gamma^*\omega = (-1)^n \Gamma(s) \int_{\mathbb{R}^n-\{s\}} \omega \cup \partial 1_s$$

In particular with $\omega = \omega_s$ we get by 7.3

$$I(\gamma; s) = (-1)^n \Gamma(s) \int_{\mathbb{R}^n-\{s\}} \omega_s \cup \partial 1_s$$

From formula 5.7, 6.2 and 7.1 we deduce

$$(-1)^n \int_{\mathbb{R}^n - \{s\}} \omega_s \cup \partial 1_s = (-1)^n \int_{\mathbb{R}^n} j_s! \, (\omega_s \cup \partial 1_s) = \int_{\mathbb{R}^n} \partial \omega_s \cup 1_s = 1$$

and consequently $\Gamma(s) = I(\gamma;s)$. This proves the first part of the theorem. From 7.5 we get with $\theta = (-1)^n \partial \Gamma$

$$\int \gamma^*\omega = (-1)^n \int_U \omega \cap \partial \Gamma = \int_{\mathbb{R}^n} \partial \omega \cup \Gamma$$

where we have used 5.7 and 6.2.

Q.E.D.

<u>Theorem 7.6</u>. Let U be an open set of $\mathbb{R}^n$, and S a discrete subset of U, which is closed relative to U. Given a continuous map $\gamma: S^{n-1} \to U-S$ such that

$$I(\gamma;z) = 0 \quad \text{for all} \quad z \in \mathbb{R}^n - U$$

Then for any $\omega \in H^{n-1}(U-S, \mathbb{C})$

$$\boxed{\int_{S^{n-1}} \gamma^*\omega = \sum_{s \in S} \mathrm{Tr}(\omega;s) \, I(\gamma;s)}$$

<u>Proof</u>. Let P denote the support of the function $s \longmapsto I(\gamma;s)$ on the closed subset $S \cup (\mathbb{R}^n - U)$. We must have $P \subseteq S$: to exclude a point $z \in \mathbb{R}^n - U$ we can remark that $I(\gamma;z) = 0$ by assumption and that our function is locally constant. In conclusion P is a compact subset of the discrete set S, i.e. P is a finite subset of S. Let $\Gamma \in H_c^0(P, \mathbb{C})$ denote the restriction of the function $s \longmapsto I(\gamma;s)$ to P.

For each $s \in P$ choose an open disc D_s with center s contained in U such that $D_s \cap S = \{s\}$. Make the discs so small that they do not intersect and let $j: D \to U$ denote the inclusion of their union into U . We have according to 7.4, 6.5 that

$$\int_{S^{n-1}} \gamma^*\omega = \int_D j^*\partial\omega \cup \Gamma = \int_D \partial j^*\omega \cup \Gamma$$

In order to evaluate the right hand side we may replace ω by $\sum_{s \in P} \mathrm{Tr}(\omega;s)\omega_s$. This gives

$$\int_U \partial j^*\omega \cup \Gamma = \sum_{s \in P} \mathrm{Tr}(\omega;s)\, I(\gamma;s) \int_{\mathbb{R}^n} \partial\omega_s \cup 1_s$$

and the result follows from 7.1.

$$\text{Q.E.D.}$$

Example 7.7. For $s \in \mathbb{C}$ we find

$$\mathrm{Tr}\left(\frac{1}{2\pi i} \frac{1}{z-s};s\right) = 1$$

as it follows from IV.7.16. From this we find

$$\mathrm{Tr}(f(z)\,dz) = 2\pi i\, \mathrm{Res}(f(z)\,dz;s)$$

for a meromorphic function f defined in a neighbourhood of s .

Example 7.8. The closed differential form

$$\theta = r^{-n} \sum_{j=1}^{n} (-1)^{i-1} x_i \, dx_1 .. \wedge \hat{dx_i} .. \wedge dx_n$$

discussed in IV.7.17 have local trace

$$\mathrm{Tr}(\theta;0) = \sigma_{n-1}$$

This follows from 6.4. and loc.cit.

VI. Poincare Duality with General Coefficients

VI.1 Verdier duality

In this section we consider a locally compact space X of $\underline{\text{finite dimension}}$ and a $\underline{\text{noetherian}}$ commutative ring k. Recall that $D^+(X,k)$ denotes the homotopy category of bounded below injective complexes of k-sheaves on X and $D^+(k)$ the homotopy category of bounded below complexes of injective k-modules. We are going to prove that the derived functor

$$R\Gamma_c(X,-): \ D^+(X,k) \ \to \ D^+(k)$$

has a right adjoint.

$\underline{\text{Verdier duality 1.1}}$. There exists a functor

$$G^{\cdot} \longmapsto G^{\cdot\,!}, \quad D^+(k) \to D^+(X,k)$$

and a natural isomorphism

$$[R\Gamma_c(X,I^{\cdot}),G^{\cdot}] = [I^{\cdot},G^{\cdot\,!}]$$

as $I^{\cdot}$ varies through $D^+(X,k)$ and $G^{\cdot}$ varies through $D^+(k)$.

290

<u>Proof</u>. 1) For a soft, flat k-sheaf S and a k-module G
we are going to define a k-sheaf $D(S,G)$ on X: For an open
subset U of X we put

$$\Gamma(U,D(S,G)) = \mathrm{Hom}(\Gamma_c(U,S),G)$$

The inclusion $j: V \to U$ of two open subsets will induce

$$j_!: \Gamma_c(V,S) \to \Gamma_c(U,S)$$

The transform of this by the functor $\mathrm{Hom}_k(-,G)$ is restriction
from U to V in the presheaf $D(S,G)$. The proof of V.1.2
shows that $D(S,G)$ is a <u>sheaf</u>.

2) The next thing to do is to notice that the tensor product $S \otimes F$ of a soft and flat k-sheaf S with an arbitary
k-sheaf F, is a <u>soft</u> k-sheaf by the proof of V.1.3, see II.11.

3) For an open subset U of X, the composite

$$\Gamma(U,F) \otimes \Gamma_c(U,S) \to \Gamma_c(U,F \otimes S) \to \Gamma_c(X,F \otimes S)$$

transforms by $\mathrm{Hom}_k(-,G)$ into a map

$$\mathrm{Hom}(\Gamma_c(X,F \otimes S),G) \to \mathrm{Hom}(\Gamma(U,F),\mathrm{Hom}(\Gamma_c(U,S),G))$$

and by variation of U in fact a map

$$\mathrm{Hom}(\Gamma_c(X,F \otimes S),G) \to \mathrm{Hom}(F,D(S,G))$$

which is seen to be an <u>isomorphism</u> still provided S is soft and
flat: use the proof of V.1.5.

4) If G is an injective k-module, then the sheaf $D(S,G)$
is <u>injective</u> in $\mathrm{Sh}(X,k)$: According to the isomorphism above it

suffices to prove that the functor

$$F \longmapsto \mathrm{Hom}(\Gamma_c(X, F \otimes S), G)$$

is exact on $\mathrm{Sh}(X,k)$. This follows from 2).

5) Let $\underset{\sim}{k} \to S^{\boldsymbol{\cdot}}$ be a bounded resolution of $\underset{\sim}{k}$ by soft and flat sheaves, see 1.3 below. An object $I^{\boldsymbol{\cdot}}$ of $D^+(k)$ gives rise to a complex $D^{\boldsymbol{\cdot}}(S^{\boldsymbol{\cdot}}, I^{\boldsymbol{\cdot}})$ by general principles, I.11. In fact a bounded below complex of injective k-modules as it follows from 4). The isomorphism from 3) extends to an isomorphism

$$\mathrm{Hom}^{\boldsymbol{\cdot}}(\Gamma_c(X, F^{\boldsymbol{\cdot}} \otimes S^{\boldsymbol{\cdot}}), G^{\boldsymbol{\cdot}}) \overset{\sim}{\to} \mathrm{Hom}^{\boldsymbol{\cdot}}(F^{\boldsymbol{\cdot}}, D^{\boldsymbol{\cdot}}(S^{\boldsymbol{\cdot}}, G^{\boldsymbol{\cdot}}))$$

6) In case $I^{\boldsymbol{\cdot}}$ is an object of $D^+(X,k)$ we successively deduce quasi-isomorphisms

$$I^{\boldsymbol{\cdot}} \to I^{\boldsymbol{\cdot}} \otimes k \to I^{\boldsymbol{\cdot}} \otimes S^{\boldsymbol{\cdot}} \qquad \text{by II.11.7}$$

$$\Gamma_c(X, I^{\boldsymbol{\cdot}}) \to \Gamma_c(X, I^{\boldsymbol{\cdot}} \otimes S^{\boldsymbol{\cdot}}) \quad \text{by 2) and I.7.5.}$$

$$\mathrm{Hom}^{\boldsymbol{\cdot}}(\Gamma_c(X, I^{\boldsymbol{\cdot}} \otimes S^{\boldsymbol{\cdot}}), G^{\boldsymbol{\cdot}}) \to \mathrm{Hom}^{\boldsymbol{\cdot}}(\Gamma_c(X, I^{\boldsymbol{\cdot}}), G^{\boldsymbol{\cdot}}) \quad \text{by I.6.2}$$

7) A combination of 5) and 6) yields a <u>quasi-isomorphism</u>

$$1.2 \qquad \mathrm{Hom}^{\boldsymbol{\cdot}}(I^{\boldsymbol{\cdot}}, D^{\boldsymbol{\cdot}}(S^{\boldsymbol{\cdot}}, G^{\boldsymbol{\cdot}})) \to \mathrm{Hom}^{\boldsymbol{\cdot}}(\Gamma_c(X, I^{\boldsymbol{\cdot}}), G^{\boldsymbol{\cdot}})$$

Put $G^{\boldsymbol{\cdot}!} = D^{\boldsymbol{\cdot}}(S^{\boldsymbol{\cdot}}, G^{\boldsymbol{\cdot}})$ and apply H^0 to 1.2 to get

$$[I^{\boldsymbol{\cdot}}, G^{\boldsymbol{\cdot}!}] \underset{\sim}{\to} [\Gamma_c(X, I^{\boldsymbol{\cdot}}), G^{\boldsymbol{\cdot}}]$$

and our journey ends. $\hspace{4cm}$ Q.E.D.

Proposition 1.3. Let X be a locally compact space of finite dimension and k a noetherian ring. There exists a bounded resolution $\underset{\sim}{k} \to S^{\cdot}$ of $\underset{\sim}{k}$ in $Sh(X,k)$ by soft and k-flat sheaves.

Proof. More generally let F be a flat k-sheaf and

$$0 \longrightarrow C^0F \xrightarrow{\partial^0} C^1F \xrightarrow{\partial^1} \ldots \longrightarrow C^nF \xrightarrow{\partial^n} C^{n+1}F$$

the Godement resolution of F, II.3.6. Let us prove that C^nF and $Im\partial^n$ are flat sheaves for all $n \in \mathbb{N}$. By the iterative nature of the Godement resolution it suffices to treat the case $n = 0$. For an open subset U of X we have

$$\Gamma(U,C^0F) = \prod_{x \in U} F_x$$

and it follows from Lemma 1.4 below that $\Gamma(U,C^0F)$ is a flat k-module: since a direct limit of flat k-modules is flat it follows that the stalks of C^0F are flat k-modules. Noticing that $F \to C^0F$ has local retractions it follows that $Im\,\partial^0$ is a flat k-sheaf.

Let the dimension of X be n. The truncation $S^{\cdot} = \tau_{\leq n}C^{\cdot}F$ is flat and soft, III.9.9.

Q.E.D.

$\underline{\text{Lemma 1.4}}$. Let A be a noetherian ring and $(F_i)_{i \in I}$ a family of flat A-modules. The product $F = \prod_{i \in I} F_i$ is a flat A-module.

$\underline{\text{Proof}}$. Let M be a finitely generated A-module. Let us prove that the canonical map

$$M \otimes F \to \prod_{i \in I} M \otimes F_i$$

is an isomorphism. This is clear in the case where M is finitely generated and free: check the case $M = A$. In the general case choose an exact sequence of the form $A^p \to A^q \to M \to 0$ and consider the exact commutative diagram

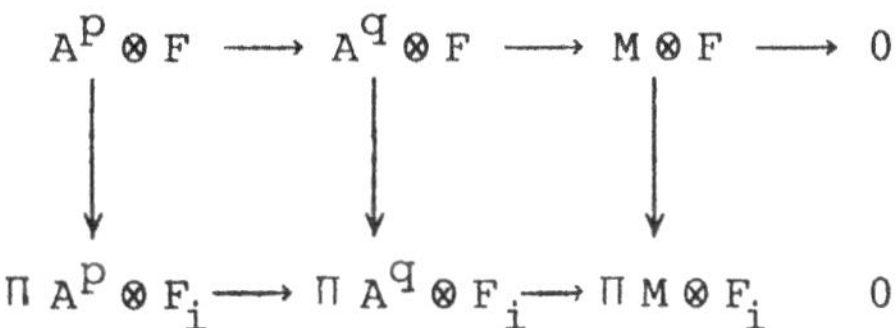

to conclude that the vertical arrow to the right is an isomorphism. As a consequence, the functor $\otimes F$ is exact on the category of finitely generated k-modules. Let us now prove that

$$\text{Tor}_1^A(M, F) = 0$$

for any A-module M. The case where M is finitely generated follows by considering an exact sequence of the form $0 \to N \to A^r \to M \to 0$. The general case follows by considering M as the direct limit of its finitely generated submodules: notice that it follows from II.8.1 that $\text{Tor}_A^p(-, F)$ commutes with direct limits.

Q.E.D.

VI.2 The dualizing complex $\mathcal{D}^{\cdot}$.

Let X denote a locally compact space of finite dimension and k a noetherian ring. We choose an injective resolution $k \to K^{\cdot}$ and define the __dualizing complex__ $\mathcal{D}^{\cdot} = K^{\cdot!}$ with the notation of 1.1. Recall that we have a natural isomorphism

$$2.1 \qquad\qquad [I^{\cdot}, \mathcal{D}^{\cdot}] \xrightarrow{\sim} [\Gamma_c(X, I^{\cdot}), K^{\cdot}]$$

as $I^{\cdot}$ varies through $D^{+}(X, k)$. Put $I^{\cdot} = \mathcal{D}^{\cdot}$ and let the image of $1 \in [\mathcal{D}^{\cdot}, \mathcal{D}^{\cdot}]$ be denoted

$$2.2 \qquad\qquad \int_X : \Gamma_c(X, \mathcal{D}^{\cdot}) \longrightarrow K^{\cdot}$$

This is called the __trace map__. According to the Yoneda principle for representable functors the duality isomorphism can be recovered from the trace map by the formula

$$2.3 \qquad\qquad \varphi \longmapsto \int_X \circ \, \Gamma_c(X, \varphi) \qquad ; \quad \varphi \in [I^{\cdot}, \mathcal{D}^{\cdot}]$$

It is useful to give this a slightly broader formulation

__Proposition 2.4.__ For any bounded below complex $S^{\cdot}$ of __soft__ k-sheaves on X, the trace map induces an isomorphism

$$[S^{\cdot}, \mathcal{D}^{\cdot}] \xrightarrow{\sim} [\Gamma_c(X, S^{\cdot}), K^{\cdot}]$$

__Proof.__ Analogous to the proof of V.5.4.

Q.E.D.

<u>Proposition 2.5</u>. Let X be a locally compact space of dimension n. The dualizing complex may be represented by a bounded complex $\mathcal{D}^{\cdot}$ of injectives with $\mathcal{D}^p = 0$ for $p < -n$. Moreover

$$\Gamma(U, H^{-n}\mathcal{D}^{\cdot}) = \mathrm{Hom}(H_c^n(U,k),k)$$

for any open subset U of X.

<u>Proof</u>. Let $\underset{\sim}{k} \to S^{\cdot}$ be a soft resolution of $\underset{\sim}{k}$ of length n. Consider the inclusion $j: U \to X$ of an open subset. By 2.4 applied to the soft complex $j_! j^* S^{\cdot}[p]$, $p \in \mathbb{Z}$, we get

$$[j_! j^* S^{\cdot}, \mathcal{D}^{\cdot}[-p]] = [\Gamma_c(U, S^{\cdot}[p]), K^{\cdot}]$$

where we have used III.7.1. The left hand side can be identified by means of II.6.6 and II.7.2 obtaining

2.6 $$H^{-p}\Gamma(U,\mathcal{D}^{\cdot}) = [\Gamma_c(U,S^{\cdot}[p]),K^{\cdot}]$$

For $p > n$, the complex $\Gamma_c(U,S^{\cdot}[p])$ is concentrated in strictly negative degrees, thus the right hand side is 0. By II.2.7 we conclude that

$$H^{-p}\mathcal{D}^{\cdot} = 0 \qquad \text{for}\ \ p > n$$

Thus we may replace $\mathcal{D}^{\cdot}$ by $\tau_{\geq -n}\mathcal{D}^{\cdot}$. In the sequel we shall assume that $\mathcal{D}^p = 0$ for $p < -n$. This gives rise to an exact sequence of sheaves on X

$$0 \to H^{-n}\mathcal{D}^{\cdot} \to \mathcal{D}^{-n} \to \mathcal{D}^{-n+1} \to \quad \cdots$$

and isomorphisms

$$\Gamma(U,H^{-n}\mathcal{D}^{\cdot}) = H^{-n}\Gamma(U,\mathcal{D}^{\cdot}) = [\Gamma_c(U,S^{\cdot}[n]),K^{\cdot}]$$

The right hand side can be evaluated by noticing that the first
of the two complexes is located in negative degrees and the
second complex is located in positive degrees. It follows that

$$[\Gamma_c(U,S^{\cdot}[n]),K^{\cdot}] = \mathrm{Hom}(H^0\Gamma_c(U,S^{\cdot}[n]),H^0(K^{\cdot})) = \mathrm{Hom}(H_c^n(U,k),k)$$

$$\text{Q.E.D.}$$

Let us consider the inclusion h: W $\to$ X of a locally closed
subspace W. The adjunction morphism, II.6.6

$$h_! h^! \mathcal{D}^{\cdot} \to \mathcal{D}^{\cdot}$$

will induce a chain map, compare III.7.1

$$\Gamma_c(W,h^!\mathcal{D}^{\cdot}) \longrightarrow \Gamma_c(X,\mathcal{D}^{\cdot})$$

which we compose with $\int_X$ to obtain a chain map

$$2.7 \qquad \int_W : \Gamma_c(W,h^!\mathcal{D}^{\cdot}) \to K^{\cdot}$$

The proof of V.5.6 shows that $h^!\mathcal{D}^{\cdot}$ is a dualizing complex
for W with trace map $\int_W$.

VI.3 Lefschetz duality

In this section we consider an n-dimensional topological manifold X with boundary ∂X, compare III.8.14. We let k denote a noetherian ring and $o\kappa_X$, the orientation sheaf relative to k. The sections over an open subset U of X are given by

3.1
$$\Gamma(U,o\kappa_X) = \operatorname{Hom}_k(H^n_C(U,k),k)$$

The dualizing complex $\mathcal{D}^\bullet$ and the orientation sheaf are related through a quasi-isomorphism

3.2
$$o\kappa_X[n] \overset{\sim}{\to} \mathcal{D}^\bullet$$

Proof. Let $\underset{\sim}{k} \to S^\bullet$ be a soft resolution of $\underset{\sim}{k}$. Recall from 2.6 that for each open subset U of X we have an isomorphism

$$H^{-p}\Gamma(U,\mathcal{D}^\bullet) = [\Gamma_C(U,S^\bullet[p]),K^\bullet] \qquad ; p \in \mathbb{Z}$$

In case U is isomorphic to $\mathbb{R}^n$ or $\mathbb{R}^{n-1} \times \mathbb{R}_+$, the complex $\Gamma_C(U,S^\bullet[p])$ is quasi-isomorphic to $H^n_C(U,k)[p-n]$ and we find that

$$H^{-p}\Gamma(U,\mathcal{D}^\bullet) = \operatorname{Ext}^{n-p}(H^n_C(U,k),k)$$

Using that $H^n_C(U,k)$ is a free module (in fact isomorphic to k or 0 respectively) we find

$$H^{-p}\Gamma(U,\mathcal{D}^\bullet) = 0 \qquad ; p \neq n$$

From this we can calculate the stalks of $H^{-p}\mathcal{D}^\bullet$ to conclude that $H^{-p}\mathcal{D}^\bullet = 0$ for $p \neq n$. The result now follows from 2.5.

Q.E.D.

298

Proposition 3.3. Let X denote a manifold with boundary ∂X and $j: \dot{X} \to X$ the inclusion of the interior $\dot{X} = X - \partial X$. The orientation sheaf or_X for X is given by $or_X = j_! or$ where or is the orientation sheaf for $\dot{X}$, the interior of X.

Proof. Let us first remark that the stalk of or_X at a boundary point is zero, as it follows from 3.1. Next remark that or_X and $j_! or$ have the same restriction to $\dot{X}$. It follows from II.6.4 that these two sheaves are identical.

Q.E.D.

Lefschetz duality 3.4. Let X be a manifold with boundary ∂X. If $\dot{X} = X - \partial X$ is oriented relative to k, we have an isomorphism

$$H^{n-p}(X, \partial X; k) \xrightarrow{\sim} [R^\bullet \Gamma_c(X, k)[p], K^\bullet]$$

where $K^\bullet$ is an injective resolution of the k-module k.

Proof. Let $\underset{\sim}{k} \to I^\bullet$ be an injective resolution. Consider the duality isomorphism $\quad [I^\bullet[p], \mathcal{D}^\bullet] = [\Gamma_c(X, I^\bullet[p]), K^\bullet]$

$$[I^\bullet[p], \mathcal{D}^\bullet] = [I^\bullet[p-n], \mathcal{D}^\bullet[-n] = [\underset{\sim}{k}[p-n], \mathcal{D}^\bullet[-n]] = H^{n-p}(X, \mathcal{D}^\bullet[-n])$$

Using the orientation $\mathcal{D}^\bullet[-n] \simeq or_X = j_! \underset{\sim}{k}$ we get

$$H^{n-p}(X, j_! \underset{\sim}{k}) \xrightarrow{\sim} [\Gamma_c(X, I^\bullet)[p], K^\bullet]$$

Recall the definition of the relative group, IV.8.1:

$H^\bullet(X, \partial X; k) = H^\bullet(X, j_! \underset{\sim}{k})$ and the result follows.

Q.E.D.

VI.4 Algebraic duality

Let $X^\cdot$ and $E^\cdot$ be complexes over a fixed commutative ring A. For $n \in \mathbb{Z}$ consider the map

$$X^n \to \mathrm{Hom}^n(\mathrm{Hom}^\cdot(X^\cdot,E^\cdot),E^\cdot) = \prod_{i \in \mathbb{Z}} \mathrm{Hom}^i(X^\cdot,E^\cdot),E^{i+n})$$

which to $x^n \in X^n$ assigns the product over $i \in \mathbb{Z}$ of the maps

$$\mathrm{Hom}^i(X^\cdot,E^\cdot) \longrightarrow E^{i+n} \; ; \qquad f \longmapsto (-1)^{in} f(x_n)$$

It is left to the reader to show that this defines a chain map

4.1
$$\mathrm{ev}\colon X^\cdot \to \mathrm{Hom}^\cdot(\mathrm{Hom}^\cdot(X^\cdot,E^\cdot),E^\cdot)$$

which we call the _evaluation_ map.

<u>Definition 4.2</u>. Let A denote a noetherian ring. A <u>dualizing complex</u> for A is a bounded complex $D^\cdot$ of injective modules with finitely generated cohomology modules, such that

$$\mathrm{ev}\colon X^\cdot \to \mathrm{Hom}^\cdot(\mathrm{Hom}^\cdot(X^\cdot,D^\cdot),D^\cdot)$$

is a quasi-isomorphism for any bounded complex $X^\cdot$ with finitely generated cohomology modules.

<u>Proposition 4.3</u>. A bounded complex $D^\cdot$ of injective modules with finitely generated cohomology modules is a dualizing complex for A if and only if

$$\mathrm{ev}\colon A \longrightarrow \mathrm{Hom}^\cdot(D^\cdot,D^\cdot)$$

is a quasi-isomorphism.

    <u>Proof</u>. Assume that the evaluation map is an isomorphism for $X^{\cdot} = A$. Consider a finitely generated module M. Recall that we consider M a complex, placing M in degree zero and placing zero's elsewhere. Let us prove that

$$\text{ev: } M \longrightarrow \text{Hom}^{\cdot}(\text{Hom}^{\cdot}(M,D^{\cdot}),D^{\cdot})$$

is a quasi-isomorphism. For $p \in \mathbb{Z}$ put

$$U^{p}(M) = H^{p} \text{Hom}^{\cdot}(\text{Hom}^{\cdot}(M,D^{\cdot}),D^{\cdot})$$

The evaluation map gives rise to morphisms

$$\text{ev}^{p}(M): H^{p}(M) \longrightarrow U^{p}(M) \qquad ; \quad p \in \mathbb{Z}$$

which we proceed to prove are isomorphisms. Notice that this is the case for $M = A^{m}$: by additivity of the functors involved it suffices to consider M = A and use the assumption.

    We will now prove the statement by decreasing induction on $p \in \mathbb{Z}$. Notice, that the result is true for large values of p since $D^{\cdot}$ is bounded. Assume the result is true for p+1. Choose a short exact sequence $\qquad 0 \longrightarrow N \longrightarrow A^{m} \longrightarrow M \longrightarrow 0$ and consider the resulting long exact sequence

$$
\begin{array}{ccccccccc}
H^{p}(N) & \longrightarrow & H^{p}(A^{m}) & \longrightarrow & H^{p}(M) & \longrightarrow & H^{p+1}(N) & \longrightarrow & H^{p+1}(A^{m}) \\
\downarrow & & \downarrow & & \downarrow & & \downarrow & & \downarrow \\
U(N) & \longrightarrow & U^{p}(A^{m}) & \longrightarrow & U^{p}(M) & \longrightarrow & U^{p+1}(N) & \longrightarrow & U^{p+1}(A^{m}) \\
\downarrow & & & & & & & & \\
0 & & & & & & & &
\end{array}
$$

Using I.1.4 we deduce an exact sequence

$$\text{Cok } ev^p(A^m) \to \text{Cok } ev^p(M) \to \text{Cok } ev^{p+1}(N)$$

from which we deduce that $ev^p(M)$ is surjective for any finite-ly generated module. In particular we can insert the dotted arrow in the diagram and conclude by I.1.3.

We shall now treat the general case of a bounded complex $X^{\cdot}$ by induction on the "length" of $X^{\cdot}$. In case $X^{\cdot}$ is concentrated in a single degree the result follows by decalage from the first part of the proof. The general case follows by considering the exact sequence I.5.7

$$0 \to \tau_{\leq n} X^{\cdot} \to X^{\cdot} \to X^{\cdot}/\tau_{\leq n} X^{\cdot} \to 0$$

of complexes.

Q.E.D.

Corollary 4.4. Let A be a noetherian ring with the proper-ty that A admits a bounded injective resolution $A \to D^{\cdot}$ (a Gorenstein ring). Then $D^{\cdot}$ is a dualizing complex.

Proof. The quasi-isomorphism $A \to D^{\cdot}$ gives rise to a quasi-isomorphism, compare I.6.2 $\quad \text{Hom}^{\cdot}(D^{\cdot},D^{\cdot}) \to \text{Hom}^{\cdot}(A,D^{\cdot}) = D^{\cdot}$ from which we deduce that $ev: A \to \text{Hom}^{\cdot}(D^{\cdot},D^{\cdot})$ is a quasi-isomorphism.

Q.E.D.

Example 4.5. The complex of abelian groups

$$0 \longrightarrow \mathbb{Q} \longrightarrow \mathbb{Q}/\mathbb{Z} \longrightarrow 0$$

is a dualizing complex for the ring $\mathbb{Z}$.

Proposition 4.6. Let $C^\cdot$ denote a bounded complex of abelian groups. Put

$$DC^\cdot = \text{Hom}^\cdot(C^\cdot, K^\cdot)$$

where $K^\cdot$ is the complex 4.5. Then there is a natural exact sequence

$$0 \to \text{Ext}(H^{i+1}C^\cdot; \mathbb{Z}) \to H^{-i}DC^\cdot \to \text{Hom}(H^i C^\cdot; \mathbb{Z}) \to 0$$

Proof. A combination of the exact sequence

$$0 \to \tau_{\le i}C^\cdot \to C^\cdot \to C^\cdot/\tau_{\le i}C^\cdot \to 0$$

and the quasi-isomorphism I.5.9

$$C^\cdot/\tau_{\le i}C^\cdot \longrightarrow \tau_{\ge i+1}C^\cdot$$

yields the exact sequence , see also I.6.14

$$H^{-i-1}D\tau_{\le i}C^\cdot \to H^{-i}D\tau_{\ge i+1}C^\cdot \to H^{-i}DC^\cdot \to H^{-i}D\tau_{\le i}C^\cdot \to H^{-i+1}D\tau_{\ge i+1}C^\cdot$$

The two extreme groups are zero for rather trivial reasons. We leave it to the reader to identify the resulting short exact sequence.

Q.E.D.

VI.5 Universal coefficients

Let X denote a locally compact space and N an abelian group. We are going to establish a natural exact sequence for each $p \in \mathbb{Z}$

$$5.1 \qquad 0 \to H_c^p(X,\mathbb{Z}) \otimes N \to H_c^p(X,N) \to \mathrm{Tor}_1(H_c^{p+1}(X,\mathbb{Z}),N) \to 0$$

The derivation is based on the following

Lemma 5.2. Let X be a locally compact space. There exists a positive complex $L^{\cdot}$ of torsion free abelian groups and a natural quasi-isomorphism

$$L^{\cdot} \otimes N \xrightarrow{\sim} R^{\cdot}\Gamma_c(X,N)$$

as N varies through the category of abelian groups.

Proof. Let S be a torsion free sheaf on X, i.e. a sheaf whose stalks are all torsion free abelian groups. An integer $m \neq 0$ will induce a monomorphism $0 \to S \xrightarrow{m} S$ which transforms into a monomorphism $\Gamma_c(X,S) \xrightarrow{m} \Gamma_c(X,S)$. This proves that $\Gamma_c(X,S)$ is a torsion free abelian group. Suppose in addition that S is a soft sheaf. Let us prove that for any abelian group N the canonical map

$$\Gamma_c(X,S) \otimes N \to \Gamma_c(X,S \otimes N)$$

is an isomorphism. Observe that N is direct limit of its finitely generated subgroups. Thus we can use III.5.1 and II.8.1

304

to reduce to the case where N is finitely generated. In
that case we can choose a presentation $0 \to \mathbb{Z}^p \to \mathbb{Z}^q \to N \to 0$
and deduce an exact, commutative diagram

$$
\begin{array}{ccccccccc}
0 & \longrightarrow & \Gamma_c(X,S) \otimes \mathbb{Z}^p & \longrightarrow & \Gamma_c(X,S) \otimes \mathbb{Z}^q & \longrightarrow & \Gamma_c(X,S) \otimes N & \longrightarrow & 0 \\
& & \downarrow & & \downarrow & & \downarrow & & \\
0 & \longrightarrow & \Gamma_c(X,S \otimes \mathbb{Z}^p) & \longrightarrow & \Gamma_c(X,S \otimes \mathbb{Z}^q) & \longrightarrow & \Gamma_c(X,S \otimes N) & \longrightarrow & 0
\end{array}
$$

from which the result follows. - Let us now prove that $S \otimes N$
is a soft sheaf on X: Let $i: K \to X$ denote the inclusion of a
compact subspace. We have

$$\Gamma(K, i^*(S \otimes N)) = \Gamma(K, i^*S \otimes N) = \Gamma(K,S) \otimes N$$

which implies that $\Gamma(X, S \otimes N) \to \Gamma(K, S \otimes N)$ is surjective.

Let us now consider the Godement resolution $\mathbb{Z} \to C^{\cdot}$, II.3.6.
This is a resolution of $\mathbb{Z}$ by soft, torsion free sheaves. Moreover, at each stalk this is a homotopy equivalence, II.3.6. For
an abelian group N we deduce the soft resolution $C^{\cdot} \otimes N$ of
N. In conclusion

$$\Gamma_c(X, C^{\cdot}) \otimes N = \Gamma_c(X, C^{\cdot} \otimes N) = R^{\cdot}\Gamma_c(X, N)$$

i.e. we can use $L^{\cdot} = \Gamma_c(X, C^{\cdot})$.

Q.E.D.

<u>Proof of 5.1</u>. Consider the canonical exact sequence

$$0 \longrightarrow \tau_{\leq p} L^{\boldsymbol{\cdot}} \longrightarrow L^{\boldsymbol{\cdot}} \longrightarrow L^{\boldsymbol{\cdot}}/\tau_{\leq p} L^{\boldsymbol{\cdot}} \longrightarrow 0$$

and the quasi-isomorphism, I.5.9

$$L^{\boldsymbol{\cdot}}/\tau_{\leq p} L^{\boldsymbol{\cdot}} \longrightarrow \tau_{\geq p+1} L^{\boldsymbol{\cdot}}$$

Consider at the same time a free resolution $F^{\boldsymbol{\cdot}} \to N$ of length 1.
From these data we deduce an exact sequence

$$H^{p-1}(\tau_{\geq p+1} L^{\boldsymbol{\cdot}} \otimes F^{\boldsymbol{\cdot}}) \to H^{p}(\tau_{\leq p} L^{\boldsymbol{\cdot}} \otimes F^{\boldsymbol{\cdot}}) \to H^{p}(L^{\boldsymbol{\cdot}} \otimes F^{\boldsymbol{\cdot}}) \to H^{p}(\tau_{\geq p+1} L^{\boldsymbol{\cdot}} \otimes F^{\boldsymbol{\cdot}}) \to H^{p+1}(\tau_{\leq p} L^{\boldsymbol{\cdot}} \otimes F^{\boldsymbol{\cdot}})$$

The two extreme groups are zero for rather simple reasons. We leave
it to the reader to identify the resulting short exact sequence,
using the following diagram

$$
\begin{array}{ccccccc}
0 & \longrightarrow & H^{p+1} \otimes F_1 & \longrightarrow & \mathrm{Cok}\,\partial^p \otimes F_1 & \longrightarrow & L^{p+1} \otimes F_1 \\
& & \downarrow & & \downarrow & & \downarrow \\
0 & \longrightarrow & H^{p+1} \otimes F_0 & \longrightarrow & \mathrm{Cok}\,\partial^p \otimes F_0 & \longrightarrow & L^{p+1} \otimes F_0
\end{array}
$$

Q.E.D.

Let us add some information on the lowest cohomology groups
with compact support.

<u>Proposition 5.3</u>. For a locally compact space X the group
$H_c^0(X,\mathbb{Z})$ is a free abelian group and the group $H_c^1(X,\mathbb{Z})$ is
torsion free.

<u>Proof</u>. Let F denote the group of all $\mathbb{Z}$-valued functions on the set X which are bounded. This is a free abelian group according to a theorem of Nöbeling (1). The group $H_c^0(X,\mathbb{Z})$ is a subgroup of F, which makes it a free group.

To prove that $H_c^1(X,\mathbb{Z})$ is torsion free we consider an integer $m \neq 0$ and the long exact sequence

$$\to H_c^0(X,\mathbb{Z}) \to H_c^0(X,\mathbb{Z}/m) \to H_c^1(X,\mathbb{Z}) \xrightarrow{m} H_c^1(X,\mathbb{Z}) \to$$

Choose a set theoretical section $s: \mathbb{Z}/m \to \mathbb{Z}$ to the projection $\mathbb{Z} \to \mathbb{Z}/m$ such that $s(0) = 0$. This induces a set theoretical section to the first arrow in the exact sequence above. This proves that multiplication with m on $H_c^1(X,\mathbb{Z})$ is injective.

Q.E.D.

VI.6 Alexander duality

Let X denote an n-dimensional oriented topological manifold. For a closed subset Z of X there is a quasi-isomorphism,
Alexander duality

$$6.1 \qquad R^{\cdot}\Gamma_Z(X,\mathbb{Z}) \overset{\sim}{\to} DR^{\cdot}\Gamma_C(Z,\mathbb{Z})[-n]$$

where we have used the notation of 4.6.

Proof. As usual we let $\mathcal{D}^{\cdot}$ denote the dualizing complex
and $K^{\cdot}$ denote the complex $\mathbb{Q} \to \mathbb{Q}/\mathbb{Z}$. For an object $I^{\cdot}$ of
$D^+(X,\mathbb{Z})$ we have a quasi-isomorphism 2.1

$$6.2 \qquad [I^{\cdot},\mathcal{D}^{\cdot}] \overset{\sim}{\to} D\Gamma_C(X,I^{\cdot})$$

Let $i: Z \to X$ denote the inclusion and $I^{\cdot}$ and injective resolution of $i_*\mathbb{Z}$. This gives a quasi-isomorphism, II.9.13

$$\Gamma_Z(X,\mathcal{D}^{\cdot}) \overset{\sim}{\to} DR^{\cdot}\Gamma_C(Z,\mathbb{Z})$$

The orientation of X provides us with a quasi-isomorphism
$\mathbb{Z}[n] \overset{\sim}{\to} \mathcal{D}^{\cdot}$, and the result follows.

Q.E.D.

Proposition 6.3. Let X denote an n-dimensional topological
manifold with orientation sheaf $o\hbar$ relative to $\mathbb{Z}$. For any
noetherian ring k the orientation sheaf relative to k is
$o\hbar \otimes_{\mathbb{Z}} k$.

308

<u>Proof</u>. The sheaf $\mathit{or} \otimes_{\mathbb{Z}} k$ is generated by the presheaf

$$U \longmapsto \mathrm{Hom}_{\mathbb{Z}} (H_c^n(U,\mathbb{Z}),\mathbb{Z}) \otimes_{\mathbb{Z}} k$$

while the orientation sheaf relative to k is given by, 5.1

$$U \longmapsto \mathrm{Hom}_{\mathbb{Z}} (H_c^n(U,\mathbb{Z}),k)$$

These explicit formulas define a morphism of sheaves from $\mathit{or} \otimes_{\mathbb{Z}} k$ to the orientation sheaf relative to k. This is an isomorphism as one checks by localization.

Q.E.D.

<u>Theorem 6.4</u>. Let X denote a connected n-dimensional manifold. The following conditions are equivalent

1) $H_c^n(X,\mathbb{Z}) \xrightarrow{\sim} \mathbb{Z}$

2) X is orientable relative to $\mathbb{Z}$.

3) X is orientable relative to $\mathbb{Z}/p$ for any prime p.

4) $H_c^n(X,\mathbb{Z}/p) \xrightarrow{\sim} \mathbb{Z}/p$ for any prime p.

<u>Proof</u>. 1) $\Rightarrow$ 4) Follows from universal coefficients 5.1.

4) $\Rightarrow$ 3) This has already been noted in V.3.6.

1) $\Rightarrow$ 2) Given $x \in X$ and D an open neighbourhood of x homeomorphic to a disc. Let us show that $r: H_c^n(D,\mathbb{Z}) \to H_c^n(X,\mathbb{Z})$ is an isomorphism. For any prime p consider the commuatative diagram

$$
\begin{array}{ccc}
H_c^n(D,\mathbb{Z}) \otimes \mathbb{Z}/p & \xrightarrow{\ r\otimes 1\ } & H_c^n(X,\mathbb{Z}) \otimes \mathbb{Z}/p \\
\downarrow & & \downarrow \\
H_c^n(D,\mathbb{Z}/p) & \xrightarrow{\hspace{2cm}} & H_c^n(X,\mathbb{Z}/p)
\end{array}
$$

The two vertical arrows are isomorphisms according to universal coefficient, 5.1. The horizontal map at the bottom is an isomorphism since X is oriented relative to $\mathbb{Z}/p$. Thus we have proved that r is an isomorphism mod p for all primes. Since $H_c^n(D,\mathbb{Z})$ and $H_c^n(X,\mathbb{Z})$ are finitely generated and free, it follows that r is an isomorphism.

2) $\Rightarrow$ 3) This is an immediate consequence of 6.3.

3) $\Rightarrow$ 1) Let U denote an open connected subset of X which is the union of finitely many discs, i.e. open subsets of X homeomorphic to $\mathbb{R}^n$. Let us remark that $H_c^n(U,\mathbb{Z})$ is finitely generated as it follows by a simple argument based on the Mayer-Vietoris sequence III.7.5

$$\to H_c^n(V,\mathbb{Z}) \oplus H_c^n(W,\mathbb{Z}) \to H_c^n(V \cup W,\mathbb{Z}) \to 0$$

From universal coefficients 5.1 and Poincaré duality III.3.2 and III.4.1 follows that for any finite field k

$$\dim_k H_c^n(U,\mathbb{Z}) \otimes k = \dim_k H_c^n(U,k) = \#\ c.c.(U) = 1$$

Using the structure theorem for finitely generated abelian groups we conclude that $H_c^n(U,\mathbb{Z}) \xrightarrow{\sim} \mathbb{Z}$, and thereby that U is oriented. - Let $\mathcal{U}$ be the set of all non empty open subsets of the type just described. Let us notice that any two sets U and V from $\mathcal{U}$ is contained in a third set W from X. This follows from the fact that for any two points x and y in X there exists a chain of discs $D_0,\ldots,D_s$, with $x \in D_0$ and $y \in D_s$ such that $D_{i-1} \cap D_i$ is non empty for $i = 1,\ldots,s$. We can now conclude from 6.5 below that

310

$$H_c^n(X, \mathbb{Z}) = \varinjlim_{\mathcal{U}} H_c^n(U, \mathbb{Z})$$

Let $j: U \to V$ denote the inclusion between two sets from $\mathcal{U}$.
Let us prove that $j_!: H_c^n(U, \mathbb{Z}) \longrightarrow H_c^n(V, \mathbb{Z})$ is an isomorphism.
From the fact that V is oriented we conclude that

$$j_!^{\vee}: \mathrm{Hom}(H_c^n(V, \mathbb{Z}), \mathbb{Z}) \to \mathrm{Hom}(H_c^n(U, \mathbb{Z}), \mathbb{Z})$$

is an isomorphism. Since $H_c^n(U, \mathbb{Z})$ and $H_c^n(V, \mathbb{Z})$ are isomorphic
to $\mathbb{Z}$ we conclude that $j_!$ is an isomorphism. - The result now
follows by a simple argument involving the direct limit description
of $H_c^n(X, \mathbb{Z})$.

Q.E.D.

Lemma 6.5. Let X denote a locally compact space and
$\mathcal{U}$ a covering of X by open subsets such that any two sets U
and V from $\mathcal{U}$ is contained in a third set W from $\mathcal{U}$.
Then for any sheaf F on X

$$H_c^p(X, F) = \varinjlim H_c^p(U, F)$$

where the limit is taken over all U in $\mathcal{U}$.

$\underline{Proof}$. For an open set U put $F_U = j_! j^*F$ where $j: U \to X$ is the inclusion. It follows from II.2.8 that $F = \varinjlim F_U$. Using III.5.1 we deduce that

$$H_c^P(X,F) = \varinjlim H_c^P(X,F_U)$$

and the result follows from III.7.3.

$$Q.E.D.$$

$\underline{Remark\ 6.6}$. Let us notice an interesting consequence of 6.5 compare the proof of V.1.2: Let X denote a locally compact space of dimension n. Consider a commutative ring k and a k-sheaf F on X. The presheaf

$$U \longmapsto \mathrm{Hom}_k(H_c^n(U,F),k)$$

is a sheaf, proceeding as in the proof of V.1.2.

$\underline{Proposition\ 6.7}$. Let X denote an oriented n-dimensional connected manifold. For any point $x \in X$, the restriction map

$$H_{\{x\}}^n(X,\mathbb{Z}) \to H_c^n(X,\mathbb{Z})$$

is an isomorphism.

$\underline{Proof}$. Let $j: U \to X$ denote the inclusion of an open connected neighbourhood of x. The map

$$j_!^\vee: \mathrm{Hom}(H_c^n(X,\mathbb{Z}),\mathbb{Z}) \to \mathrm{Hom}(H_c^n(U,\mathbb{Z}),\mathbb{Z})$$

312

is an isomorphism according to the theory of orientation 3.1
It follows from 6.4 that

$$j_! : H_c^n(U,\mathbb{Z}) \to H_c^n(X,\mathbb{Z})$$

is an isomorphism. Thus by excision the question is local,
and we may replace X by S^n. This case is left to the reader.

Q.E.D.

<u>Theorem 6.8</u>. Let X denote an oriented compact manifold
of dimension n. For any proper open subset U of X we have

$$H^n(U,\mathbb{Z}) = 0$$

<u>Proof</u>. Put $Z = X-U$ and consider the exact sequence

$$\to H^n(X,\mathbb{Z}) \to H^n(U,\mathbb{Z}) \to H_Z^{n+1}(X,\mathbb{Z}) \to$$

The first map is zero: Pick $x \in X-U$ and notice that the map
can be factored through $H^n(X-\{x\},\mathbb{Z})$ which is zero as it
follows from 6.7. By Alexander duality 6.1

$$H_Z^{n+1}(X,\mathbb{Z}) = H^1 DR \cdot \Gamma_c(Z,\mathbb{Z})$$

From 4.6 we deduce an isomorphism

$$H_Z^{n+1}(X,\mathbb{Z}) = \text{Ext}(H^0(Z,\mathbb{Z}),\mathbb{Z})$$

From this the result follows since $H^0(Z,\mathbb{Z})$ is a free abelian
group, 5.3.

Q.E.D.

VII. Direct Image with Proper Support

<u>VII.1 The functor $f_!$</u>

The inclusion $h: W \to X$ of a locally subspace of a
locally compact space gives rise to a very useful functor
$h_!: Sh(W) \to Sh(X)$, compare II.6 and III.7. In order to generalize
this notion let us observe that h is a proper map if and only
if W is a closed subspace of X, III.6.

<u>Definition 1.1</u>. Let $f: X \to Y$ be a continuous map between
locally compact spaces. For a sheaf F on X and an open subset V of Y put

$$\Gamma(V, f_! F) = \left\{ s \in \Gamma(f^{-1}(V), F) \;\middle|\; \begin{array}{c} \text{Supp}(s) \xrightarrow{f} Y \\ \text{is a proper map} \end{array} \right\}$$

We consider $f_! F$ as a subpresheaf of $f_* F$.

<u>Proposition 1.2</u>. The presheaf $f_! F$ is a sheaf on Y.

<u>Proof</u>. It suffices to prove that the map $f: X \to Y$ has the
following property:

Given a family $(V_i)_{i \in I}$ of open subsets of Y with union V and a closed subset S of $f^{-1}(V)$. If the restriction $f: S \cap V_i \to V_i$ is proper for all i, then the restriction $f: S \to V$ is proper.

Let K be a compact subset of V. We are going to construct a family $(K_i)_{i \in I}$ of compact subsets of K, with $K_i \subseteq V_i$ for each $i \in I$, whose union is K and such that K_i is empty except for finitely many $i \in I$.

For the construction of the K_i's we may assume that I is a finite set by Borel-Heine. For $x \in K$ choose a compact neighbourhood K_x of x contained in some V_i, $i \in I$. Using Borel-Heine we obtain a finite covering of K by compact subsets of K each of which is contained in some V_i, $i \in I$. Let K_i be the union at those of the components of the covering which are contained in U_i. This gives the desired family $(K_i)_{i \in I}$. − To conclude the proof notice that

$$S \cap f^{-1}(K) = S \cap f^{-1}(\cup K_i) = \cup (S \cap f^{-1}(K_i))$$

which shows that $S \cap f^{-1}(K)$ is compact.

Q.E.D.

Let us consider the left exact functor

1.3 $$f_!: Sh(X) \longrightarrow Sh(Y)$$

The i'th derived functor evaluated on the sheaf F will be denoted $R^i f_! F$.

<u>Theorem 1.4</u>. Let $f: X \to Y$ denote a continuous map between locally compact spaces. For $y \in Y$ we have a natural isomorphism

$$(R^i f_! F)_y = H_c^i(f^{-1}(y), F) \qquad ; \ i \in \mathbb{Z}$$

as F varies through the category of sheaves on X.

<u>Proof</u>. Let us first treat the case $i = 0$. Consider an open neighbourhood V of y in Y and the restriction

$$\Gamma(V, f_! F) \longrightarrow \Gamma(f^{-1}(y), F)$$

Consider a $s \in \Gamma(V, f_! F)$ which maps to zero. This means that $\operatorname{Supp}(s) \cap f^{-1}(y)$ is empty or otherwise expressed $y \notin f(\operatorname{Supp}(s))$. Let W denote the complement of $\operatorname{Supp}(s)$ in V, this is an open neighbourhood of y to which the restriction of s is zero. Thus we have proved that the restriction

$$(f_! F)_y \to \Gamma_c(f^{-1}(y), F)$$

is injective. In case F is soft the restriction map is surjective as it follows by remarking that $\Gamma_c(X, F)$ is a subgroup of $\Gamma(Y, f_! F)$ and that restriction from $\Gamma_c(X, F)$ to $\Gamma_c(f^{-1}(y), F)$ is surjective III.2.6. - In the general case consider an exact sequence $0 \to F \to S \to T$ with S and T soft and use the resulting diagram

316

$$0 \longrightarrow (f_!F)_y \longrightarrow (f_!S)_y \longrightarrow (f_!F)_y$$
$$\downarrow \qquad\qquad \downarrow \qquad\qquad \downarrow$$
$$0 \longrightarrow \Gamma_c(f^{-1}(y),F) \to \Gamma_c(f^{-1}(y),F) \to \Gamma_c(f^{-1}(y),F)$$

to conclude that restriction is an isomorphism. - Let us now choose an injective resolution $F \to I^{\cdot}$. We have

$$(R^i f_!F)_y = (H^i f_! I^{\cdot})_y = H^i(f_! I^{\cdot})_y = H^i \Gamma_c(f^{-1}(y),I^{\cdot})$$

The result follows by noticing that the restriction of $I^{\cdot}$ to $f^{-1}(y)$ is a soft resolution of the restriction of F to $f^{-1}(y)$, III.2.5.

$$\text{Q.E.D.}$$

A diagram as below is called <u>cartesian</u> if $(q,h): A \to X \times B$ induces an isomorphism between A and the subspace $\{(x,b) \in X \times B \mid f(x) = p(b)\}$ of $X \times B$.

<u>Corollary 1.5</u>. Consider a cartesian square of locally compact spaces. For any sheaf F on X we have

$$p^* R^i f_! F \xrightarrow{\sim} R^i h_! q^* F \quad ; \ i \in \mathbb{Z}$$

$$\begin{array}{ccc} A & \xrightarrow{q} & X \\ \downarrow h & & \downarrow f \\ B & \xrightarrow{p} & Y \end{array}$$

<u>Proof</u>. The canonical morphism

$$p^* f_! F \longrightarrow h_! q^* F$$

__Corollary 1.6.__ Let $f: X \to Y$ denote a continuous map between locally compact spaces. A soft sheaf S on X is transformed into a soft sheaf $f_!S$ on Y. Moreover $R^i f_! S = 0$ for $i > 0$.

__Proof.__ Let us prove that for any compact subset K of Y and any sheaf S on X

$$1.7 \qquad \Gamma(K, f_! S) = \Gamma_c(f^{-1}(K), S)$$

To this end let $p: K \to Y$ and $q: f^{-1}(K) \to X$ denote the inclusions and $h: f^{-1}(K) \to K$ the restriction of f. According to 1.5 we have $p^* f_! S = h_! q^* S$. Apply $\Gamma(K,-)$ to this identity to get 1.7.

In case S is soft we shall prove that $f_! S$ is soft, i.e. that the restriction map

$$\Gamma(Y, f_! S) \longrightarrow \Gamma(K, f_! S)$$

is surjective. Notice that $\Gamma(Y, f_! S)$ contains $\Gamma_c(X, S)$ as a subgroup. Restriction from this group to $\Gamma_c(f^{-1}(K), S)$ is surjective according to III.2.6 and the result follows from 1.7.

The second part follows from 1.4 by localization using the fact that S induces soft sheaves on any of the fibres of f, III.2.5.

Q.E.D.

318

is an isomorphism as one sees by localization using 1.5 and
identification of the fibre of h over $b \in B$ with the fiber
of f over p(b). Let us remark that in case S is soft,
then q*S is acyclic for $h_!$: Notice that q*S induces soft
sheaves on the fibres of h and apply 1.4. For a soft resolution
S' of F we have

$$p*H^i f_! S' = H^i p*f_! S' = H^i h_! q*S'$$

from which the result follows using I.7.5.

Q.E.D.

Let us now consider a commutative ring k and interprete
$f_!$ as a functor

$$f_!: \mathrm{Sh}(X,k) \longrightarrow \mathrm{Sh}(Y,k)$$

and similarly for the derived functor

1.8 $$Rf_!: D^+(X,k) \longrightarrow D^+(Y,k)$$

Given two continuous maps between locally compact spaces
f: $X \to Y$ and g: $Y \to Z$ then it is easily seen that

1.9 $$(f \circ g)_! = g_! \circ f_!$$

and as a consequence of 1.6 and I.7.15

1.10 $$R(g \circ f)_! = Rg_! \circ Rf_!$$

VII.2 The Künneth formula

Let k denote a noetherian ring and $f: X \to Y$ a continuous map of locally compact spaces both of finite dimension. We are first going to construct with the notation of XI.2 a functor

$$2.1 \qquad\qquad Rf_! : D^-(X,k) \longrightarrow D^-(Y,k)$$

on the basis of the following

Lemma 2.2. Let X denote a locally compact space of finite dimension and k a commutative ring. Given a soft sheaf S and a sheaf F on X. If either S or F is flat then $S \otimes_k F$ is a soft sheaf.

Proof. In case S is flat we can use the proof of V.1.3. In case F is flat, that proof needs a small modification: notice that $K = \mathrm{Ker}\, \partial_{n-1}$ is a flat sheaf II.11.1.

$$\qquad\qquad\qquad\qquad\qquad\qquad\qquad\qquad Q.E.D.$$

Lemma 2.3. Let X denote a locally compact space of finite dimension and k a noetherian ring. For any $F^\cdot$ in $K^-(X,k)$ there exists a quasi-isomorphism $F^\cdot \to T^\cdot$ in $K^-(X,k)$ where $T^\cdot$ is a complex of soft sheaves (<u>a soft resulution of $F^\cdot$</u>)

Proof. Choose according to VI.1.3 a quasi-isomorphism $k \to S^\cdot$ where $S^\cdot$ is a bounded complex of soft and flat sheaves.

320

According to II.11.2 this gives a quasi-isomorphism
$F^{\cdot} \to S^{\cdot} \otimes F^{\cdot}$. The complex $S^{\cdot} \otimes F^{\cdot}$ is a bounded above complex
of soft sheaves by Lemma 2.2.

$$Q.E.D.$$

Construction of $Rf_!: D^-(X,k) \to D^-(Y,k)$

Let $S^-(X,k)$ denote the homotopy category of bounded
above complexes of soft k-sheaves. It follows from 2.2 and
XI.2 that we may identify $D^-(X,k)$ with the category obtained
by inverting all quasi-isomorphisms in $S^-(X,k)$. - The functor
$f_!$ extends to a functor

$$f_!: S^-(X,k) \longrightarrow S^-(Y,k)$$

according to 1.6. It remains to establish that $f_!$ _transforms
quasi-isomorphisms into quasi-isomorphisms_. To see this remark
that $R^n f_! = 0$ for $n > \dim X$ and use I.7.6.

Let us remark that the functor $f^*: Sh(Y,k) \to Sh(X,k)$ being
exact extends immediately to a functor

$$f^*: D^-(Y,k) \longrightarrow D^-(X,k)$$

With the notations above and the notation of II.11.10 we
have the fundamental

Projection formula 2.4.

$$Rf_!(F^{\cdot} \overset{L}{\otimes} f^*G^{\cdot}) = (Rf_! F^{\cdot}) \overset{L}{\otimes} G^{\cdot}$$

for $F^{\cdot}$ in $D^-(X,k)$ and G in $D^-(Y^{\cdot},k)$.

$\underline{\text{Proof}}$. Let us first prove that for a sheaf F on X and a flat sheaf G on Y there is a canonical isomorphism

$$2.5 \qquad f_!(F) \otimes G \xrightarrow{\sim} f_!(F \otimes f^*G)$$

Let us first notice that there is a natural transformation from the left hand side of the formula to the right hand side. Thus it follows from 1.4 that we may assume that Y is a point. Next, let us remark that both sides of the formula are left exact functors in F. Thus we may assume that F is soft.

It is now time to vary the k-module G. The case where G is a free module follows from the fact that Γ_c preserves direct sums, III.5.1. In general consider an exact sequence

$$0 \to L_2 \to L_1 \to L_0 \to G \to 0$$

where L_1 and L_0 are free modules. Since G is flat we can conclude that L_2 is flat. There results an exact sequence of sheaves on X

$$0 \to S \otimes L_2 \to S \otimes L_1 \to S \otimes L_0 \to S \otimes G \to 0$$

This is a sequence of soft sheaves on X by 2.2. It follows from I.7.5 that the sequence

$$0 \to \Gamma_c(X, S \otimes L_2) \to \Gamma_c(X, S \otimes L_1) \to \Gamma_c(X, S \otimes L_0) \to \Gamma_c(X, S \otimes G) \to 0$$

is exact, in particular we deduce a commutative exact diagram

$$\Gamma_c(X,F) \otimes L_1 \longrightarrow \Gamma_c(X,F) \otimes L_0 \longrightarrow \Gamma_c(X,F) \otimes G \longrightarrow 0$$

$$\Gamma_c(X,F \otimes L_1) \longrightarrow \Gamma_c(X,F \otimes L_0) \longrightarrow \Gamma_c(X,F \otimes G) \longrightarrow 0$$

from which we conclude that the vertical arrow to the right is an isomorphism. This proves 2.5.

Let us notice that in case F is soft, then $F \otimes f^*G$ is soft as it follows from 2.2 and the fact that f^*G is flat. With this remark in hand it is a simple matter to extend the formula 2.5 to the derived categories.

Q.E.D.

<u>Base change 2.6</u>. Consider a cartesian square of locally compact spaces of finite dimension. Then

$$p^*Rf_!F^{\cdot} \xrightarrow{\sim} Rh_!q^*F^{\cdot}$$

for all $F^{\cdot}$ in $D^-(X,k)$.

$$\begin{array}{ccc} A & \xrightarrow{q} & X \\ \downarrow h & & \downarrow f \\ B & \xrightarrow{p} & Y \end{array}$$

<u>Proof</u>. Let us remark that a soft sheaf S on X transforms into a sheaf q^*S on A which is acyclic for $h_!$ as it follows from 1.5 and 1.6. For a complex $S^{\cdot}$ in $K^-(X,k)$ choose a soft resolution $q^*S^{\cdot} \to T^{\cdot}$ in $K^-(A,k)$. According to 1.6 we have

$$p^*f_!S^{\cdot} \xrightarrow{\sim} h_!q^*S^{\cdot} \to h_!T^{\cdot}$$

The second arrow is a quasi-isomorphism by I.7.7.

Q.E.D.

With the notation of 2.6 put $c = fq = ph$.

<u>Künneth formula 2.7</u>.

$$Rc_!(h*E^{\cdot} \overset{L}{\otimes} q*F^{\cdot}) = (Rp_!E^{\cdot}) \overset{L}{\otimes} (Rf_!F^{\cdot})$$

for $E^{\cdot}$ in $D^-(B,k)$ and $F^{\cdot}$ in $D^-(X,k)$.

<u>Proof</u>. Use first the projection formula 2.4 and next the base change formula 2.6 to get

$$Rq_!(h*E^{\cdot} \overset{L}{\otimes} q*F^{\cdot}) = Rq_!(h*E^{\cdot}) \overset{L}{\otimes} F = f*(Rp_!E^{\cdot}) \overset{L}{\otimes} F^{\cdot}$$

Apply $Rf_!$ to this and use $Rc_! = Rf_! \circ Rq_!$ to get

$$Rc_!(h*E^{\cdot} \overset{L}{\otimes} q*F) = Rf_!(f*(Rp_!E^{\cdot}) \overset{L}{\otimes} F^{\cdot})$$

Apply the projection formula once more to get the result.

Q.E.D.

In particular for two locally compact spaces X and Y of finite dimension and a noetherian ring k

$$2.8 \qquad R\Gamma_c^{\cdot}(X \times Y,k) = R\Gamma_c^{\cdot}(X,k) \overset{L}{\otimes}_k R\Gamma_c^{\cdot}(Y,k)$$

VIII.3 Global form of Verdier duality

Let k be a noetherian ring and $f: X \to Y$ a continuous
map between locally compact spaces of finite dimension.

Theorem 3.1. There exists an additive functor

$$f^!: D^+(Y,k) \to D^+(X,k)$$

and a natural isomorphism

$$[Rf_!F^\cdot,G^\cdot] \xrightarrow{\sim} [F^\cdot,f^!G^\cdot]$$

as $F^\cdot$ varies through $D^+(X,k)$ and $G^\cdot$ varies through $D^+(Y,k)$.

Proof. We shall follow the proof of VI.1.1 closely and give
the needed modifications. For a soft and flat sheaf S on X and
a sheaf G on Y the functor

$$F \longrightarrow \mathrm{Hom}(f_!(S \otimes F),G)$$

from $\mathrm{Sh}(X,k)$ to the category of k-modules transforms kernels
into cokernels and direct sums into direct products: To see
this notice that the functor $F \mapsto f_!(S \otimes F)$ is exact as it follows
from 2.2 and 1.6. The same functor preserves direct sums or
more generally direct limits as it follows from 1.4 and III.5.

1) With S and G as above, the presehaf on X

$$U \longmapsto \mathrm{Hom}(f_!(S_U),G)$$

is a sheaf on X, here $S_U = j_! j^* S = S \otimes j_! k$ where $j: U \to X$

denotes the inclusion: For open subsets U and V of X

consider the transform of the exact sequence

$$0 \longrightarrow \underset{\sim}{k}_{U \cap V} \longrightarrow \underset{\sim}{k}_U \oplus \underset{\sim}{k}_V \longrightarrow \underset{\sim}{k}_{U \cap V} \longrightarrow 0$$

by the functor $F \longmapsto \mathrm{Hom}(f_! (S \otimes F), G)$.

    2) Let $f^!(S,G)$ denote the above sheaf on X. Here S

is a k-flat and soft sheaf on X and G is any sheaf on Y.

    3) There is a canonical isomoprphism

$$3.2 \qquad\qquad \mathrm{Hom}(f_!(F \otimes S),G) \overset{\sim}{\longrightarrow} \mathrm{Hom}(F,f^!(S,G))$$

as F varies through the category $\mathrm{Sh}(X,k)$. - Let us first

establish the identification

$$3.3 \qquad\qquad f_* f^!(S,G) = \mathcal{H}om(f_! S,G)$$

To this end consider the inclusion $j: V \to Y$ of an open subset

of Y and notice that

$$\mathrm{Hom}(j^* f_! S, j^* G) = \mathrm{Hom}(j_! j^* f_! S, G) = \mathrm{Hom}(f_!(S_{f^{-1}(V)}),G)$$

as it follows by the base change property 1.5. -

By adjunction we deduce from 3.3 a morphism of sheaves

$$f^* \mathcal{H}om(f_! S,G) \to f^!(S,G)$$

Let us now depart from the morphism $\quad f_* F \otimes f_! S \to f_!(F \otimes S)$

and deduce first a morphism, II.12

$$\mathrm{Hom}(f_!(F \otimes S),G) \to \mathrm{Hom}(f_*F, \mathit{Hom}(f_!S,G))$$

and by adjunction a morphism

$$\mathrm{Hom}(f_!(F \otimes S),G) \to \mathrm{Hom}(F, f_* \mathit{Hom}(f_!S,G))$$

Combine this with the morphism above to get

$$\mathrm{Hom}(f_!(F \otimes S),G) \to \mathrm{Hom}(F, f^!(S,G))$$

To prove that this is an isomorphism we shall vary F. It suffices to check the case $F = j_! \underset{\sim}{k}$ where j is the inclusion of an open subset of X.

4-7) Needs no essential modifications.

$$\mathrm{Q.E.D.}$$

$\underline{\text{Example 3.4}}$. Let $h \colon W \to X$ denote the inclusion of a locally closed subspace. With the notation of II.6 the functor

$$h^! \colon \mathrm{Sh}(X,k) \to \mathrm{Sh}(W,k)$$

is left exact and transforms injectives into injectives. Thus it extends to a functor $\quad h^! \colon D^+(X,k) \to D^+(W,k)$

which is a right adjoint to the functor $h_!$ described in II.6. This provides an extension of Theorem 3.1 for this kind of immersions beyond the framwork of locally compact spaces. The case of a closed subspace will be explored in Chapter VIII.

VII.4   Covering spaces

Let $f: X \to Y$ be a finite covering space of degree n, i.e. for each point $y \in Y$ there exists an open neighbourhood V of y such that $f: f^{-1}(V) \to V$ is isomorphic to the projection $V \times [1,n] \to V$. For a sheaf F on X and $y \in Y$ we have a canonical isomorphism

$$4.1 \qquad (f_*F)_y \xrightarrow{\sim} \bigoplus_{x \in f^{-1}(y)} F_x$$

from which we conclude that the functor

$$4.2 \qquad f_*: Sh(X,k) \to Sh(Y,k) \quad \text{is exact}$$

For a sheaf G on Y we deduce from 4.1 an isomorphism on the stalks at $y \in Y$

$$(f_*f^*G)_y \xrightarrow{\sim} \bigoplus_{x \in f^{-1}(y)} G_y$$

Compose this with "summation" to get the <u>trace map</u>

$$4.3 \qquad tr_y: (f^*f_*G)_y \longrightarrow G_y \qquad ; y \in Y$$

There exists a unique morphism of sheaves

$$4.4 \qquad tr: f_*f^*G \longrightarrow G$$

whose stalks are those recorded in 4.3: Uniqueness follows from II.2.2.i. By II.12 the problem is local on Y. Thus it suffices to treat the case where the covering is trivial which

328

is left to the reader. - For a sheaf F on X the trace map
induces an isomorphism

$$4.5 \qquad \boxed{\operatorname{Hom}(F,f^*G) \xrightarrow{\sim} \operatorname{Hom}(f_*F,G)}$$

Proof. The construction above yields more generally a
morphism of sheaves on Y

$$4.6 \qquad f_*\mathcal{H}om(F,f^*G) \to \mathcal{H}om(f_*F,G)$$

We shall in fact prove that the morphism 4.6 is an isomorphism:
The new problem is local on Y. Thus it suffices to treat the
case of a trivial covering which is left to the reader.

Q.E.D.

Corollary 4.7. The pull back functor

$$f^*\colon \operatorname{Sh}(Y,k) \to \operatorname{Sh}(X,k)$$

transforms injectives into injectives.

Proof. Follows formally from the presence of a left
adjoint which is exact by 4.2.

Q.E.D.

From our discussion follows that in the case where X
and Y are locally compact we have $\underline{f^! = f^* \quad in \quad this}$
$\underline{particular\ case\ of\ Verdier\ duality\ 3.1.}$

For a sheaf G on Y let res: $G \to f_* f^* G$ denote the standard adjunction morphism, II.4.9. By localization on Y follows
that

4.8
$$\boxed{tr \circ res = n}$$

Let us consider an injective resolution $\underset{\sim}{k} \to G^{\cdot}$. Apply the
functor $H^{\cdot}\Gamma(Y,-)$ to the morphisms

4.9
$$G^{\cdot} \xrightarrow{res} f_* f^* G^{\cdot} \xrightarrow{tr} G^{\cdot}$$

to obtain morphisms on cohomology $f^* = res$

4.10
$$H^{\cdot}(Y,k) \xrightarrow{res} H^{\cdot}(X,k) \xrightarrow{tr} H^{\cdot}(Y,k)$$

which satisfies the relation 4.8. Notice that

4.11
$$tr(f^*\eta \cup \xi) = \eta \cup tr(\xi) \qquad ;\ \eta \in H^{\cdot}(Y,k),\ \xi \in H^{\cdot}(X,k)$$

as it follows from the construction above.

VII.5 Local form of Verdier duality

In this section we shall give a local version of Verdier duality. To do so we shall use the general theory of derived categories as exposed in Chapter XI.

Given a topological space and a commutative ring k . For $F^\bullet$ in $D^-(X,k)$ and $F^\bullet$ in $D^+(X,k)$ we shall define $R\,Hom^\bullet(E,F)$ in $D^+(X,k)$: choose an injective resolution $F^\bullet \to J^\bullet$ and put

$$5.1 \qquad\qquad R\,Hom^\bullet(E^\bullet,F^\bullet) = Hom^\bullet(E^\bullet,J^\bullet).$$

This takes a particularly simple form if we represent the derived categories as follows $D^-(X,k)$: The category obtained from the homotopy category of bounded above complexes of flat sheaves by inverting all quasi-isomorphisms. $D^+(X,k)$: The homotopy category of bounded below complexes of injective sheaves. In particular, with $E^\bullet$ and $F^\bullet$ such represented, the complex $Hom^\bullet(E^\bullet,F^\bullet)$ is automatically a bounded below complex of injective sheaves, II. 12.3.

<u>Theorem 5.2.</u> Let k be a noetherian ring and $f: X \to Y$ a continuous map of locally compact spaces of finite dimension. There is a natural isomorphism in $D^+(X,k)$

$$R\,Hom^\bullet(Rf_! E^\bullet,F^\bullet) \xrightarrow{\sim} Rf_* \, R\,Hom^\bullet(E^\bullet,f^! F^\bullet)$$

as $E^\bullet$ varies through $D^-(X,k)$ and $F^\bullet$ through $D^+(X,k)$.

Proof. Let S be a soft and flat sheaf on X. The iso-morphism 3.2 extends easily to an isomorphism

$$5.3 \qquad Hom(f_!(E \otimes S),F) = f_* Hom(E,f^!(S,F))$$

as E varies through Sh(X,k) and F through Sh(Y,k). Let us choose a fixed bounded flat and soft resolution $\underset{\sim}{k} \to S^{\cdot}$ on X, VI.1.3. For a bounded above complex $E^{\cdot}$ of <u>flat</u> sheaves on X and a bounded below complex $F^{\cdot}$ of <u>injective</u> sheaves on Y we deduce from 5.3 an isomorphism.

$$5.4 \qquad Hom^{\cdot}(f_!(E^{\cdot} \otimes S^{\cdot}),F^{\cdot}) \overset{\sim}{\to} f_* Hom^{\cdot}(E^{\cdot},f^!(S^{\cdot},F^{\cdot}))$$

We can conclude the proof by the following four references: The morphism $E^{\cdot} \to E^{\cdot} \otimes S^{\cdot}$ is a soft resolution of $E^{\cdot}$ by 2.2. The complex $f^!(S^{\cdot},F^{\cdot})$ is a bounded below complex of injective sheaves on X as it follows from the proof of 3.1. The complex $Hom^{\cdot}(E^{\cdot},f^!(S^{\cdot},F^{\cdot}))$ is a bounded below complex of injective sheaves as it follows from II.12.3. Any bounded above complex admits a flat resolution by II.11.8.

Q.E.D.

<u>Example 5.5</u>. Consider a finite covering $f: X \to Y$. Verdier duality is simply represented by the isomorphism 4.6.

VIII. Characteristic Classes

<u>VIII.1 Local duality</u>

In this section we consider a fixed commutative ring k.
Let us recall that the inclusion $i: Z \to X$ of a closed subspace gives rise to two functors

$$Sh(Z,k) \underset{i_*}{\overset{i^!}{\rightleftarrows}} Sh(X,k)$$

which are mutually adjoints

$$1.1 \qquad \text{Hom}(i_*E,F) = \text{Hom}(F,i^!F)$$

The functor $i^!$ is left exact and carries injectives into
injectives. We shall study the derived functor $R^p i^!$, $p \in \mathbb{Z}$.

For a sheaf F on X we shall give an explicit description of the sheaf $i^!F$ on Z or rather the sheaf $i_*i^!F$ on
X. It follows from the formula II.6.7 and II.9.1 that the
sections of $i_*i^!F$ over the open subset V of X are

$$1.2 \qquad \Gamma(V,i_*i^!F) = \Gamma_{Z \cap V}(V,F)$$

For an integer $p \in \mathbb{Z}$ we shall describe a presheaf on X:
An open subset U of the open subset V of X gives rise to
a restriction map

$$H_{V \cap Z}^{p}(V,F) \longrightarrow H_{U \cap Z}^{p}(U,F)$$

and thereby a presheaf $U \longmapsto H_{U \cap Z}^{p}(U,F)$.

Proposition 1.3. The sheaf associated to the presehaf

$$U \longmapsto H_{U \cap Z}^{d}(U,F) \qquad ; \ d \in \mathbb{Z}$$

is $i_{*}R^{d}i^{!}F$. If $R^{p}i^{!}F = 0$ for all $p < d$, then

$$\Gamma(U, i_{*}R^{d}i^{!}F) = H_{Z \cap U}^{d}(U,F)$$

for any open subset U of X.

Proof. Let $I^{\cdot}$ denote an injective resolution of the sheaf
F on W. We have

$$i_{*}R^{p}i^{!}F = i_{*}H^{p}i^{!}I^{\cdot} = H^{p}i_{*}i^{!}I^{\cdot}$$

Thus it follows from II.2.7 that $i_{*}R^{p}i^{!}F$ is the sheaf associated
to the presheaf of

$$U \longmapsto H^{p}\Gamma(U, i_{*}i^{!}I^{\cdot})$$

From the description 1.2 we deduce that

$$H^p \, \Gamma(U, i_* i^! I^{\boldsymbol{\cdot}}) = H^p \, \Gamma_{U \cap Z}(U, I^{\boldsymbol{\cdot}}) = H_{U \cap Z}^{\;\;\;p}(U, F)$$

which proves the first part of the statement.

If $R^p i^! F = 0$ for $p < d$ then the complex $i^! I^{\boldsymbol{\cdot}}$ is homotopic to $\tau_{\geq d} \, i^! I^{\boldsymbol{\cdot}}$ and we deduce from the formula above that

$$H_{U \cap Z}^p(U, F) = H^p \Gamma(U, i_* \tau_{\geq d} i^! I^{\boldsymbol{\cdot}}) = \Gamma(U, i_* H^p i^! I^{\boldsymbol{\cdot}})$$

from which the result follows since $R^p i^! F = H^p i^! I^{\boldsymbol{\cdot}}$.

Q.E.D.

Most of our calculations will be based on the following proposition

Proposition 1.4. Let $p: X \to Z$ be a retraction to the inclusion $i: Z \to X$ of a closed subspace. For a sheaf F on X, the sheaf associated to the presheaf

$$V \longmapsto H_V^{\;n}(p^{-1}(V), F) \qquad\qquad ;\quad n \in \mathbb{Z}$$

is $R^n i^! F$. If $R^n i^! F = 0$ for all $n < d$, then

$$\Gamma(V, R^d i^! F) = H_{i(V)}^{\;\;\;d}(p^{-1}(V), F)$$

for any open subset V of Z.

$\underline{\text{Proof}}$. Let $I^{\cdot}$ denote an injective resolution of F. $R^n i^! F$ is the sheaf associated to the presheaf

$$V \longmapsto H^n \Gamma(V, i^! I^{\cdot}) \qquad ; \ V \subseteq Z$$

Using formula 1.2 we deduce that

$$H^n \Gamma(V, i^! I^{\cdot}) = H^n \Gamma(p^{-1}(V), i_* i^! I^{\cdot}) = H^n \Gamma_{i(V)}(p^{-1}(V), I^{\cdot})$$

from which the first result follows.

If $R^n i^! F = 0$ for $n < d$ we deduce that $i^! I^{\cdot}$ is homotopic to $\tau_{\geq d} i^! I^{\cdot}$ and get that

$$H_{i(V)}^d(p^{-1}(V), F) = H^d \Gamma(V, \tau_{\geq d} i^! I^{\cdot}) = \Gamma(V, H^d i^! I^{\cdot})$$

and the second result follows since $H^d i^! I^{\cdot} = R^d i^! F$.

$$\text{Q.E.D.}$$

$\underline{\text{Local duality theorem 1.5}}$. Let $i: Z \to X$ denote the inclusion of a closed subspace. We have

$$[i_* I^{\cdot}, J^{\cdot}] = [I^{\cdot}, i^! J^{\cdot}]$$

for $I^{\cdot}$ in $D^+(Z, k)$ and $J^{\cdot}$ in $D^+(X, k)$.

$\underline{\text{Proof}}$. The identity 1.1 gives an isomorphism

$$\text{Hom}^{\cdot}(i_* I^{\cdot}, J) = \text{Hom}^{\cdot}(I^{\cdot}, i^! J^{\cdot})$$

from which the formula results by applying the functor H^0.

$$\text{Q.E.D.}$$

336

1.6
$$H_Z^{\ d}(X,k) = [\, \underset{\sim}{k}[-d], R^{\cdot}i^{\,!}\underset{\sim}{k}\,]$$

__Proof__. Apply 1.5 with $I^{\cdot}$ equal to the constant sheaf $\underset{\sim}{k}[p]$
on Z and $J^{\cdot}$ an injective resolution of $\underset{\sim}{k}$ on X.

Q.E.D.

__Proposition 1.7__. Let X denote a locally compact space
of finite dimension, Z a closed subspace of X and k a
noetherian ring. If $\mathcal{D}_X^{\cdot}$ denotes the dualizing complex for X,
then the dualizing complex $\mathcal{D}_Z^{\cdot}$ for Z is given by

$$\mathcal{D}_Z^{\cdot} \ = \ R\,i^{\,!}\mathcal{D}_X^{\cdot}$$

__Proof__. Let $k \to K^{\cdot}$ denote an injective resolution in the
category of k-modules. For $I^{\cdot}$ in $D^{+}(Z,k)$ we have

$$[I^{\cdot}, i^{\,!}\mathcal{D}_X^{\cdot}] = [i_{*}I^{\cdot}, \mathcal{D}_X^{\cdot}] = [\Gamma_c(X, i_{*}I^{\cdot}), K^{\cdot}] = [\Gamma_c(Z, I^{\cdot}), K^{\cdot}]$$

where we have used Verdier duality VI.2.1.

Q.E.D.

__Corollary 1.8__. Let X denote an n-dimensional topological
manifold and $i: Z \to X$ the inclusion of a closed submanifold of
dimension m. If both manifolds are oriented relative to the noetherian ring k, then

$$R^{\cdot}i^{\,!}\underset{\sim}{k} \ \xrightarrow{\sim} \ \underset{\sim}{k}[m-n]$$

__Proof__. We have $\mathcal{D}_X^{\cdot} \xrightarrow{\sim} \underset{\sim}{k}[n]$ and $\mathcal{D}_Z^{\cdot} \xrightarrow{\sim} \underset{\sim}{k}[m]$ by VI.3.2.

Q.E.D.

VIII.2   Thom Class

In this section we consider the inclusion $i: Z \to X$ of a closed subspace and a commutative ring k. We shall study a fixed quasi-isomorphism

$$2.1 \qquad\qquad \tau: \underset{\sim}{k}[-d] \xrightarrow{\sim} R^{\cdot}i^{!}\underset{\sim}{k}$$

where $d \in \mathbb{N}$ is a fixed integer. The basic examples are the case where Z and X are oriented manifolds 1.8 and the case where i is the zero section in an oriented vector bundle or more generally an oriented microbundle VIII.3. Our basic objective is to establish some important cup product formulas arising in this situation.

By local duality, 1.6 we may interprete τ as a local cohomology class, the so called <u>Thom class</u>

$$2.2 \qquad\qquad \tau \in H_{Z}^{d}(X,k)$$

The extraordinary cup product II.9.14 gives rise to the

<u>Thom isomorphism 2.3</u>. The cup product

$$\alpha \to \tau \cup \alpha \qquad ; \quad H^{p}(Z,k) \to H^{d+p}_{Z}(X,k)$$

is an isomorphism for all $p \in \mathbb{Z}$.

<u>Proof</u>. Let $I^{\cdot}$ be an injective resolution of the constant sheaf $\underset{\sim}{k}$ on Z and $J^{\cdot}$ an injective resolution of $\underset{\sim}{k}$ on X. Using $\tau[d]^{-1}: i^{!}J^{\cdot}[d] \xrightarrow{\sim} I^{\cdot}$ we get from the local duality that

$$H_Z^{d+p}(X,k) = [i_*I^{\cdot},J^{\cdot}[d+p]] = [I^{\cdot},i^!J^{\cdot}[d+p]] \overset{\sim}{\to} [I^{\cdot},I^{\cdot}[p]]$$

from which the Thom isomorphism can be read off.

Q.E.D.

The Thom isomorphism can be composed with restriction
$r: H_Z^{\cdot}(X,k) \to H^{\cdot}(X,k)$ to obtain the <u>Gysin map</u>

$$i_*: H^p(Z,k) \to H^{d+p}(X,k) \qquad ; \qquad p \in \mathbb{Z}$$

2.4

$$i_*(\alpha) = r(\tau \cup \alpha) \qquad ; \qquad \alpha \in H^p(Z,k)$$

The Gysin map satisfies the

<u>Projection formula 2.5</u>

$$i_*(\alpha \cup i^*\beta) = i_*(\alpha) \cup \beta \qquad ; \qquad \alpha \in H^p(Z,k), \beta \in H^q(X,k)$$

<u>Proof</u>. Using the formula II.9.17 twice we get

$$r(\tau \cup \alpha \cup i^*\beta) = (-1)^{(p+d)q} r(\beta \cup \tau \cup \alpha)$$

$$(-1)^{(d+p)q} \beta \cup r(\tau \cup \alpha) = r(\tau \cup \alpha) \cup \beta$$

from which the projection formula follows.

Q.E.D.

<u>Definition 2.6</u>. The <u>Euler class</u> $e \in H^d(Z,k)$ is given by $e = i^*r(\tau)$ where $\tau \in H_Z^d(X,k)$ is the Thom class.

<u>Formula 2.7</u>. $i^*i_*\alpha = e \cup \alpha$; $\alpha \in H^p(Z,k)$. If d is odd, then $2e = 0$. In particular $e \cup \alpha = \alpha \cup e$ for all $\alpha \in H^\cdot(Z,k)$.

<u>Proof</u>. For the first part it suffices to prove that $\tau \cup i^*i_*\alpha = \tau \cup e \cup \alpha$. Using the formulas II.9.15-16 we get

$$\tau \cup i^*i_*\alpha = \tau \cup i^*r(\tau \cup \alpha) = r(\tau) \cup (\tau \cup \alpha)$$

$$\tau \cup (e \cup \alpha) = \tau \cup i^* r(\tau) \cup \alpha = r(\tau) \cup \tau \cup \alpha$$

Let us finally prove the formula $e = (-1)^p e$ or what amounts to the same $\tau \cup e = (-1)^p \tau \cup e$. The formula 2 lines above with $\alpha = 1$ gives $\tau \cup e = r(\tau) \cup \tau$, on the other hand by II.9.17

$$\tau \cup e = \tau \cup i^*r(\tau) = (-1)^{d^2} r(\tau) \cup \tau$$

and the result follows since $d^2 \equiv d \bmod 2$.

Q.E.D.

VIII.3   Oriented microbundles

Let X denote a topological space. By a <u>microbundle</u> of fibre dimension d we understand a triple (E,p,i) where E is a topological space, p: E → X a continuous map and i: X → E a continuous section to p such that for each part x of X there exists an open neighbourhood V of i(x) in E and an open neighbourhood U of x in X with p(V) ⊆ U and i(U) ⊆ V and a homeomorphism a: V → U × $\mathbb{R}^d$ which makes the following diagram commutative

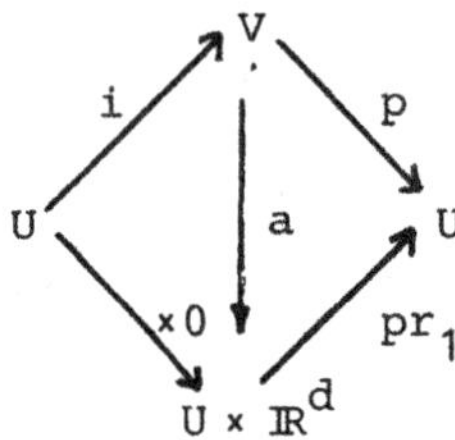

Microbundles (E,p,i) and (F,q,j) are said to be isomorphic, if there exists an open neighbourhood V of i(X) in E and an open neighbourhood W of j(X) in F such that the triples (V,p,i) and (W,q,j) are isomorphic. The concept of microbundles was introduced by Milnor (1). It is known that any microbundle over a paracompact base space is isomorphic to an $\mathbb{R}^d$-bundle, Kister (1), Holm (1).

<u>Example 3.1</u>. Let E be a d dimensional vector bundle over the space X and V and open neighbourhood of the zero section in F. The projection of V onto X and the zero section makes V into a microbundle.

<u>Example 3.2.</u> Let M denote a d-dimensional topological mani-
fold. The projection $p_1 \colon M \times M \to M$ and the diagonal $\Delta \colon M \to M \times M$
constitutes a microbundle: For $x \in M$ choose an open neighbourhood
U of x in M and a homeomorphism $f \colon U \xrightarrow{\sim} \mathbb{R}^d$: The neighbourhood $U \times U$ of (x,x) is mapped homeomorphically onto $U \times \mathbb{R}^n$
by the map $(x,y) \longmapsto (x, f(y) - f(x))$. This microbundle is called
the <u>tangent microbundle</u>. - In case of a smooth manifold the
tangent microbundle is isomorphic to the tangent bundle, Milnor (1).

<u>Theorem 3.3.</u> Let (E,p,i) be a microbundle of fibre dimension
d over the topological space X.   The presheaf

$$U \longmapsto H_{i(U)}^d (p^{-1}(U), k)$$

is a locally constant k-sheaf. The stalk of $x \in X$ is

$$H_{\{i(x)\}}^d (p^{-1}(x), k)$$

<u>Proof.</u> Let us calculate the stalk at $x \in X$ of the sheaf
$R^s i^! \underset{\sim}{k}, \ s \in \mathbb{Z}.$ According to 1.4 we have

$$R^s i^! \underset{\sim}{k}_x = \varinjlim H_{i(U)}^s (p^{-1}(U), k)$$

where the limit is taken over all open neighbourhoods U of x
in X.   Consider the long exact commutative ladder

$$H^{s-1}(p^{-1}(U)) - i(U), k) \longrightarrow H_{i(U)}^s (p^{-1}(U), k) \longrightarrow H^s (p^{-1}(U), k)$$
$$\downarrow \qquad\qquad\qquad\qquad \downarrow \qquad\qquad\qquad\qquad \downarrow$$
$$H^{s-1}(p^{-1}(x) - \{i(x)\}, k) \longrightarrow H_{\{i(x)\}}^s (p^{-1}(x), k) \longrightarrow H^s (p^{-1}(x), k)$$

By excision we may assume that $E = X \times \mathbb{R}^d$. Pass to the direct limit over all open neighbourhoods U of x to obtain a new exact commutative ladder. Apply theorem IV.1.6 to the fibrations E and $E-i(X)$ over X and use the five lemma to conclude that the stalk of $R^s i_! \underset{\sim}{k}$ at the point x is as described.

From this we conclude that $R^s i_! \underset{\sim}{k} = 0$ for $s \neq d$. It follows from 1.4 that the sections of $R^d i^! k$ over the open set U of X are given by

$$\Gamma(U, R^d i^! k) = H^d_{i(U)}(p^{-1}(U), k)$$

which shows that our presheaf is indeed a sheaf.

To prove that our sheaf is locally constant we may assume that the bundle is the trivial bundle $\mathrm{pr}_1 : X \times \mathbb{R}^d \to X$. Let us pick a generator $\tau \in H^d_{\{0\}}(\mathbb{R}^d, k)$ and pull this back along $\mathrm{pr}_2 : X \times \mathbb{R}^d \to \mathbb{R}^d$ to obtain a class

$$\mathrm{pr}_2^* \tau \in H^d_{X \times \{0\}}(X \times \mathbb{R}^d, k)$$

From the first part of the proof follows that $\mathrm{pr}_2^* \tau$ generates our k-sheaf.

Q.E.D.

The microbundle (E, p, i) is said to be <u>orientable</u> relative to k if the k-sheaf of the theorem is isomorphic to the constant k-sheaf $\underset{\sim}{k}$. Expressed otherwise

Definition 3.4. By an _orientation_ relative to k of the microbundle (E,p,i) of fibre dimension d we understand a cohomology class

$$\tau \in H_{i(X)}{}^{d}(E,k)$$

which for each $x \in X$ restricts to a generator of the k-mdoule $H_{\{i(x)\}}{}^{d}(p^{-1}\{x\},k)$.

If (E,p,i,τ) is an oriented microbundle of fiber dimension d, τ will also be called the _Thom class_ of the bundle. The image of τ by the composite

$$H_X{}^{n}(E,k) \xrightarrow{\ r\ } H^n(E,k) \xrightarrow{\ i^*\ } H^n(X,k)$$

is called the _Euler class_, $e(E)$ of the oriented microbundle.

Pull back of microbundles

Let $f: X \to Y$ be a continuous map and (F,j,q) a microbundle on Y of fibre dimension d. Let

$$f^*F = \{ (x,v) \in X \times F \mid f(x) = q(v) \}$$

and let $p: f^*F \to X$ denote the restriction of the first projection and let $i: X \to f^*E$ denote the map given by $i(x) = (x,jf(x))$. In this way we have constructed a microbundle (f^*F,i,p), the _pull back of_ (F,j,q) along f.

If $\tau \in H_{j(Y)}{}^{d}(F,k)$ is an orientation of F then $f^*\tau \in H_{i(X)}{}^{d}(f^*F,k)$ is an orientation of f^*E which follows by identifying the fibers $p^{-1}(x)$ and $q^{-1}(x)$, $x \in X$. If $e(F)$ is the Euler class of F then the <u>Euler class</u> $e(f^*F)$ is given by

$$3.5 \qquad\qquad e(f^*F) = f^*e(F)$$

Products of microbundles

Let (E,p,i) be a microbundle on X of fibre dimension m and (F,q,j) a microbundle on Y of fibre dimension n. Then $(E \times F, p \times q, i \times j)$ is a fibre bundle on $X \times Y$ of fibre dimension $m+n$ as one easily verifies.

Given orientations $\sigma \in H_X^m(E,k)$ and $\tau \in H_Y^n(F,k)$ respectively. Then the <u>cross product</u>, II.10.7

$$\sigma \times \tau \in H_{X \times Y}^{m+n}(E \times F,k)$$

is an orientation of $(E \times F, p \times q, i \times j)$ as it follows from Corollary 3.9 below. If $e(E) \in H^m(X,k)$ and $e(F) \in H^n(Y,k)$ denotes the Euler class of E and F respectively, then the Euler class $e(E \times F)$ of the product is given by

$$3.6 \qquad\qquad e(E \times F) = e(E) \times e(F)$$

Whitney sum of microbundles

Let E and F be microbundles on X and $E \times F$ the product bundle on $X \times X$. Consider the diagonal map $\Delta: X \to X \times X$ and let the <u>Whitney</u> sum $E \oplus F$ be defined by

$$E \oplus F = \Delta^* E \times F$$

In case E and F are oriented microbundles the Euler class $e(E \oplus F)$ of $E \oplus F$ is given by

3.7
$$e(E \oplus F) = e(E) \cup e(F)$$

as it follows from 3.5, 3.6 and II.10.8.

<u>Proposition 3.8</u>. Let D denote a d-dimensional real vector-space and $\tau \in H_0^{\,d}(D,k)$ an orientation. For any pair X,Z consisting of a topological space X and a closed subset Z, the cross product with τ will induce an isomorphism

$$\times \tau \; : \; H_Z^{\,p}(X,k) \longrightarrow H_{Z \times 0}^{\,p+d}(X \times D,k) \qquad ; \quad p \in \mathbb{Z}$$

<u>Proof</u>. Let us first remark that in case Z is empty the result follows from 3.3 and 2.3. In the general case consider the commutative exact ladder II.10.2

$$\begin{array}{ccccccc}
\longrightarrow & H^{p-1}(X-Z,k) & \longrightarrow & H_Z^{\,p}(X,k) & \longrightarrow & H^p(X,k) & \longrightarrow \\
& \downarrow {\scriptstyle \times \tau} & & \downarrow {\scriptstyle \times \tau} & & \downarrow {\scriptstyle \times \tau} & \\
\longrightarrow & H_{(X-Z) \times 0}^{\,p-1+d}((X-Z) \times D,k) & \rightarrow & H_{Z \times 0}^{\,p+d}(X \times D,k) & \longrightarrow & H_{X \times 0}^{\,p+d}(X \times D,k) &
\end{array}$$

Finally apply the 5-lemma.

Q.E.D.

<u>Corollary 3.9</u>. Let E and F be finite dimensional real vector spaces. Cross product induces an isomorphism

$$H_{\{0\}}^{\;\bullet}(E,k) \otimes H_{\{0\}}^{\;\bullet}(F,k) \xrightarrow{\sim} H_{\{0\}}^{\;\bullet}(E \times F,k)$$

346

<u>Example 3.10</u>. Let E denote an arbitrary microbundle
of fibre dimension d. The Whitney sum $E \oplus E$ has a <u>canonical
orientation</u> relative to $\mathbb{Z}$: For $x \in X$ let τ_x generate
$H_{\{x\}}^{d}(E_x, \mathbb{Z})$. Then $\tau_x \times \tau_x$ is a local orientation of $E \oplus E$,
which is independent of τ_x since $(-\tau_x) \times (-\tau_x) = \tau_x \times \tau_x$.
These local data fit together to give a global orientation of
$E \oplus E$, 3.3.

In the following two sections we shall consider the case
where the coefficient ring k equals $\mathbb{F}_2$. Any microbundle has
a unique orientation relative to $\mathbb{F}_2$ as it follows from the fact
that a locally constant $\mathbb{F}_2$-sheaf of rank 1 is uniquely isomorphic
to $\underset{\sim}{\mathbb{F}}_2$, compare V. 3.8.

<u>3.11. The Möbius band</u> is the classical microbundle M of
fibre dimension 1 over S^1. Let us investigate the Euler class
of M in cohomology with coefficients in $\mathbb{F}_2$. In this case
the Thom class is unique V.3.8. - From the fact that $M-S^1$ is
connected follows that restriction from $H_{S^1}^{1}(M, \mathbb{F}_2) \to H^1(M, \mathbb{F}_2)$
is a monomorphism and consequently by dimension reasons, re-
striction is an isomorphism. It follows that the Euler class of
the Möbius band is the non trivial-element of $H^1(S^1, \mathbb{F}_2)$.

VIII.4 Cohomology of real projective space

The set of lines in $\mathbb{R}^{n+1}$ constitute $P^n(\mathbb{R})$, real projective n-space. The multiplicative group $\mathbb{R}^*$ of real numbers $\neq 0$ acts on $\mathbb{R}^{n+1}-(0)$ and we can represent $P^n(\mathbb{R})$ as the orbit space

$$P^n(\mathbb{R}) = \mathbb{R}^{n+1}-0 \,/\, \mathbb{R}^*$$

Projective space carries a <u>canonical line bundle L</u> given as the orbit space

$$L = (\mathbb{R}^{n+1}-0) \times \mathbb{R} \,/\, \mathbb{R}^*$$

where the action of $\mathbb{R}^*$ is given by

$$r(v,x) = (rv,rx) \quad ; \quad r \in \mathbb{R}^*,\ v \in \mathbb{R}^{n+1}-0, \quad x \in \mathbb{R}$$

The projection $p\colon L \to P^n$ is induced by the projection of $(\mathbb{R}^n-0) \times \mathbb{R}$ onto its first factor.

<u>Theorem 4.1.</u> Let $e \in H^1(P^n(\mathbb{R}),\mathbb{F}_2)$ denote the Euler class of the canonical line bundle L on $P^n(\mathbb{R})$. The powers

$$1,e,e^2,\ldots,e^n$$

are a basis for the $\mathbb{F}_2$-vector space $H^\cdot(P^n(\mathbb{R}),\mathbb{F}_2)$.

<u>Proof.</u> The Thom class $\tau \in H^1_{P^n}(L,\mathbb{F}_2)$ of L induces an isomorphism for each $p \in \mathbb{Z}$

$$H^p(P^n,\mathbb{F}_2) \xrightarrow{\ \sim\ } H^{p+1}_{P^n}(L,\mathbb{F}_2) \quad\quad ; \quad \alpha \longmapsto \tau \cup \alpha$$

The complement of the zero section in L may be identified with $\mathbb{R}^{n+1}-0$ by mapping v into the orbit of $(v,1)$ in $(\mathbb{R}^{n+1}-0) \times \mathbb{R}$. Thus the long exact sequence of local cohomology may be written, compare 2.7

$$
\begin{array}{ccccccc}
H^p(L-0,\mathbb{F}_2) & \longrightarrow & H_p^{p+1}(L,\mathbb{F}_2) & \longrightarrow & H^{p+1}(L,\mathbb{F}_2) & \longrightarrow & H^{p+1}(L-0,\mathbb{F}_2) \\
\downarrow & & \downarrow{\scriptstyle \tau^{-1}} & & \downarrow{\scriptstyle i^*} & & \downarrow \\
H^p(\mathbb{R}^{n+1}-0,\mathbb{F}_2) & \rightarrow & H^p(P,\mathbb{F}_2) & \xrightarrow{\cup e} & H^{p+1}(P,\mathbb{F}_2) & \xrightarrow{p^*} & H^{p+1}(\mathbb{R}^{n+1}-0,\mathbb{F}_2)
\end{array}
$$

The fact that $\mathbb{R}^{n+1}-0$ is connected and $H^p(\mathbb{R}^{n+1}-0,\mathbb{F}_2) = 0$ for $p = 1,\ldots,n-1$ implies that

$$
H^p(P^n,\mathbb{F}_2) \xrightarrow{\cup e} H^{p+1}(P^n,\mathbb{F}_2)
$$

is an isomorphism for $p = 0,\ldots,n-2$ and a monomorphism for $p = n-1$. Using that $H^{n+1}(P^n,\mathbb{F}_2) = 0$ we find that

$$
H^n(\mathbb{R}^{n+1}-0,\mathbb{F}_2) \longrightarrow H^n(P^n,\mathbb{F}_2)
$$

is an epimorphism. From this first fact we find that $H^n(P^n,\mathbb{F}_2) \neq 0$ and from the second fact we conclude that $H^n(P^n,\mathbb{F}_2)$ is one dimensional over $\mathbb{F}_2$.

$$\text{Q.E.D.}$$

<u>Remark 4.2.</u> The canonical projection $p: S^n \to P^n$ induces the <u>zero</u> map in mod 2 cohomology

$$p*: H^n(P^n, \mathbb{F}_2) \longrightarrow H^n(S^n, \mathbb{F}_2)$$

as it follows from the previous proof.

As an example of application of the idea of characteristic class let us consider a theorem of <u>K. Borzuk</u> proved in 1933 on the basis of a conjecture of <u>St.Ulam</u>.

<u>Borzuk-Ulam theorem 4.3.</u> Any continuous map $f: S^n \to \mathbb{R}^n$ must identify a pair of antipodal points.

<u>Proof.</u> Suppose to the contrary that

$$f(-x) \neq f(x) \qquad ; \quad x \in S^n$$

Then the map $g: S^n \to S^{n-1}$ given by

$$g(x) = \frac{f(x) - f(-x)}{|f(x) - f(-x)|}$$

will satisfy $g(-x) = -g(x)$ for all $x \in S^n$. The formula

$$f(x) = |x|g\left(\frac{x}{|x|}\right) \qquad ; \quad x \in \mathbb{R}^{n+1}-0$$

defines a continuous map $f: \mathbb{R}^{n+1}-0 \to \mathbb{R}^n-0$, which satisfies

$$f(rx) = rf(x) \qquad\qquad ; \quad x \in \mathbb{R}^{n+1}-0$$

From this formula follows that f induces a continuous map $F: P^n \to P^{n-1}$ with the property $F*L = L$. The last fact is deduced from the commutative diagram

$$
\begin{array}{ccc}
(\mathbb{R}^{n+1}-0) \times \mathbb{R} & \xrightarrow{\ f \times 1\ } & (\mathbb{R}^n-0) \times \mathbb{R} \\
\downarrow & & \downarrow \\
\mathbb{R}^{n+1}-0 & \xrightarrow{\quad f \quad} & \mathbb{R}^n-0
\end{array}
$$

which is $\mathbb{R}*$-equivariant. Let us now remark that $F*e_{n-1} = e_n$, where e_{n-1} and e_n denote the Euler classes of the canonical bundles on $\mathbb{P}^{n-1}$ and P^n. This gives

$$e_n{}^n = F*e_{n-1}{}^n = F*0 = 0$$

which is a contradiction.

Q.E.D.

We shall present two applications of the Borzuk-Ulam theorem; the first result goes back to about 1930.

<u>Lusternik-Snirelmann theorem 4.4</u>. If S^n is covered by $n+1$ closed sets, then one of the sets contains a pair of antipodal points.

<u>Proof</u>. Let the closed sets $Z_0, Z_1, \ldots, Z_n$ cover S^n. For $i = 1, \ldots, n$ let $d(x, Z_i)$ denote the distance from x to Z_i. These n functions define a continuous map $d \colon S^n \to \mathbb{R}^n$. By 4.4 we can find $x \in S^n$ with $d(x) = d(-x)$. If $d(x, Z_i) > 0$ for $i = 1, \ldots, n$, then x and $-x$ lie in Z_0. On the other hand, if $d(x, Z_i) = 0$ for some $i > 0$, then x and $-x$ lie in Z_i, since Z_i is a closed set.

Q.E.D.

<u>Theorem 4.5</u>. Let $K_1, \ldots, K_n$ be compact subsets of $\mathbb{R}^n$. There exists a hyperplane in $\mathbb{R}^n$ that divides each of $K_1, \ldots, K_n$ into two parts of equal volume.

<u>Proof</u>. We shall identify $\mathbb{R}^n$ with the hyperplane $\mathbb{R}^n \times 0$ in $\mathbb{R}^{n+1}$. For $x \in S^n$ let $P(x)$ denote hyperplane in $\mathbb{R}^{n+1}$ through $(0, \ldots, 0, \tfrac{1}{2})$ orthogonal to x. Notice that $P(x) = P(-x)$. For $i = 1, \ldots, n$ and $x \in S^n$ let $\mu_i(x)$ denote the volume of the part of K_i which is on the same side of $P(x)$ as x. Let us now apply the Borsuk-Ulam theorem to the map

$$x \longmapsto (\mu_1(x), \ldots, \mu_n(x)) \qquad ; \quad x \in S^n$$

to find an $x \in S^n$ such that $\mu_i(x) = \mu_i(-x)$ for $i = 1, \ldots, n$. It is left to the reader to prove that $P(x) \cap \mathbb{R}^n$ solves our problem.

Q.E.D.

VIII.5   Stiefel-Whitney classes

This section is intended to give a brief introduction to Stiefel-Whitney classes. The general reference is Milnor and Stasheff (1). Recall from Bredon (1) II.21:

Steenrod squares

For a closed subspace Z of the topological space X and $i \in \mathbb{N}$ there is given a linear map "square upper i"

$$Sq^i: H_Z^p(X, \mathbb{F}_2) \longrightarrow H_Z^{p+i}(X, \mathbb{F}_2) \qquad ; \quad p \in \mathbb{Z}$$

with the following properties: For $a \in H_Z^n(X, \mathbb{F}_2)$

5.1 $\qquad\qquad Sq^0(a) = a, \quad Sq^n(a) = a \cup a, \quad Sq^i(a) = 0 \quad \text{for} \quad i > n$

As a consequence of this we write

$$Sq(a) = Sq^0(a) + Sq^1(a) + \ldots + Sq^n(a)$$

Given a continuous map $f: X \to Y$ and closed sets Z and W of X and Y with $f(X-Z) \subsetneq Y-W$,   then

5.2 $\qquad\qquad Sq(f^*a) = f^*Sq(a) \qquad\qquad ; \quad a \in H_W^\bullet(Y, \mathbb{F}_2)$

For closed subsets Z and W of X we have

5.3 $\qquad\qquad Sq(\alpha \cup \beta) = Sq(\alpha) \cup Sq(\beta) \qquad ; \quad \beta \in H_Z^\bullet(X, \mathbb{F}_2), \quad \alpha \in H_W^\bullet(X, \mathbb{F}_2)$

Stiefel Whitney classes

Let E denote a rank n vector bundle on the topological space X and $\tau \in H_X^n(E, \mathbb{F}_2)$ the Thom class. Recall that the Thom isomorphism is given by

$$\alpha \longmapsto \tau \cup \alpha \quad ; \quad H^p(X, \mathbb{F}_2) \to H^{p+n}(X, \mathbb{F}_2) \qquad ; \ p \in \mathbb{Z}$$

The total <u>Stiefel-Whitney</u> class $w(E) \in H(X, \mathbb{F}_2)$ of E is given as the image of $Sq(\tau)$ by the inverse of the Thom isomorphism, i.e.

<u>Thoms formula 5.4.</u>

$$Sq(\tau) = \tau \cup w(E)$$

From the formula 5.1 follows that we can write

$$5.5 \qquad w(E) = 1 + w_1(E) + \ldots + w_n(E) \qquad ; \ w_i(E) \in H^i(X, \mathbb{F}_2)$$

From the proof of 2.7 we see that the Euler class e satisfies $\tau \cup \tau = \tau \cup e$, consequently by 5.1

$$5.6 \qquad w_n(E) \text{ is the Euler class of the rank } n \text{ bundle } E$$

For a continuous map $f: Z \to X$ we have

$$5.7 \qquad w(f^*E) = f^*w(E)$$

as it follows from Thoms formula 5.4. - For bundles E and F on the space X we have

$$5.8 \qquad w(E \oplus F) = w(E) \cup w(F)$$

354

<u>Proof</u>. Consider the bundle $E \times F$ on $X \times X$. It suffices to prove that

$$w(E \times F) = w(E) \times w(F)$$

To see this let $\sigma \in H^m_X(E, \mathbb{F}_2)$ and $\tau \in H^n_X(F, \mathbb{F}_2)$ denote the Thom classes for E and F respectively. The Thom class of $E \times F$ is $\sigma \times \tau$ as it follows from 3.9. Using Thoms formula 5.4 and the formulas 5.2, 5.3 we get

$$Sq(\sigma \times \tau) = Sq(\sigma) \times Sq(\tau) = (\sigma \cup w(E)) \times (\tau \cup w(F)) = (\sigma \times \tau) \cup (v(E) \times w(F))$$

from which the result follows.

Q.E.D.

The formula 5.8 has the following variant

<u>Proposition 5.9</u>. Let X denote a locally compact space, countable at infinity.[*] Given an exact sequence of vector bundles

$$0 \longrightarrow E \longrightarrow F \longrightarrow G \longrightarrow 0$$

Then $w(F) = w(E) \cup w(G)$.

<u>Proof</u>. Introduce a Riemannian metric on F to obtain an iso-morphism $F \xrightarrow{\sim} E \oplus G$.

Q.E.D.

---

[*] It suffices that X is a paracompact topological space.

<u>Normal bundles</u>

Let us consider a smooth manfold X and the inclusion
i: Z → X of a closed smooth submanifold of codimension d.
We have

$$5.10 \qquad R^p i^! \, \underset{\sim}{\mathbb{F}}_2 = \begin{cases} \mathbb{F}_2 & \text{for } p = d \\ 0 & \text{for } p \neq d \end{cases}$$

as it follows by noticing that the result is in fact local, which
allows us to view i as the zero section of a microbundle. With
the notation of VIII.2 we deduce a Gysin map

$$i_* : H^p(Z, \mathbb{F}_2) \longrightarrow H^{p+d}(X, \mathbb{F}_2) \quad ; \ p \in \mathbb{Z}$$

in case X is countable at infinity we can find a diffeomorphism
of the <u>normal bundle N</u> of Z in X onto an open neighbourhood
of Z in X which transforms the zero section into i, Milnor
and Stasheff (1). It follows that i: Z → X and the zero section
Z → N have the same Euler class. In particular we get from 2.7

$$5.11 \qquad \boxed{i^* i_* \alpha = w_d(N) \cup \alpha} \qquad ; \qquad \alpha \in H^\bullet(Z, \mathbb{F}_2)$$

From this formula we can deduce a useful non imbedding criterion.

<u>Proposition 5.12</u>. Let X be an n-dimensional smooth mani-
fold which admits a closed, smooth imbedding into $\mathbb{R}^{n+d}$, d > 0.
Then

$$[w(X)^{-1}]_d = 0$$

i.e. the d'th homogeneous component of the inverse of the total

Stiefel-Whitney class of the tangent bundle of X is zero.

$\underline{Proof}$. Let N denote the normal bundle for $i: X \to \mathbb{R}^{n+k}$. From the canonical exact sequence

$$0 \longrightarrow T(X) \longrightarrow i*T(\mathbb{R}^{n+k}) \longrightarrow N \longrightarrow 0$$

we deduce from 5.9 with $w(X) = w(TX)$ that

$$w(X) \cup w(N) = 1$$

From 5.11 and the fact that $H^d(\mathbb{R}^{n+d}, \mathbb{F}_2) = 0$ we deduce that

$$w_d(N) = i*i_*1 = 0$$

Combine these two formulas to get $[w(X)^{-1}]_d = 0$.

$$Q.E.D.$$

Imbeddings of projective n-space

Recall from VIII.3 that P^n comes equipped with a canonical line bundle L. The bundle $L \otimes \mathbb{R}^{n+1}$ has a canonical section s: We can represent $L \otimes \mathbb{R}^{n+1}$ as the orbit space

$$(\mathbb{R}^{n+1}-0) \times \mathbb{R}^{n+1}/\mathbb{R}*$$

under the action $r(v,z) = (rv, rz)$. The section $s: P^n \to L \otimes \mathbb{R}^{n+1}$ is then induced by the $\mathbb{R}*$-equivariant map

$$\mathbb{R}^{n+1}-0 \longrightarrow (\mathbb{R}^{n+1}-0) \times \mathbb{R}^{n+1} ; \qquad v \longmapsto (v,v)$$

There results an exact sequence of bundles

$$5.13 \qquad 0 \longrightarrow P \times \mathbb{R} \xrightarrow{\ s\ } L \otimes \mathbb{R}^{n+1} \longrightarrow T \longrightarrow 0$$

where T is the tangent bundle on P^n. We shall calculate $w(P) = w(T)$ from this sequence using 5.9 and the fact that $w(L) = 1+e$ where e is the non trivial element of $H^1(P^n, \mathbb{F}_2)$. The result is

$$5.14 \qquad w(P^n) = (1+e)^{n+1} \qquad ; \ e \in H^1(P^n, \mathbb{F}_2), \quad e \neq 0$$

<u>Example 5.15</u>. For $n = 4$ we find that

$$w(P^n) = 1 + e + e^4$$

A simple calculation yields

$$w(P^4)^{-1} = 1 + e + e^2 + e^3$$

We deduce from 5.12 that P^n cannot be imbedded in $\mathbb{R}^7$. It is known that P^5 can be imbedded in $\mathbb{R}^8$. For additional information see Milnor and Stasheff (1).

VIII.6   Chern classes

Let V be a real d-dimensional vector space. Recall that we have a canonical isomorphism

$$H_{\{0\}}{}^d(V,\mathbb{Z}) = H_c{}^d(V,\mathbb{Z})$$

It follows from III.8.10 that a linear automorphism f of V acts on $H_{\{0\}}{}^d(V,\mathbb{Z})$ as multiplication by $\operatorname{sign}(\det(f))$.

An ordered basis $a_1,\ldots,a_d$ for V will determine a generator for $H_{\{0\}}{}^d(V,E)$ as it follows by decomposing $V = \mathbb{R}\,a_1 \oplus \ldots \oplus \mathbb{R}\,a_d$ and using 3.9. A second basis $b_1,\ldots,b_d$ will determine the same generator if and only if the transision matrix from the first basis to the second basis has positive determinant. Accordingly we shall identify an <u>orientation of V</u> with a generator for $H_{\{0\}}{}^d(V,\mathbb{Z})$.

Let us consider a complex vector space E . An ordered complex basis $e_1,\ldots,e_n$ for E give rise to an ordered real basis

$$e_1 \,,\, ie_1 \,,\, e_2 \,,\, ie_2 \,,\, \ldots \,,\, e_n \,,\, ie_n \ .$$

The resulting orientation of E is independent of the complex basis considered as it follows from the fact that a complex linear automorphism of E , considered as an automorphism of the underlying space, has positive determinant.

Let E denote a complex vector bundle of fibre dimension n. According to the discussion above E has a canonical orientation, i.e. a <u>Thom class</u>

$$6.1 \qquad\qquad \tau(E) \in H_X{}^{2n}(E,\mathbb{Z})$$

which for each $x \in X$ restricts to the complex orientation class of E_x, compare 3.3. According to 2.6 this gives rise to a characteristic class

$$6.2 \qquad\qquad e(E) \in H^{2n}(X, \mathbb{Z})$$

the <u>Euler class</u> of the complex vector bundle E.

<u>Proposition 6.3</u>. The Euler class of the tensor product of two complex line bundles K and L on the topological space X is given by

$$e(K \otimes L) = e(K) + e(L)$$

<u>Proof</u>. We shall first prove that the projections of the Whitney sum $K \oplus L$ onto its factors induce an isomorphism

$$H_X^{2}(K, \mathbb{Z}) \oplus H_X^{2}(L, \mathbb{Z}) \to H_{K \cup L}^{2}(K \oplus L, \mathbb{Z})$$

We leave it to the reader to show that the projection of $K \oplus L$ onto its factors induce isomorphisms

$$H_X^{\cdot}(K, \mathbb{Z}) \stackrel{\sim}{\to} H_L^{\cdot}(K \oplus L, \mathbb{Z}) \qquad H_X^{\cdot}(L, \mathbb{Z}) \stackrel{\sim}{\to} H_K^{\cdot}(K \oplus L, \mathbb{Z})$$

The Mayer-Vietoris sequence II.9.11 for the subspaces $K = K \times 0$ and $L = 0 \times L$ of $K \oplus L$ gives us an exact sequence

$$\to H_X^{2}(K \oplus L, \mathbb{Z}) \to H_K^{2}(K \oplus L, \mathbb{Z}) \oplus H_L^{2}(K \oplus L, \mathbb{Z}) \to H_{K \cup L}^{2}(K \oplus L, \mathbb{Z})$$

and the result follows from the fact that $H_X^{i}(K \oplus L, \mathbb{Z}) = 0$ for $i = 0, 1, 2, 3$.

The map $K \oplus L \to K \otimes L$ given by $(x,y) \mapsto x \otimes y$ induces a map on local cohomology

$$H_X^2(K \otimes L, \mathbb{Z}) \to H_{K \cup L}^2(K \oplus L, \mathbb{Z})$$

Compose this with the inverse of the isomorphism established above to get a map

$$H_X^2(K \otimes L, \mathbb{Z}) \to H_X^2(K, \mathbb{Z}) \oplus H_X^2(L, \mathbb{Z})$$

We are going to prove that this map transforms the Thom class of $K \otimes L$ into the direct sum of the Thom class of K and the Thom class of L.

According to 3.3 it suffices to treat the case where X is a point. Thus we may assume that $K = L = \mathbb{C}$ in which case we may identify three relevant local cohomology groups with $H^1(\mathbb{C}^*, \mathbb{Z})$. It suffices to prove the following.

Let p and q denote two projections of $\mathbb{C}^* \times \mathbb{C}^*$ onto its factors and $\pi: \mathbb{C}^* \times \mathbb{C}^* \to \mathbb{C}^*$ the product morphism $\pi(z,w) = zw$; $z,w \in \mathbb{C}^*$. Then

$$\pi^*\alpha = p^*\alpha + q^*\alpha \qquad ; \qquad \alpha \in H^1(\mathbb{C}^*, \mathbb{Z})$$

To prove this formula, notice that the two canonical injections $i_1, i_2 = \mathbb{C}^* \to \mathbb{C}^* \times \mathbb{C}^*$ give rise to an isomorphism

$$H^1(\mathbb{C}^* \times \mathbb{C}^*, \mathbb{Z}) \xrightarrow{\sim} H^1(\mathbb{C}^*, \mathbb{Z}) \oplus H^1(\mathbb{C}^*, \mathbb{Z})$$

as it follows from the Künneth formula. This completes the proof of the statement about the Thom class. From this it is easy to deduce the statement about the Euler class.

Let $p: E \to X$ be a complex vector bundle of rank n on the topological space X. Define

$$P(E) = E-X/\mathbb{C}^*$$
$$L(E) = (E-X) \times \mathbb{C}/\mathbb{C}^*$$

i.e. the quotient spaces under the action of the complex multiplicative group $\mathbb{C}^*$. In the latter case the action is given by

$$z(e,x) = (ze,zx) \qquad ; \ z \in \mathbb{C}^*, \ e \in E-X, \ z \in X$$

The projection of $(E-X) \in \mathbb{C}$ onto its first factor induces a map $L(E) \to P(E)$ which in fact makes $L(E)$ a line bundle on $P(E)$, the <u>canonical line bundle</u> on $P(E)$. If we let $\pi: P(E) \to X$ denote the projection and $E^\vee$ the dual bundle, evaluation defines a $\mathbb{C}^*$-equivariant map

$$(E-X) \times E^\vee \to (E-X) \times \mathbb{C} \quad ; \ (e,f) \rightarrow (e,f(e))$$

and consequently a map of bundles

6.4 $$\pi^*E^\vee \longrightarrow L(E) \longrightarrow 0$$

A different way of describing this is to start with the $\mathbb{C}^*$-equivariant map

$$(E-0) \to (E-0) \times E \quad ; \ e \longmapsto (e,e)$$

where the action on the second space is given by $r(e,v) = (re,rv)$. This defines a non vanishing section

6.5 $$s: P(E) \longrightarrow L(E) \otimes \pi^*E$$

<u>Theorem 6.6</u>. Let $p: E \to X$ denote a complex vector bundle of rank n, and $\xi \in H^2(P(E), \mathbb{Z})$ the Euler class of the canonical line bundle $L(E) \to P(E)$. Then $H^{\cdot}(P(E), \mathbb{Z})$, is a free $H^{\cdot}(X, \mathbb{Z})$-module with basis

$$1, \xi, \ldots, \xi^{n-1}$$

<u>Proof</u>. The fibre of $P(E)$ at $x \in X$ is $P(E_x)$ and the restriction of $L(E)$ to $P(E_x)$ is the canonical line bundle $L(E_x)$ on $P(E_x)$. Thus the image $\xi_x \in H^2(P(E_x), \mathbb{Z})$ of ξ is the Euler class of $L(E_x)$. It follows by a proof similar to that of 4.1 that $H^{\cdot}(P(E_x), \mathbb{Z})$ is a free $\mathbb{Z}$-module with basis $1, \xi_x, \ldots, \xi_x^{n-1}$. The theorem is now a consequence of the following Lemma.

Q.E.D.

<u>Lemma 6.7</u>. Let $f: P \to X$ be a fibre bundle whose fibres have the homotopy type of a compact space and let k denote a commutative ring. Given homogeneous elements $\alpha_1, \ldots, \alpha_n$ of $H^{\cdot}(P, \underset{\sim}{k})$ which for each $x \in X$ restricts to a basis for the k-module $H^{\cdot}(P_x, k)$. Then the $H^{\cdot}(X, k)$-module $H^{\cdot}(P, k)$ is free with basis $\alpha_1, \ldots, \alpha_n$.

<u>Proof</u>. Let $I^{\cdot}$ denote an injective resolution of $\underset{\sim}{k}$ in $\mathrm{Sh}(P, k)$. For $i = 1, \ldots, n$ let d_i denote the degree of α_i and interprete $\alpha_i \in H^{d_i}(P, k)$ as a morphism

$$\alpha_i: \underset{\sim}{k}[-d_i] \to I^{\cdot} \qquad \text{in } K^+(P, k)$$

Using that $f^*\underset{\sim}{k} = \underset{\sim}{k}$, we may by adjunction interprete α_i as a morphism

$$\alpha_i : \underset{\sim}{k}[-d_i] \to f_* I^{\cdot} \qquad \text{in} \quad K^+(X,k)$$

Let us form the direct sum of these morphisms

$$\oplus_i \underset{\sim}{k}[-d_i] \longrightarrow f_* I^{\cdot}$$

By Theorem IV.1.6 this is a quasi-isomorphism of complexes of k-sheaves on X. Now apply $R\Gamma(X,-)$ to this quasi-isomorphism to get a quasi-isomorphism

$$\oplus_i R^{\cdot}\Gamma(X,k)[-d_i] \longrightarrow \Gamma(P,I^{\cdot})$$

and the result follows by passing to cohomology.

$$Q.E.D.$$

With the notation of the previous theorem there is a unique sequence of cohomology classes

$$c_1(E),\ldots,c_n(E) \qquad ; \quad c_i(E) \in H^{2i}(X,\mathbb{Z})$$

such that the following relation holds in $H^{\cdot}(P(E),\mathbb{Z})$, here and in the following we omit the $\cup$-sign,

6.8
$$\xi^n + c_1(E)\,\xi^{n-1} + \ldots + c_{n-1}(E)\,\xi + c_n(E) = 0$$

With the conventions

$$c_0(E) = 1, \quad c_i(E) = 0 \quad \text{for} \quad i > n = \mathrm{rk}\,E$$

we have defined the __i'th Chern class__ of E

$$c_i(E) \in H^{2i}(X, \mathbb{Z}) \qquad ; \ i \in \mathbb{N}$$

and the __total Chern class__ $c.(E) \in H^{\cdot}(X, \mathbb{Z})$ of E

$$c.(E) = c_0(E) + c_1(E) + \ldots$$

Given a continuous map $f: W \to X$, then

$$6.9 \qquad\qquad c.(f^*E) = f^*c.(E)$$

as it follows rather immediately from the construction of $c.(E)$.

__Proposition 6.10.__ Given an exact sequence of bundles

$$0 \longrightarrow E \longrightarrow F \longrightarrow G \longrightarrow 0$$

on the topological space X. Then

$$c.(F) = c.(E) \cup c.(G)$$

__Proof.__ Let $\xi \in H^2(P(F), \mathbb{Z})$ denote the Euler class of $L(F)$ and consider the cohomology classes α and β in $H^{\cdot}(P(F), \mathbb{Z})$

$$\alpha = \xi^m + c_1(E)\xi^{m-1} + \ldots + c_m(E) \qquad ; \ m = \mathrm{rk}\, E$$

$$\beta = \xi^n + c_1(F)\xi^{n-1} + \ldots + c_n(F) \qquad ; \ n = \mathrm{rk}\, G$$

According to the relation 6.8 the sought for formula is equivalent to $\alpha \cup \beta = 0$. To see this, remark first that the restriction of α to $P(E)$ is zero, since $L(F)$ restricts to $L(E)$ on $P(E)$.

Second let us remark that the restriction of β to $P(F) - P(E)$ is zero: the pull back of $L(G)$ along the projection

$$P(F) - P(E) \to P(G)$$

equals the restriction of $L(F)$ to $P(F) - P(E)$. Conclusion by lemma 6.12 below.

Q.E.D.

Corollary 6.11. Let $p: F \to X$ be a rank n bundle on X. If p has a continuous section $s: X \to F$ with $s(x) \neq 0$ for all $x \in X$, then

$$c_n(F) = 0$$

Proof. This implies the existence of a short exact sequence of bundles as in 6.10 where E is the trivial bundle. Hence $c.(F) = c.(G)$.

Q.E.D.

Lemma 6.12. Let Z denote a closed subspace of the topological space X and k a commutative ring. Given $\alpha \in H^p(X,k)$ whose image in $H^p(Z,k)$ is zero and $\beta \in H^q(X,k)$ whose image in $H^q(X-Z,k)$ is zero. Then $\alpha \cup \beta = 0$ in $H^{p+q}(X,k)$.

Proof. If we let $i: Z \to X$ denote the inclusion we have $i*\alpha = 0$. In virtue of the exact sequence

$$H_Z^p(X,k) \xrightarrow{\ r\ } H^p(X,k) \longrightarrow H^p(X-Z,k)$$

we can write $\beta = r(\gamma)$ where $\gamma \in H_Z^p(X,k)$. Thus

$$\alpha \cup \beta = \alpha \cup r(\gamma) = r(\alpha \cup \gamma) = (-1)^{pq}(\gamma \cup i^*\alpha) = 0$$

where we have used the formula II.9.17.

Q.E.D.

Proposition 6.13. For a rank n vector bundle E on the topological space X, the Euler class of E equals $c_n(E)$, the top Chern class of E.

Proof. Let us first treat the case $n = 1$. From the definition 6.8 and 6.4 we find that $c_1(E) = -e(E^\vee)$, which according to 6.3 equals $e(E)$, using that $E \otimes E^\vee$ is the trivial bundle.

In the general case we can refer to the proof of 6.14 which establishes that

$$c_n(E) = \theta_n \, e(E)$$

where θ_n is a constant depending only on n. To determine this constant consider the canonical line bundle L on P^n and apply this formula to $E = L^{\oplus n}$. We get

$$e(E) = e(L)^n, \quad c_n(E) = c_1(L)^n$$

as it follows from 3.7 and 6.10 respectively. Thus we can conclude that $\theta_n = 1$.

Q.E.D.

A Chern class formula for the Thom class

Let E denote a rank n vector bundle on the topological space X. Put $\widetilde{E} = E \oplus \mathbb{C}$ and consider $P(\widetilde{E})$. The canonical map $E \to E \oplus \mathbb{C} - 0$, $e \mapsto (e,1)$ induces an imbedding $j: E \to P(\widetilde{E})$ of E onto an open subset of $P(\widetilde{E})$. We shall consider X as a closed subset of E and $P(\widetilde{E})$ through the zero section of E. This gives an excision isomorphism

$$j^*: \; H_X^{\,\cdot}(P(\widetilde{E}),\mathbb{Z}) \xrightarrow{\;\sim\;} H_X^{\,\cdot}(E,\mathbb{Z})$$

which allows us to view the Thom class of E as

$$\tau(E) \in H_X^{\,\cdot}(P(\widetilde{E}),\mathbb{Z})$$

The restriction map

$$r: \; H_X^{\,\cdot}(P(\widetilde{E}),\mathbb{Z}) \longrightarrow H^{\,\cdot}(P(\widetilde{E}),\mathbb{Z})$$

is <u>injective</u> and the following formula holds

6.14 $$r(\tau(E)) = c_n(E \otimes L(E))$$

Proof. Let us first prove that the cohomology class

$$c_n(E \otimes L(E)) \in H^{2n}(P(\widetilde{E}),\mathbb{Z})$$

has image zero in $H^{2n}(P(\widetilde{E}) - X,\mathbb{Z})$. According to 6.11 it suffices to prove that $E \otimes L(\widetilde{E})$ has a section over $P(\widetilde{E})$ which is non vanishing over $P(\widetilde{E})-X$. In fact the standard section, 6.7 of $\widetilde{E} \otimes L(\widetilde{E})$ has a projection onto $E \otimes L(\widetilde{E})$ of the required sort. This section is induced by the $\mathbb{C}^*$-equivariant map

$$E \oplus \mathbb{C}-0 \longrightarrow (E \oplus \mathbb{C}-0) \times E \qquad ; \quad (e,z) \longmapsto (e,z,e)$$

Let us exploit this in case X is a point. We conclude that $c_1(L)^n$ the free generator of $H^{2n}(P^n, \mathbb{Z})$ can be lifted back to $H_0^{2n}(P^n, \mathbb{Z})$. This group we have identified with $H_0^{2n}(\mathbb{C}^n, \mathbb{Z})$ so let $\tau \in H_0^{2n}(P^n, \mathbb{Z})$ denote the Thom class. Thus we can write

$$r(\tau) = \theta_n c_1(L) \qquad ; \quad \theta_n = \pm 1$$

The constant θ_n will be determined indirectly.

Returning to the global situation we can choose a cohomology class $\alpha \in H_X^{2n}(P(\widetilde{E}), \mathbb{Z})$ with $r(\alpha) = \theta_n c_n(E \otimes L(\widetilde{E}))$. We shall now prove that $\alpha = \tau(E)$. Using 3.3 we see that it suffies to treat the case where X is a point, which has been accomplished above. Hence

$$r(\tau(E)) = \theta_n c_n(E \otimes L(\widetilde{E}))$$

where $\theta_n = \pm 1$ is a constant which depends only on $n \in \mathbb{N}$.

If we pull the formula above back to X we get $e(E) = \theta_n c_n(E)$.

It is established during the proof of 6.13 that $\theta_n = 1$.

Let us now prove that the restriction map r mentioned in the line before 6.14 is injective. Put $c_1(L(E)) = \xi$. Then $1, \xi, \ldots, \xi^n$ form a basis for the $H^\cdot(X, \mathbb{Z})$-module $H^\cdot(P(\widetilde{E}), \mathbb{Z})$ according to 6.6, while the restriction of $1, \xi, \ldots, \xi^{n-1}$ to $P(E)-X$ form a basis for the $H^\cdot(X, \mathbb{Z})$-module $H^\cdot(P(E) - X, \mathbb{Z})$ as it follows from 6.7 and our investigations above of the case where X is a point. This proves that the map

$$H^\cdot(P(\widetilde{E}), \mathbb{Z}) \longrightarrow H^\cdot(P(\widetilde{E})-X, \mathbb{Z})$$

is surjective. The long exact sequence of local cohomology proves that r is injective.

Q.E.D.

The splitting principle 6.15. Let E be a complex vector bundle on the topological space X. There exists a continuous map $f: W \to X$ such that

1) $f^*: H^\cdot(X, \mathbb{Z}) \longrightarrow H^\cdot(W, \mathbb{Z})$ is injective.

2) f^*E admits a finite filtration by sub-bundles whose quotients are line bundles.

Proof. Repeated applications of 6.4 and 6.6.

Q.E.D.

Let us give some applications of the splitting principle. For a complex vector bundle E, we have for the dual bundle $E^\vee$

$$6.16 \qquad c_i(E^\vee) = (-1)^i c_i(E)$$

370

Proof. Let us look at the even part of $H^{\cdot}(X,\mathbb{Z})$

$$H^{ev}(X,\mathbb{Z}) = \bigoplus_{i\in\mathbb{N}} H^{2i}(X,\mathbb{Z})$$

Notice that this is a commutative ring. For this reason we shall omit the cup product sign. This ring has an automorphism $\alpha \longmapsto \bar{\alpha}$ given by

$$\bar{\alpha} = (-1)^{P}\alpha \quad ; \quad \alpha \in H^{2P}(X,\mathbb{Z})$$

In order to prove the formula 6.16 we may assume that E admits a filtration whose quotients are line bundles $L_1,\ldots,L_n$. It follows from 6.3 that $c.(L_s{}^{\vee}) = \bar{c}.(L_s)$. Using 6.10 we get

$$c.(E^{\vee}) = c.(L_1{}^{\vee})\ldots c.(L_n{}^{\vee}) = \bar{c}.(L_1)\ldots\bar{c}.(L_n) = \bar{c}.(E)$$

from which the formula 6.16 follows.

Q.E.D.

For a rank n bundle E and a line bundle L we have

6.17 $\qquad c_n(E \otimes L) = c_n(E) + c_{n-1}(E)c_1(L) + c_{n-2}(E)c_1(L)^2 + \ldots$

Proof. This will be done by induction on $n \in \mathbb{N}$. By the splitting principle we may assume that E has a sub bundle F such that $E/F = K$ is a line bundle. Notice that

$$c_i(E) = c_i(F) + c_{i-1}(F)c_1(K) \qquad ; \ i \in \mathbb{N}$$

By the induction hypothesis we have

$$c_{n-1}(F \otimes L) = c_{n-1}(F) + c_{n-2}(F)c_1(L) + \dots$$

Consequently we get

$$c_n(E \otimes L) = c_{n-1}(F \otimes L)c_1(K \otimes L) =$$

$$(c_{n-1}(F) + c_{n-2}(F)c_2(L) + \dots)(c_1(K) + c_1(L))$$

from which the formula follows.

Q.E.D.

For a complex vector bundle E of rank n, let $\bar{E}$ denote the conjugate bundle: The underlying real bundle is the same as that of E but the action of $i = \sqrt{-1}$ on $\bar{E}$ is that of $-i$ on E. We have

6.18
$$c_i(\bar{E}) = (-i)^i c_i(E) \qquad ; \ z \in \mathbb{N}$$

<u>Proof</u>. Let us notice in general that

6.19
$$\tau(\bar{E}) = (-1)^n \tau(E) \qquad ; \ n = \mathrm{rk}\, E$$

as it follows by inspection of the case where X is a point, compare the introductory remarks of this section.

Q.E.D.

VIII.7   Pontrjagin classes

Consider a real vector bundle E of rank n on the topological space X . To this we can assign the complexified
bundle $F \otimes_{\mathbb{R}} \mathbb{C}$. Let us remark that the Chern classes of this
bundle satisfy

7.1
$$2c_i(E \otimes_{\mathbb{R}} \mathbb{C}) = 0 \quad \text{for } i \text{ odd}$$

as it follows from 6.19 by noticing that the bundle $E \otimes_{\mathbb{R}} \mathbb{C}$ is
isomorphic to the conjugate bundle

$$E \otimes_{\mathbb{R}} \mathbb{C} \xrightarrow{\sim} E \otimes_{\mathbb{R}} \mathbb{C} \quad ; \ e \otimes z \longmapsto e \otimes \bar{z}$$

In consequence of 7.1 we define the <u>i'th Pontrjagin class</u> of
the real bundle E on the space X to be

7.2
$$p_i(E) = (-1)^i c_{2i}(E \otimes_{\mathbb{R}} \mathbb{C})$$

which is considered as a class in rational cohomology

7.3
$$p_i(E) \in H^{4i}(X, \mathbb{Q})$$

The <u>total Pontrjagin class</u> $p.(E) \in H^{\cdot}(X, \mathbb{Q})$ is given by

$$p(E) = p_0(E) + p_1(E) + \ldots .$$

The following properties of Pontrjagin classes follow immediately
from the properties of the Chern class

7.4
$$p_0(E) = 1, \quad p_i(E) = 0 \quad \text{for} \quad i > \tfrac{1}{2} + \text{rk } E$$

For a continuous map $f: W \to X$ we have

7.5
$$p.(f*E) = f*p.(E)$$

For an exact sequence of bundles on X

$$0 \longrightarrow E \longrightarrow F \longrightarrow G \longrightarrow 0$$

we have the product formula

7.6
$$p.(F) = p(E) \cup p.(G)$$

We shall draw attention to

<u>Hirzebruch's signature theorem 7.7</u>. Let X denote a compact oriented 4n-dimensional smooth manifold. The pairing

$$\int_M \alpha \cup \beta \quad ; \quad \alpha, \beta \in H^{2n}(M, \mathbb{R})$$

is symmetric and non-degenerated by Poincaré duality. The <u>signature of M, sign(M)</u> is by definition the signature of this quadratic form. The <u>Pontrjagin classes of M</u>, $p_1(M)$, $p_2(M)$, ... are by definition the Pontrjagin classes of the tangent bundle of M. According to Hirzebruch one has

$$n = 1: \quad \text{sign}(M) = \int_M \tfrac{1}{3} p_1(M)$$

$$n = 2: \quad \text{sign}(M) = \int_M \tfrac{1}{45} (7p_2(M) - p_2(M)^2)$$

In general there is a polynomium $L_n \in \mathbb{Q}[T_1, \ldots, T_n]$ such that

$$\text{sign}(M) = \int_M L_n(p_1, \ldots, p_n)$$

IX. Borel Moore Homology

<u>IX.1   Proper homotopy invariance</u>

Throughout this section we consider a commutative no-etherian ring k and let $K^{\cdot}$ denote an injective resolution of k in the category of k-modules. For a complex $C^{\cdot}$ of k-modules we put

$$DC^{\cdot} = \operatorname{Hom}_k^{\cdot}(C^{\cdot}, K^{\cdot})$$

For a locally compact space X we define

1.1
$$H_i(X,k) = H^{-i}DR^{\cdot}\Gamma_c(X,k) \qquad ; \ i \in \mathbb{Z}$$

the i'th <u>Borel-Moore homology group</u> with coefficients in k. A proper continuous map $f: X \to Y$ between locally compact spaces induces a morphism of complexes of k-modules

$$R^{\cdot}\Gamma_c(Y,k) \longrightarrow R^{\cdot}\Gamma_c(X,k)$$

and consequently a morphism on homology

1.2
$$f_*: H_i(X,k) \longrightarrow H_i(Y,k) \qquad ; \ i \in \mathbb{Z}$$

A second proper map $g: Y \to W$ will induce a homology map g_* subjected to the rule

1.3
$$(g \circ f)_* = g_* \circ f_*$$

i.e. Borel-Moore homology is a covariant functor from the category of locally compact spaces and proper map to the category of k-modules.

Vietoris Begle theorem 1.4. Let $f: X \to Y$ be a proper map with acyclic fibers, III.5.3. Then

$$f_*: H_i(X,k) \longrightarrow H_i(Y,k) \qquad ; \quad i \in \mathbb{Z}$$

is an isomorphism.

Proof. The morphism of complexes

$$R^{\cdot}\Gamma_c(Y,k) \longrightarrow R^{\cdot}\Gamma_c(X,k)$$

is a quasi-isomorphism by III.6.4 and the functor D preserves quasi-isomorphisms.

Q.E.D.

Homotopy invariance 1.5. Properly homotopic continuous maps $f,g: X \to Y$ induce the same maps in homology

$$f_* = g_*: H_i(X,k) \to H_i(Y,k) \qquad ; \quad i \in \mathbb{Z}$$

Proof. The same as that of III.6.6 using 1.4.

Q.E.D.

376

Proposition 1.6. Let X be a locally compact space of finite dimension n. Then

$$H_i(X,k) = 0 \qquad ; \ i > n$$

For homology with coefficients in $\mathbb{Z}$ we have

$$H_i(X,\mathbb{Z}) = 0 \qquad ; \ i < 0$$

Proof. The complex $R^{\cdot}\Gamma_c(X,k)$ may be represented by a complex $P^{\cdot}$ of projective modules with $P^i = 0$ for $i > n$. It follows that

$$DR^{\cdot}\Gamma_c(X,k) \xrightarrow{\sim} DP^{\cdot} \xrightarrow{\sim} \mathrm{Hom}(P^{\cdot},k)$$

from which the first result follows.

In case $k = \mathbb{Z}$ we get from VI.4.6 a short exact sequence of abelian groups

$$1.7 \qquad 0 \to \mathrm{Ext}(H_c^{i+1}(X,\mathbb{Z}),\mathbb{Z}) \to H_i(X,\mathbb{Z}) \to \mathrm{Hom}(H_c^{i}(X,\mathbb{Z}),\mathbb{Z}) \to 0$$

In particular for $i = -1$ we deduce that $H_{-1}(X,\mathbb{Z}) = 0$ from the fact that $H_c^{0}(X,\mathbb{Z})$ is a free abelian group VI.5.3.

Q.E.D.

IX.2   Restriction maps

Let X denote a locally compact space, $i: Z \to X$ the inclusion of a closed subspace and $j: U \to X$ the inclusion of the complement of Z in X. A soft resolution $\underset{\sim}{k} \to S^{\cdot}$ gives rise to a short exact sequence of k-sheaves

$$0 \to j_! j^* S^{\cdot} \to S^{\cdot} \to i_* i^* S^{\cdot} \to 0$$

and consequently a short exact sequence of k-modules

$$0 \to R^{\cdot}\Gamma_c(Z,k) \to R^{\cdot}\Gamma_c(X,k) \to R^{\cdot}\Gamma_c(U,k) \to 0$$

from which we deduce the exact sequence

2.1 $$\longrightarrow H_p(Z,k) \xrightarrow{\ i_*\ } H_p(X,k) \xrightarrow{\ j^{\#}\ } H_p(U,k) \longrightarrow H_{p-1}(Z,k) \longrightarrow$$

The morphism $j^{\#}$ is called the <u>restriction map</u>. Notice that if $h: O \to U$ is the inclusion of a second open subset then

2.2 $$(h \circ j)^{\#} = j^{\#} \circ h^{\#}$$

in particular $U \mapsto H_p(U,F)$ carries the structure of a <u>pre-sheaf</u> on X.

From the exact sequences in II.6 can be deduced a large number of interrelated exact sequences. Let us given an example. Given open subsets U and V of X then there is a <u>Mayer Vietoris sequence 2.3</u>.

$$H_p(U \cup V,F) \longrightarrow H_p(U,F) \oplus H_p(V,A) \longrightarrow H_p(U \cap V,F) \longrightarrow H_{p-1}(U \cup V,F)$$

IX.3   Cap products

Let Z denote a closed subspace of the locally compact space X.   We are going to introduce a <u>cap product</u>

3.1
$$\alpha \cap \xi \in H_{a-p}(Z,k) \qquad ; \ \alpha \in H_a(X,k), \xi \in H_Z^{\ p}(X,k)$$

<u>Construction</u>. Let $i: Z \to X$ denote the inclusion. We may interprete ξ as a morphism in the derived category
$\xi: i_*\underset{\sim}{k} \to \underset{\sim}{k}[p]$ or $\xi: i_*\underset{\sim}{k}[-p] \to \underset{\sim}{k}$.   From this   we derive

$$R^\cdot\Gamma_c(Z,k)[-p] \longrightarrow R^\cdot\Gamma_c(X,k)$$

Apply the functor D to this to get, compare I.11.10

$$DR^\cdot\Gamma_c(X,k) \longrightarrow D(R^\cdot\Gamma_c(Z,k)[-p]) \overset{\overset{\nu_p}{}}{\longrightarrow} DR^\cdot\Gamma_c(Z,k)[p]$$

Homology of the composed arrow is cap product with ξ

$$\cap \xi \ : \ H_a(X,k) \longrightarrow H_{a-p}(Z,k) \qquad ; \qquad \xi \in H_Z^{\ p}(X,k)$$

Let us first remark that the cap product makes $H.(X,k)$ into a <u>right module</u> over $H^\cdot(X,k)$.   Let us make a list over formulas whose proofs are left to the reader, compare II.11.11. Consider the inclusion $i: Z \to X$ of a closed subspace

3.2
$$\alpha \cap \xi \cup \eta = (\alpha \cup \xi) \cap \eta \qquad ;$$
$$\alpha \in H_a(X,k), \quad \xi \in H^p(X,k), \eta \in H_Z^{\ q}(X,k)$$

3.3
$$\alpha \cap (\xi \cup \zeta) = (\alpha \cap \eta) \cap \zeta \ ;$$
$$\alpha \in H_a(X,k), \ \eta \in H_Z^{\ q}(X,k), \ \xi \in H^r(Z,k)$$

Given a second closed subspace A of X

$$3.4 \qquad \alpha \cap (\xi \cap \eta) = (\alpha \cap \xi) \cap i^*\eta$$

$$\alpha \in H_a(X,k), \quad \xi \in H_Z^p(X,k), \quad \eta \in H_A^q(X,k)$$

Let $j: U \to X$ denote the inclusion of the complement of Z in X. Then we have two commutative diagrams

3.5 where $\alpha \in H_a(X,k)$

$$
\begin{array}{ccccccc}
H_Z^p(X,k) & \longrightarrow & H^p(X,k) & \xrightarrow{\;j^*\;} & H^p(U,k) & \longrightarrow & H_Z^{p+1}(X,k) \\
\downarrow{\scriptstyle \alpha\cap} & & \downarrow{\scriptstyle \alpha\cap} & & \downarrow{\scriptstyle \alpha\cap} & & \downarrow{\scriptstyle \alpha\cap} \\
H_{a-p}(Z,k) & \longrightarrow & H_{a-p}(X,k) & \xrightarrow{\;j^{\#}\;} & H_{a-p}(U,k) & \longrightarrow & H_{a-p-1}(Z,k)
\end{array}
$$

3.6 where $\zeta \in H^p(X,k)$

$$
\begin{array}{ccccccc}
H_a(Z,k) & \longrightarrow & H_a(X,k) & \xrightarrow{\;j^{\#}\;} & H_a(U,k) & \longrightarrow & H_{a-1}(Z,k) \\
\downarrow{\scriptstyle \cap\, i^*\xi} & & \downarrow{\scriptstyle \cap\, \xi} & & \downarrow{\scriptstyle \cap\, j^*\xi} & & \downarrow{\scriptstyle \cap\, \xi(-1)^p} \\
H_{a-p}(Z,k) & \longrightarrow & H_{a-p}(X,k) & \xrightarrow{\;j^*\;} & H_{a-p}(U,k) & \longrightarrow & H_{a-p-1}(Z,k)
\end{array}
$$

<u>Projection formula</u>. Let $f: X \to Y$ denote a proper map. For a closed subspace B of Y and $A = f^{-1}(B)$ we have

$$3.7 \qquad f_*(\alpha \cap f^*\eta) = f_*(\alpha) \cap \eta \qquad ;\, \alpha \in H_a(X,k), \quad \eta \in H_B^q(Y,k)$$

a formula valid in $H_{a-q}(B,k)$.

IX.4   Poincaré duality

Let k denote a noetherian ring and X a locally compact space of finite dimension n . If we let $\mathcal{D}^{\cdot}$ denote the dualizing complex on X we get from Poincaré duality VI.2.1 that

$$4.1 \qquad H_p(X,k) \xrightarrow{\sim} H^{-p}\Gamma(X,\mathcal{D}^{\cdot})$$

For the inclusion $j: U \to X$ of an open subset of X the restriction $j^*\mathcal{D}^{\cdot}$ of $\mathcal{D}^{\cdot}$ to U is a dualizing complex for U , VI.2.7, and the following diagram is commutative

$$4.2 \qquad
\begin{array}{ccc}
H_p(X,k) & \xrightarrow{\ \sim\ } & H^{-p}\Gamma(X,\mathcal{D}^{\cdot}) \\
\downarrow{\scriptstyle j^{\#}} & & \downarrow{\scriptstyle r_{U,X}} \\
H_p(U,k) & \xrightarrow{\ \sim\ } & H^{-p}\Gamma(U,\mathcal{D}^{\cdot})
\end{array}$$

In particular for $p = n$ we get

$$4.3 \qquad H_n(U,k) = \Gamma(U, H^{-n}\mathcal{D}^{\cdot})$$

which shows that the presheaf $U \mapsto H_n(U,k)$ is a __sheaf__. In case X is a topological manifold with orientation sheaf $o\hbar_X$

$$4.4 \qquad H_n(U,k) = \Gamma(U, o\hbar_X)$$

In case X is an oriented topological n-manifold with orientation $\mu_X: \underset{\sim}{k} \xrightarrow{\sim} o\hbar_X$ the formula 4.4 with $U = X$ gives us

$$4.5 \qquad H_n(X,k) = \operatorname{Hom}_k(\underset{\sim}{k}, o\hbar_X)$$

which allows us

to interpret the orientation as a <u>fundamental homology class</u>

4.6
$$\mu_X \in H_n(X,k)$$

<u>Theorem 4.7</u>. Let X be an oriented topological n-manifold with orientation class μ_X. For any closed subset Z of X cap product induces an isomorphism

$$\mu_X \cap : H_Z^p(X,k) \to H_{n-p}(Z,k) \qquad ; \; p \in \mathbb{Z}$$

<u>Proof</u>. Alexander duality VI.6.1 formulated there for integral cohomology has an obvious generalization.

$$\text{Q.E.D.}$$

<u>Thom class 4.8</u>. With the notation of 4.7 let $j: U \to X$ denote the inclusion of an open subset of X. The restriction of μ_X to U defines an orientation of U with fundamental homology class $j^{\#}\mu_X.$, VI.2.7. It follows that cap product induces an isomorphism

$$j^{\#}\mu_X \cap : H_{Z \cap U}^p(U,k) \to H_{n-p}(U \cap Z)$$

In particular if Z is a closed submanifold of X of codimension p we find that the sheaf VIII.1.3 $\quad U \longmapsto H_{Z \cap U}^p(U,k)$ is constant with <u>Thom class</u> $\tau_Z \in H_Z^p(X,k)$ if and only if Z is orientable with fundamental homology class $\mu_Z \in H_{n-p}(Z,k)$ given by

4.9
$$\boxed{\mu_Z = \mu_X \cap \tau_Z}$$

IX.5 Cross products

In this section we consider exclusively locally compact spaces of <u>finite dimension</u>. The commutative ring k is still assumed to be noetherian.

Let us first work out a convenient way of representing the complex $DR^{\cdot}\Gamma_c(X,k)$. Choose a quasi-isomorphism $P^{\cdot} \to R^{\cdot}\Gamma_c(X,k)$ where $P^{\cdot}$ is a bounded above complex of projective k-modules, and put

$$5.1 \qquad C.(X,k) = \mathrm{Hom}_k(P^{\cdot},k) = P^{\cdot V}$$

Since k is noetherian, VI.1.4, this is a <u>bounded below complex of flat k-modules</u>, uniquely determined up to homotopy with the property

$$5.2 \qquad H_p C.(X,k) = H_p(X,k) \qquad ; \; p \in \mathbb{Z}$$

as it follows from the quasi-isomorphism

$$\mathrm{Hom}^{\cdot}(P^{\cdot},k) \xrightarrow{\sim} \mathrm{Hom}^{\cdot}(P^{\cdot},I^{\cdot}) \xleftarrow{\sim} \mathrm{Hom}^{\cdot}(R^{\cdot}\Gamma_c(X,k),I^{\cdot})$$

<u>Remark 5.3</u>. In case $H_c^p(X,k)$ is finitely generated for each $p \in \mathbb{Z}$, the complex $P^{\cdot}$ in 5.1 may be chosen to be a bounded above complex of finitely generated free k-modules, as it follows by a procedure dual to the one used in I.6.

As a consequence, in case $H_c^p(X,k)$ is finitely generated for all $p \in \mathbb{Z}$, and $k \to l$ is an extension of rings, we have

$$5.4 \qquad C.(X,k) \otimes_k l \xrightarrow{\sim} C.(X,l)$$

as it follows from VI.5.2.

Given two locally compact spaces X and Y, choose bounded above complexes of projective k-modules $P^{\boldsymbol{\cdot}}$ and $Q^{\boldsymbol{\cdot}}$ quasi-isomorphic to $R^{\boldsymbol{\cdot}}\Gamma_c(X,k)$ and $R^{\boldsymbol{\cdot}}\Gamma_c(Y,k)$ respectively. Consider the pairing

$$5.5 \qquad \lambda \colon \operatorname{Hom}^{\boldsymbol{\cdot}}(P^{\boldsymbol{\cdot}},k) \otimes \operatorname{Hom}^{\boldsymbol{\cdot}}(Q^{\boldsymbol{\cdot}},k) \to \operatorname{Hom}^{\boldsymbol{\cdot}}(P^{\boldsymbol{\cdot}} \otimes Q^{\boldsymbol{\cdot}},k)$$

given in terms of the evaluation symbol $<,>$

$$p \otimes q \longmapsto (-1)^{rs}<\alpha,p><\beta,q>$$

where $\alpha \in \operatorname{Hom}^r(P^{\boldsymbol{\cdot}},k) = \operatorname{Hom}(P^{-r},k)$ and $\beta \in \operatorname{Hom}^s(Q^{\boldsymbol{\cdot}},k) = \operatorname{Hom}(Q^{-s},k)$. According to the Künneth formula for cohomology with compact support VII.2.8 we have

$$R^{\boldsymbol{\cdot}}\Gamma_c(X \times Y,k) \overset{\sim}{\to} P^{\boldsymbol{\cdot}} \otimes Q^{\boldsymbol{\cdot}}$$

Thus the pairing 5.5 defines a map of complexes

$$5.6 \qquad C.(X,k) \otimes C.(Y,k) \longrightarrow C.(X \times Y,k)$$

<u>Definition 5.7</u>. The transformation 5.6 induces a pairing, cross product or exterior product

$$\alpha \times \beta \in H_{p+q}(X \times Y,k) \; ; \quad \alpha \in H_p(X,k), \beta \in H_q(Y,k).$$

<u>Künneth Formula 5.8</u>. Let X and Y be locally compact spaces of finite dimension. If $H_c^p(X,k)$ is finitely generated for all $p \in \mathbb{Z}$, then

$$C.(X,k) \otimes C.(Y,k) \longrightarrow C.(X \times Y,k)$$

is a homotopy equivalence.

Proof. Let us use the notation of 5.1. According to 5.3 we may assume that $P^{\cdot}$ is a bounded above complex of finitely generated free k-modules in which case the map 5.5 is an isomorphism. Conclusion by the formula preceding 5.6.

$$Q.E.D.$$

Formulas for the cross products

All spaces considered here are locally compact spaces of finite dimension. Let A and B be closed subspaces of X and Y and $\alpha \in H_a(X,k)$, $\beta \in H_b(Y,k)$, $\xi \in H_A^P(X,k)$, $\eta \in H_B^q(Y,k)$

5.9
$$(\alpha \times \beta) \cap (\xi \times \eta) = (-1)^{aq+pq} (\alpha \cap \xi) \times (\beta \cap \eta)$$

For proper continuous maps $f: X \to Y$ and $g: Z \to W$

5.10
$$(f \times g)_*(\alpha \times \beta) = (f_*\alpha) \times f_*(\beta) \qquad ; \alpha \in H.(Z,k), \alpha \in H.(Y,k)$$

For inclusions $j: U \to X$ and $h: V \to Y$ of open subsets we have

5.11
$$(j \times h)^{\#}(\alpha \times \beta) = (j^{\#}\alpha) \times (h^{\#}\beta) \qquad ; \alpha \in H.(X,k), \beta \in H.(Y,k)$$

Let $\sigma: X \times Y \to Y \times X$ be inversion of the two factors. Then

5.12
$$\sigma_*(\alpha \times \beta) = (-1)^{pq}\alpha \times \beta \qquad ; \alpha \in H_p(X,k), \beta \in H_q(Y,k)$$

$\underline{\text{Proof of 5.9}}$. With the notation above we put $X^{\cdot} = R^{\cdot}\Gamma_c(X,k)$, $Y^{\cdot} = R^{\cdot}\Gamma_c(Y,k)$, $A^{\cdot} = R^{\cdot}\Gamma_c(A,k)$, $B^{\cdot} = R^{\cdot}\Gamma_c(B,k)$. These are all thought of as bounded above complexes of projective k-modules. We let $\xi\colon A^{\cdot}[-p] \to X^{\cdot}$ and $\eta\colon B^{\cdot}[-q] \to Y^{\cdot}$ denote the morphisms induced by the two local cohomology classes. The result follows by chasing the homology class $\alpha \otimes \beta$ through the following commutative diagram

$$
\begin{array}{ccccccc}
\alpha \otimes \beta & & (\alpha\cap\xi)\otimes(\beta\cap\eta) & & (-1)^{aq}(\alpha\cap\xi)\otimes(\beta\cap\eta) & & \\[4pt]
X^{\cdot\vee}\otimes Y^{\cdot\vee} & \xrightarrow{\ \xi^{\vee}\otimes\eta^{\vee}\ } & A^{\cdot}[-p]^{\vee}\otimes B^{\cdot}[-q]^{\vee} & \xrightarrow{\ \nu_p\otimes\nu_q\ } & A^{\cdot\vee}[p]\otimes B^{\cdot\vee}[q] & \xrightarrow{\ \tau_{p,q}\ } & A^{\cdot\vee}\otimes B^{\cdot\vee}[p+q] \\[4pt]
\ \downarrow{\lambda} & & \ \downarrow{\lambda} & & & & \ \downarrow{(-1)^{p,q}\lambda\,[p+q]} \\[4pt]
(X^{\cdot}\otimes Y^{\cdot})^{\vee} & \xrightarrow{\ (\xi\otimes\eta)^{\vee}\ } & (A^{\cdot}[-p]\otimes B^{\cdot}[-q])^{\vee} & \xrightarrow{\ \tau^{\vee}_{p,q}{}^{-1}\ } & (A^{\cdot}\otimes B^{\cdot}[-p-q])^{\vee} & \xrightarrow{\ \nu^{\vee}_{p+q}\ } & (A^{\cdot}\otimes B^{\cdot})^{\vee}[p+q] \\[4pt]
\alpha \times \beta & & & & & & (\alpha\times\beta)\cap(\xi\times\eta)
\end{array}
$$

The symbols written above and below the diagram indicate the result of chasing $\alpha \otimes \beta$.

$$\text{Q.E.D.}$$

$\underline{\text{Cap product as a left module structure 5.13}}$. The general rule for converting a cap product from a right module structure to a left module structure is

$$\xi \cap \alpha = (-1)^{ap}\alpha \cap \xi \quad ; \quad \alpha \in H_a(X,k), \ \xi \in H^p(X,k)$$

Notice, this does not involve any changes of the product structure in cohomology.

IX.6 Diagonal class of an oriented manifold

In this section, we shall be concerned with topological manifolds oriented relative to a fixed noetherian ring k .

Theorem 6.1. Let X and Y be oriented topological manifolds with orientation classes μ_X and μ_Y . Then $X \times Y$ is oriented with orientation class $\mu_X \times \mu_Y$.

Proof. In virtue of the formula 5.11 the problem is local on X and Y thus we may assume $X = \mathbb{R}^p$ and $Y = \mathbb{R}^q$. Let us remark that a homology class $\mu \in H_p(\mathbb{R}^p,k)$ is an orientation if and only if we can find $\xi \in H_{\{0\}}^p(\mathbb{R}^p,k)$ such that $\mu \cap \xi = 1$. With this remark in hand the result follows from formula 5.9.

Q.E.D.

By the product of two oriented toplogical manifolds (X,μ_X) and (Y,μ_Y) we understand the oriented manifold $(X \times Y,\mu_X \times \mu_Y)$.

Let (X,μ_X) be an oriented n-dimensional toplogical manifold. The class

$$6.2 \qquad \tau_X \in H_X^n(X \times X,k)$$

given by 4.9, $\mu_X \times \mu_X \cap \tau_X = \mu_X$ is called the __Thom class__ of X . The image of τ_X by restriction

$$6.3 \qquad \Delta_X \in H^n(X \times X,k)$$

is called the __diagonal class__ of X . The pull back of Δ_X along the diagonal map $\delta_X: X \to X \times X$ is called the <u>Euler class</u> of X

6.4 $$e_X \in H^n(X,k)$$

Let us note that

6.5 $$\Delta_X \cup (1 \times \alpha) = \Delta_X \cup (\alpha \times 1) \qquad ; \ \alpha \in H^\cdot(X,k)$$

<u>Proof</u>. Using the extraordinary cup product and the formula II.9.17 it suffices to prove the more precise result

$$\tau_X \cup \delta_X^*(1 \times \alpha) = \tau_X \cup \delta_X^*(\alpha \times 1)$$

which follows from II.10.7 since $pr_1^*\alpha = \alpha \times 1$ and $pr_2^*\alpha = 1 \times \alpha$.

Q.E.D.

Compact manifolds

A compact manifold X has an <u>Euler characteristic</u> $\chi_X \in \mathbb{Z}$ given by

6.6 $$\sum_i (-1)^i \dim_k H^i(X,k) = \chi_X$$

for any field k. The expression on the left hand side is independent of the field k since there exists a bounded complex $L^\cdot$ of finitely generated free groups such that for any field k

6.7 $$H^i(X,k) = H^i(L^\cdot \otimes_{\mathbb{Z}} k) \qquad ; \ i \in \mathbb{Z}$$

as it follows from VI.5.2, compare 5.3. In case X is oriented
relative to the field k we have

6.8
$$\int_X e_X = \int_X \Delta_X \cup \Delta_X = \chi_X 1 \quad ; \quad 1 \in k$$

where the second formula follows from Lefschetz fixed point
formula 8.5 and the second formula from the more general formula

6.9
$$\int_X \delta_X^* \zeta = \int_{X \times X} \Delta_X \cup \zeta \qquad ; \qquad \zeta \in H^n(X \times X, k)$$

which is a special case of 8.4.

<u>Corollary 6.10</u>. An odd dimensional compact manifold orientable
relative to $\mathbb{Z}$ has $\chi_X = 0$.

<u>Proof</u>. If n denotes the dimension of X then $(-1)^n e_X = e_X$
according to VIII.2.7.

Q.E.D.

<u>Caution against signs 6.11</u>. Let X denote an n-dimensional
manifold and let $\sigma: X \times X \to X \times X$ reverse the two factors. Then
the action of σ on $H_X^n(X \times X, \mathbb{Z})$ is multiplication by $(-1)^n$.
It suffices to treat the case $X = \mathbb{R}^n$.   In that case it is convenient to transport the problem by means of the automorphism
$\begin{pmatrix} 1 & 1 \\ 1 & -1 \end{pmatrix}$ of $\mathbb{R}^n \times \mathbb{R}^n$. According to VIII.3.8 we have

$$H_{\mathbb{R}^n \times \{0\}}^n(\mathbb{R}^n \times \mathbb{R}^n, \mathbb{Z}) \xrightarrow{\sim} H_{\{0\}}^n(\mathbb{R}^n, \mathbb{Z})$$

The result follows from the fact that the action of $x \mapsto -x$ on
$H_{\{0\}}^n(\mathbb{R}^n, \mathbb{Z})$ is multiplication by $(-1)^n$.

<u>Local Thom class</u>. Let X denote an oriented topological manifold with orientation μ_X. For a point $x \in X$, the submanifold $\{x\}$ has a canonical orientation μ_x. In accordance with 4.9 let $\tau_x \in H_{\{x\}}^n(X,k)$ denote the <u>local Thom class</u> characterized by

$$6.12 \qquad \qquad \mu_x = \mu_X \cap \tau_x$$

Let $i_x : X \to X \times X$ denote the map given by $i_x(z) = (z,x)$. Then we have

$$6.13 \qquad \qquad \tau_x = i_x^*(\tau_X)$$

<u>Proof</u>. This follows from the computation

$$\mu_X \times \mu_x \cap \tau_X = (-1)^n (\mu_X \times \mu_x \cap 1 \times \tau_x) \cap \tau_X =$$

$$(\mu_X \times \mu_x \cap \tau_X) \cap 1 \times \tau_x = \mu_X \cap (1 \times \tau_x) = \mu_x$$

where we have used 6.12, 5.9 and 3.4.

$$\text{Q.E.D.}$$

<u>Corollary 6.14</u>. The image of $\tau_x \in H_{\{x\}}^n(X,k)$ in $H^n(X,k)$ is $i_x^* \Delta$.

IX.7   Gysin maps

In this section we consider exclusively topological manifolds oriented relative to a fixed commutative noetherian ring k. Given a proper continuous map $f: X \to Y$ where X has dimension m and Y has dimension n. Then we define

$$7.1 \qquad f_*: H^p(X,k) \longrightarrow H^{n-m+p}(Y,k) \qquad ; \; p \in \mathbb{Z}$$

by Poincaré duality via the formula

$$7.2 \qquad \mu_Y \cap f_*(\xi) = f_*(\mu_X \cap \xi) \qquad ; \qquad \xi \in H(X,k)$$

It follows immediately from 3.7 that

Projection formula 7.3.

$$f_*(\xi \cup f^*\eta) = f_*(\xi) \cup \eta$$

where $\xi \in H^\cdot(X,k)$ and $\eta \in H^\cdot(Y,k)$. Given a second proper continuous map $g: Y \to Z$ then we have

$$7.4 \qquad (g \circ f)_* = g_* \circ f_*$$

In case $i: Z \to X$ is the inclusion of an oriented closed submanifold and $\tau_Z \in H_Z^\cdot(X,k)$ denotes the Thom class, 4.8

$$7.5 \qquad i_*(\zeta) = r(\tau_Z \cap \zeta) \qquad ; \; \zeta \in H^\cdot(Z,k)$$

where $r: H_Z^\cdot(X,k) \to H^\cdot(X,k)$ denotes the restriction.

$\underline{\text{Proof}}$. The formula is equivalent to

$$\mu_X \cap (r(\tau_Z \cap \zeta)) = i_*(\mu_Z \cap \zeta)$$

which can be deduced from, compare 4.9 and 3.3

$$\mu_X \cap (\tau_Z \cap \zeta) = \mu_Z \cap \zeta$$

using the first square in 3.5.

Q.E.D.

In particular, if we let $cl^X(Z) \in H^{\cdot}(X,k)$ denote the image of the Thom class, then

7.6
$$i_*(1) = cl^X(Z)$$

$\underline{\text{Example 7.7}}$. In case f is an orientation preserving iso-morphism then $f_* = f^{*-1}$. According to 7.2 we have $f_*(1) = 1$ and the projection formula gives

$$f_*(f^*(\eta)) = f_*(1 \cup f^*(\eta)) = f_*(1) \cup \eta = \eta$$

In particular $f_*(\alpha \cup \beta) = f_*(\alpha) \cup f_*(\beta)$.

$\underline{\text{Example 7.8}}$. Suppose X is a compact oriented n-manifold. Then the map $p: X \to Pt$ is proper and we have

$$p_* \xi = \int_X \xi \qquad\qquad ; \ \xi \in H^n(X,k)$$

392

Example 7.9. Let X and Y be oriented compact manifolds
and f: X → Y a proper continuous map. Then

$$\int_Y f_*(\xi) \cup \eta = \int_X \xi \cup f^*\eta$$

where $\xi \in H^\cdot(X,k)$, $\eta \in H^\cdot(Y,k)$.

To see this let p: X → Pt and q: Y → Pt denote the
canonical maps. From 7.4 and 7.8 follows that

$$\int_Y f_*(\xi \cup f^*\eta) = \int_X \xi \cup f^*\eta$$

and finally apply the projection formula.

Example 7.10. Let f: X → Y and q: Z → W be proper
continuous maps. Then with x = dim X, z = dim Z, w = dim W

$$(f \times g)_*(\xi \times \eta) = f_*(\xi) \times f_*(\eta) (-1)^{(p+x)(z+w)}$$

where $\xi \in H^p(X,k)$ and $\eta \in H^q(Y,k)$, as it follows from 5.9.

Example 7.11. Consider the projection

$$pr: X \times Y \to Y$$

in case where X is compact. Then

$$pr_*(\xi \times \eta) = (\int_X \xi) \eta \qquad ; \quad \xi \in H^\cdot(X,k), \eta \in H^\cdot(Y,k)$$

as it follows by a combination of 7.8 and 7.10.

Proposition 7.12. Let X and Y denote oriented topological manifolds and $i: Z \to X$ the inclusion of an oriented closed submanifold. Then the following diagram is commutative

$$
\begin{array}{ccc}
H^{\boldsymbol{\cdot}}(Z \times Y, k) & \xrightarrow{\ (i \times 1)_*\ } & H^{\boldsymbol{\cdot}}(X \times Y, k) \\
\Big\uparrow{\scriptstyle p_Z{}^*} & & \Big\uparrow{\scriptstyle p_X{}^*} \\
H^{\boldsymbol{\cdot}}(Z, k) & \xrightarrow{\ i_*\ } & H^{\boldsymbol{\cdot}}(X, k)
\end{array}
$$

Proof. Let $\tau \in H_Z{}^{\boldsymbol{\cdot}}(X, k)$ denote the Thom class. Recall that $\mu_X \cap \tau = \mu_Z$ and notice that

$$
(\mu_X \times \mu_Y) \cap (\tau \times 1_Y) = \mu_Z \times \mu_Y
$$

which proves that $p_X{}^*(\tau) \in H_{Z \times Y}{}^{\boldsymbol{\cdot}}(X \times Y, k)$ is the Thom class of $Z \times Y$. The result follows from formula 8.5.

IX.8   Lefschetz fixed point formula

Let $f: X \to Y$ be a continuous map between compact topological manifolds of dimension m and n, both oriented relative to the commutative noetherian ring k. The graph of f is denoted Γ and the Thom class of Γ in $X \times Y$ is denoted

$$8.1 \qquad \tau_\Gamma \in H_\Gamma^n(X \times Y, k)$$

The image of the Thom class is denoted

$$8.2 \qquad \Gamma_f \in H^n(X \times Y, k)$$

The action of f on cohomology can be described by means of Γ_f via the formula

$$8.3 \qquad f^*(\eta) = p_*(\Gamma_f \cup q^*\eta) \qquad\qquad ; \; \eta \in H^\cdot(Y, k)$$

where $p: X \times Y \to X$ and $q: X \times Y \to Y$ denote the projections.

Proof. Let $\gamma_f: X \to X \times Y$ denote the graph map. From the formula $f = q \circ \gamma_f$ follows that it suffices to prove that

$$\gamma_f^*(\zeta) = p_*(\Gamma_f \cup \zeta) \qquad ; \; \zeta \in H^\cdot(X \times Y, k)$$

Using the projection formula we get

$$\gamma_f^*\zeta = p_* \gamma_{f*} \gamma_f^* \zeta = p_*(\gamma_{f*}(1) \cap \zeta)$$

and the result follows from $\gamma_{f*}(1) = \Gamma$.

$$\text{Q.E.D.}$$

In case X and Y are compact oriented manifolds we deduce from 7.6 and 7.9 the formula

$$8.4 \qquad \int_X \gamma_f^* \zeta = \int_{X \times Y} \Gamma_f \cup \zeta \quad ; \zeta \in H^\cdot(X \times Y, k)$$

For a continuous self map $f: X \to X$ of a compact oriented toplogical manifold we have in case k is a field

Lefschetz fixed point formula 8.5.

$$\int_{X \times X} \Gamma_f \cup \Delta = \sum (-1)^p \text{ trace } H^p(f,k)$$

Proof. Let us choose a basis (β_i) for $H^\cdot(Z,k)$ consisting of homogeneous elements. The dual basis will be denoted (α_j), i.e.

$$8.6 \qquad \int_X \alpha_i \cup \beta_j = \delta_{ij}$$

Let us prove that the diagonal class is given by

$$8.7 \qquad \Delta = \sum (-1)^{dq\beta_i} \alpha_i \times \beta_i$$

To do this let us use the Künneth theorem to write $\Delta = \sum \lambda_i \times \beta_i$, $\lambda_i \in H^\cdot(X,k)$. Let us apply q_* to the formula 6.5 to get

$$q_*(\Delta \cup q^*(\alpha)) = \sum q_*(\lambda_i \times \beta_i \cup \alpha \times 1)$$

or using the projection formula

$$q_*(\Delta) \cup \alpha = \sum (-1)^{dg\beta_i dg\alpha} q_*(\lambda_i \cup \alpha) \times \beta_i)$$

396

Notice that $q_*(\Delta) = q_*(\delta_{X*}1) = 1$. Thus by 7.11

$$\alpha = \sum (-1)^{dg\beta_i dg\alpha} \int_X (\lambda_i \cup \alpha)\beta_i$$

or with $\alpha = \beta_j$

$$(-1)^{dq\beta_i dq\beta_j} \int_X (\lambda_i \cup \beta_j) = \delta_{ij}$$

which gives $\lambda_i = (-1)^{dq\beta_i}\alpha_i$ and proves 8.7.

Let (r_{ij}) be the matrix for f with respect to the basis (β_i): $f*(\beta_i) = \sum_j r_{ij}\beta_j$. Using 8.7 we get

$$\int_X \Upsilon_f^*\Delta = \int_X \sum_i (-1)^{dg\beta_i}\Upsilon_f^*(\alpha_i \times \beta_i)$$

$$= \sum_i (-1)^{dg\beta_i} \int_X \alpha_i \cup f*\beta_i$$

$$= \sum_i (-1)^{dg\beta_i} \sum_j r_{ij} \int_X \alpha_i \cup \beta_j$$

and the result follows from 8.6 and 8.4.

Q.E.D.

<u>Corollary 8.8</u>. If $\sum (-1)^i$ trace $H^i(f,k) \neq 0$ then $f: X \to X$ has a fixed point.

<u>Proof</u>. With the notation of 6.2 and 8.1 the class $\Gamma_f \cup \Delta$, is the image of the local cohomology class

$$\tau_f \cup \tau_X \in H_{\Gamma\cap X}^A(X \times X, k)$$

which is zero in case $\Gamma \cap X = \emptyset$.

Q.E.D.

IX.9   Wu's formula

Let X denote a compact toplogical n-manifold. We shall work with $\mathbb{F}_2$-coeffcients in which case X has a unique orientation μ_X and a Thom class

$$9.1 \qquad \tau_X \in H_X{}^n(X \times X, \mathbb{F}_2)$$

Let us introduce the total Stiefel-Whitney class of X

$$9.2 \qquad w(X) \in H^\cdot(X, \mathbb{F}_2)$$

characterized by the formula

$$9.3 \qquad \tau_X \cup w(X) = Sq(\tau_X)$$

Poincaré duality allows us to introduce the <u>Wu class</u> $v(X)$ characterized by

$$9.4 \qquad \int_X v(X) \cup \xi = \int_X Sq(\xi) \quad ; \quad \xi \in H^\cdot(X, \mathbb{F}_2)$$

<u>Wu's formula 9.5.</u> $\qquad \boxed{Sq(v(X)) = w(X)}$

<u>Proof</u>. Let $p: X \times Y \to X$ and $q: X \times Y \to Y$ denote projections. Apply p_* to $\qquad \Delta \cup (v(X) \times 1) = \Delta \cup (1 \times v(X))$ and use $\Delta = \Sigma\, \alpha_i \cup \beta_i$, compare 8.7, to get

$$v(X) = \Sigma\, \alpha_i \int \beta_i \cup v(X) = \Sigma\, \alpha_i \int Sq(\beta_i)$$

From formula 9.3 we deduce using II.9.17

$$\Delta_X \cup (w(X) \times 1) = Sq(\Delta_X)$$

which by transformation by p_* yields $\quad w(X) = \Sigma\, Sq(\alpha_i) \int Sq(\beta_i)$

Compare the expressions for $w(X)$ and $v(X)$ to get the result.

$$\text{Q.E.D.}$$

IX.10 Preservation of numbers

For a compact topological space X we have a degree map

10.1
$$\deg: H_0(X,k) \longrightarrow k$$

which may be identified with p_* where $p: X \to Pt$ is the projection onto a point.

Consider a proper map $f: V \to T$ where T is an oriented connected manifold of dimension d. For $t \in T$ let $\tau_t \in H_{\{t\}}^d(T,k)$ denote the local Thom class. For $\xi \in H_d(X,k)$ put

10.2
$$\xi_t = \xi \cap \tau_t \qquad ; \ t \in T$$

Notice that $\xi_t \in H_0(f^{-1}(t),\mathbb{Z})$ and that $f^{-1}(t)$ is compact, thus we may calculate $\deg \xi_t \in k$

10.3
$$\boxed{\deg \xi_t \ \text{ is independent of } \ t \in T}$$

Proof. Let $p: f^{-1}(t) \longrightarrow \{t\}$ denote the projection. According to the projection formula 3.7

$$p_*(\xi_t) \ = \ p_*(\xi \cap \tau_t) \ = \ p_*(\xi) \cap \tau_t.$$

Since $p_*(\xi)$ is a multiple of μ_T, it suffices to prove that

10.4
$$\deg \mu_T \cap \tau_t \ \text{ is independent of } \ t \in T.$$

Since the problem is local on T, it suffices to treat the case $T = S^d$. The result follows from the fact that the image of τ_t in $H^d(S^d,k)$ is independent of t, compare 6.13.

Q.E.D.

IX.11   Trace maps in homology

Given locally compact spaces X and Y and a covering $f: X \to Y$ of degree n, compare VII.4. An injective resolution $\underset{\sim}{k} \to G^\cdot$ in $Sh(Y,k)$ gives rise to two morphisms VII.4.9

11.1
$$G^\cdot \xrightarrow{\text{res}} f*f_*G^\cdot \xrightarrow{\text{tr}} G^\cdot$$

to which we can apply $\Gamma_c(Y,-)$ to get morphisms

11.2
$$R^\cdot\Gamma_c(Y,k) \xrightarrow{\text{res}} R^\cdot\Gamma_c(X,k) \xrightarrow{\text{tr}} R^\cdot\Gamma_c(Y,k)$$

The dual of these induce homology maps

11.3
$$H.(Y,k) \xrightarrow{\text{tr}} H.(X,k) \xrightarrow{\text{res}} H.(Y,k)$$

11.4
$$\text{res} \circ \text{tr} = n \qquad ; \ f_* = \text{res}$$

For a closed subset B of Y put $A = f^{-1}(B)$ to get

11.5
$$\text{tr}(\beta \cap \eta) = \text{tr}(\beta) \cap f*(\eta) \ ; \ \beta \in H.(Y,k), \eta \in H_B^\cdot(Y,k)$$

Let $j: V \to Y$ denote the incluson of an open subspace and $h: f^{-1}(V) \to X$ the inclusion of the inverse image

11.6
$$\text{tr}(j^\#(\beta) = h^\#(\text{tr}(\beta)) \qquad ; \ \beta \in H.(Y,k)$$

<u>Proposition 11.7</u>. Let Y be a topological manifold with orientation μ_Y and $f: X \to Y$ is a finite covering. The class $\text{tr}(\mu_Y)$ is an orientation of the manifold X.

<u>Proof</u>. According to the formula 11.6 the question is local on Y. Thus it suffices to treat the case of a trivial covering. This is left to the reader.

$$\text{Q.E.D.}$$

X. Application to Algebraic Geometry

X.1   Dimension of algebraic varieties

In this chapter we shall give an introduction to the
topology of algebraic varieties over the complex numbers. For
all unexplained notation we refer to Fulton (1).

The set of geometric points of a <u>complex algebraic scheme</u> X
carries the structure of a locally compact space, which we
also denote by X. Let us recall three fundamental facts about
complex <u>algebraic varieties,</u> i.e. reduced and irreducible
schemes.

1.1. A non-singular point x of X admits an open neigh-
bourhood homeomorphic to $\mathbb{C}^n$ where $n = \dim \mathcal{O}_{X,x}$, the local
ring of X at x.

1.2. A non empty Zariski open subset of an algebraic
variety X is dense in X.

1.3. An algebraic variety is connected.

<u>Proposition 1.4.</u> For a point $x \in X$ of an algebraic scheme
X we have

$$\dim_x X = 2 \dim \mathcal{O}_{X,x}$$

where the first symbol refers to the concept of dimension introduced in III.9.

$\underline{\text{Proof}}$. Let us introduce the ad hoc notation $\cdot$ dlc X for the concept of dimension introduced in III.9 while dim X denotes the dimension as an algebraic scheme. - Let us first prove that

$$\text{dlc } X \leq 2 \dim X$$

This will be done by induction on $d = \dim X$: In case X is irreducible, let X_{ns} denote the non singular part of X. The complement $X_s = X - X_{ns}$ is an algebraic scheme of dimension $< d$. For any sheaf F on X we have the exact sequence

$$\rightarrow H_c^{2d+1}(X_{ns},F) \rightarrow H_c^{2d+1}(X,F) \rightarrow H_c^{2d+1}(X_s,F) \rightarrow$$

and the result follows from 1.1. - In the general case we shall do induction on the number s of irreducible components of X. Let Y be the union of s-1 of the irreducible components of X. For a sheaf F on X we have an exact sequence

$$\rightarrow H_c^{2d+1}(X-Y,F) \rightarrow H_c^{2d+1}(X,F) \rightarrow H_c^{2d+1}(Y,F) \rightarrow$$

and the result follows, since X-Y is irreducible of dimension $\leq$ d.

To prove the opposite inequality put $d = \dim O_{X,x}$ and choose an irreducible component Z through x of dimension d. Since $\text{dlc}_x Z \leq \text{dlc}_x X$ it suffices to treat the case where X is irreducible. - Let U be an open neighbourhood of x in X. It follows from 1.2 that $U \cap X_{ns}$ is non empty. Thus we conclude from 1.1 that U contains an open subset homeomorphic to $\mathbb{R}^{2d}$ and consequently $\text{dlc } U \geq 2d$.

$$\text{Q.E.D.}$$

402

X.2   The cohomology class of a subvariety

A non-singular algebraic variety has a natural orientation
relative to $\mathbb{Z}$. This comes from the fact that open subsets of
$\mathbb{C}^n$ inherits the natural orientation of $\mathbb{C}^n$ and the fact that
complex analytic transition functions have complex Jacobi matrices
and finally that complex linear transformations have positive
determinants considered as real linear transformations.

Theorem 2.1. Let X be a non-singular algebraic variety
and Z a closed subvariety of codimension d. Then

$$H_Z^p(X,\mathbb{Z}) = 0 \qquad \text{for} \quad p < 2d$$

Proof. By Poincaré duality with n = dim X

$$H_Z^p(X,\mathbb{Z}) \xrightarrow{\sim} H_{2n-p}(Z,\mathbb{Z})$$

Conclusion by 1.4 and IX.1.6.

Q.E.D.

Remark 2.2. With the notation of 2.1 we have

$$H^i_{Z\cap U}(U,\mathbb{Z}) = 0 \qquad ; \quad i < d$$

for any open subset of X as it follows from the proof of 2.1.
With the terminology of VIII.1, let i: $Z \to X$ denote the inclusion. We conclude from VIII.1.3 that

$$R^p i^! \underset{\sim}{\mathbb{Z}} = 0 \qquad ; \; p < 2d$$

and from VIII.1.3 that the presheaf

$$U \longmapsto H_{Z \cap U}^{2d}(U, \mathbb{Z})$$

is a sheaf on X, indeed equal to $i_* R^{2d} i^! \underset{\sim}{\mathbb{Z}}$.

<u>Corollary 2.3</u>. Let X be a non-singular variety and Z a closed subvariety of codimension d. For any Zariski open subset U of X <u>which meets Z</u>, the map

$$H_Z^{2d}(X, \mathbb{Z}) \; \to \; H_{Z \cap U}^{2d}(U, \mathbb{Z})$$

is an isomorphism.

<u>Proof</u>. The excision sequence II.9.5

$$H_{Z-U}^{2d}(X, \mathbb{Z}) \; \to \; H_Z^{2d}(X, \mathbb{Z}) \; \to \; H_{Z \cap U}^{2d}(U, \mathbb{Z}) \; \to \; H_{Z-U}^{2d+1}(X, \mathbb{Z})$$

combined with 2.1 yields the result.

Q.E.D.

<u>Definition 2.4</u>. Let X denote a non singular variety and Z a closed subvariety of codimension d. The <u>Thom class of Z in X</u>

$$\tau_Z \in H_Z^{2d}(X, \mathbb{Z}) \qquad ; \; d = \mathrm{codim}(Z, X)$$

is the class whose restriction to $X - Z_s$ is the Thom class for

the oriented submanifold Z_{ns} of the oriented manifold $X-Z_s$.
- The image of τ_Z in absolute cohomology will be denoted

$$cl^X(Z) \in H^{2d}(X,\mathbb{Z}) \qquad ; \quad d = codim(Z,X)$$

or just $cl(Z)$ when no confusion is possible.

Proposition 2.5. Let X be a non-singular variety and Z a Zariski closed subset of codimenson d. If $Z_1,\ldots,Z_s$ denote the irreducible components of X of codimension d. Then the canonical map

$$\oplus H_{Z_i}^{\,2d}(X,\mathbb{Z}) \to H_Z^{\,2d}(X,\mathbb{Z})$$

is an isomorphism.

Proof. Use the Mayer-Vietoris sequence II.9.11 in combination with 2.1.

Q.E.D.

Intersection numbers

Let V and W denote closed subvarieties of the
of the non-singular variety X. Recall that for any irreducible component Z of $V \cap W$ we have

2.6 $\qquad\qquad$ codim $Z \le$ codim $V +$ codim W

Let us assume that V and W $\underline{\text{intersects properly}}$ in X, i.e. that the inequality 2.6 is an equality for all irreducible components Z of $V \cap W$. Let V and W have codimension c and d respectively and consider the cohomology class

$$2.7 \qquad \tau_V \cup \tau_W \in H_{V \cap W}^{2(c+d)}(X, \mathbb{Z})$$

Taking the isomorphism 2.5 into account we can write

$$2.8 \qquad \tau_V \cup \tau_W = \sum_Z i(Z, V \cdot W; X) \tau_Z$$

the sum being over all irreducible components Z of $V \cap W$. The integer $i(Z, V \cdot W; X)$ is called the $\underline{\text{local intersection symbol}}$. Push the formula 2.8 into $H^{\cdot}(X, \mathbb{Z})$ to get

$$2.9 \qquad cl^X(V) \cap cl^X(W) = \sum_Z i(Z, V \cdot W; X) cl^X(Z)$$

The local intersection number can be calculated by means of Serre's alternating Tor formula

$$2.10 \qquad i(Z, V \cdot W; X) = \sum_i (-1)^i \ell(Tor_i^{\mathcal{O}}(\mathcal{O}_V, \mathcal{O}_W))$$

where $\mathcal{O}$ denotes the local ring of X of the generic point of Z. Compare Iversen (1).

X.3   Homology class of a subvariety

In this section we shall generalize some of the constructions from the previous section to singular spaces and Borel-Moore homology. The basic principle is that for a scheme X we have

$$3.1 \qquad H_i(X,\mathbb{Z}) = 0 \qquad ; \ i > 2\dim X$$

as it follows from 1.4 and IX.1.6. In Case X is a variety of dimension n and U a non-empty Zariski open subset we have a long exact sequence

$$H_{2n}(X-U,\mathbb{Z}) \to H_{2n}(X,\mathbb{Z}) \to H_{2n}(U,\mathbb{Z}) \to H_{2n-1}(X-U,\mathbb{Z})$$

and consequently an isomorphism

$$3.2 \qquad H_{2n}(X,\mathbb{Z}) \xrightarrow{\sim} H_{2n}(U,\mathbb{Z})$$

In particular if $U = X_{ns}$ we find a <u>fundamental homology class</u>

$$3.3 \qquad \mu_X \in H_{2n}(X,\mathbb{Z}) \qquad ; \qquad n = \dim X$$

characterized by requiring its restriction to X_{ns} to be the orientation class.

Theorem 3.4. Let X be an algebraic shceme of dimension n and let $X_1,\ldots,X_s$ denote the irreducible components of dimension n The images of $\mu_{X_1},\ldots,\mu_{X_s}$ in $H_{2n}(X,\mathbb{Z})$ form a basis for that group.

<u>Proof</u>. In case X is a non singular variety, this is a consequence of Poincaré duality and the fact that X is connected. In case X is a variety, the result follows from the isomorphism 3.2. In the general case we have an isomorphism

$$\underset{i}{\oplus}\; H_{2n}(X_i, \mathbb{Z}) \xrightarrow{\sim} H_{2n}(X, \mathbb{Z})$$

by an excision or Mayer-Vietoris argument.

Q.E.D.

For a scheme X let $Z_k(X)$ denote the group of <u>k-cycles</u> on X, i.e. the free abelian group based on the set of subvarieties of X of dimension k. The cycle corresponds to a subvariety V is denoted [V]. The cycle map

3.5
$$cl_X: Z_k(X) \to H_{2k}(X, \mathbb{Z})$$

is given by the formula

3.6
$$cl_X([V]) = i_*(\mu_V)$$

where i: V → X denotes the inclusion of a closed k-dimensional subvariety.

Let f: X → Y be a proper map of algebraic schemes. For a closed subvariety V of X, the image W = f(V) is a closed subvariety of Y. Define $f_*[V] = \deg(V/W)[W]$, where

$$\deg(V/W) = \begin{cases} 0 & \text{if dim } V < \text{dim } W \\ [R(V): R(W)] & \text{if dim } V = \text{dim } W \end{cases}$$

408

Here $R(-)$ denotes the function field. This defines

3.7 $f_*\colon\; Z.(X) \longrightarrow Z.(Y)$

The following diagram is commutative

$$
\begin{array}{ccc}
Z_k(X) & \xrightarrow{\;f_*\;} & Z_k(Y) \\
\downarrow & & \downarrow \\
H_{2k}(X,\mathbb{Z}) & \xrightarrow{\;f_*\;} & H_{2k}(Y,\mathbb{Z})
\end{array}
$$

3.8

<u>Proof</u>. It suffices to prove that if $f\colon V \to W$ is a surjective and proper map of algebraic varieties then

$$f_*(\mu_V) = \deg(W/V)\,\mu_W$$

In case $\dim W < \dim V = k$ then $H_{2k}(W,\mathbb{Z}) = 0$ and the result is clear. In case $\dim W = k$ we may replace W by an open subvariety and assume that $V \to W$ is a covering space of degree $\deg(V/W)$ and the result follows from IX.11.

Q.E.D.

X.4   Intersection theory

Consider a non-singular variety X of dimension n .
Poincaré duality

$$H^p(X,\mathbb{Z}) \xrightarrow{\;\mu_X\cap\;} H_{2n-p}(X,\mathbb{Z}) \qquad ; \; p \in \mathbb{Z}$$

is a isomorphism. In IX.7 we used Poincaré duality to transport stucture from homology to cohomology. In accordance
with Fulton (1) we shall here transport structure from cohomology
to homology. This has several advantages: better and more
suggestive formulas which in many cases may be generalized to
singular varities.

Let Z denote a closed subvariety of X.   Then we have the
fundamental relation, compare 2.4

4.1 $$\mu_Z = \mu_X \cap \tau_Z$$

which follows from 3.3 and IX.4.9 using the second square in
IX.3.5.

Let i: Z → X denote the inclusion. From 4.1 follows

$$\mu_Z \cap i^*\xi = (\mu_X \cap \tau_Z) \cap i^*\xi \qquad ; \quad \xi \in H^\cdot(X,\mathbb{Z})$$

Using IX.3.3 and II.9.17 and taking advantage of the fact that
τ_Z has even degree we get

4.2 $$\mu_Z \cap i^*\xi = (\mu_X \cap \xi) \cap \tau_Z$$

In case Z is non singular we can interprete the formula 4.2
as expressing that the Poincaré dual of i*: $H^\cdot(X,\mathbb{Z}) \to H^\cdot(Z,\mathbb{Z})$

410

is given by

$$4.3 \qquad \alpha \longmapsto \alpha \cap \tau_Z \quad ; \quad H.(X,\mathbb{Z}) \to H.(Z,\mathbb{Z})$$

At the same time this suggests immediately a generalization
to the case where Z is singular.

Let $f: X \to Y$ be a morphism between non-singular varieties
and $\tau_\Gamma \in H_\Gamma^{\cdot}(X \times Y, \mathbb{Z})$ the Thom class. The following formula
gives an expression for the Poincaré dual of $f^*: H^{\cdot}(Y,\mathbb{Z}) \to H^{\cdot}(X,\mathbb{Z})$
or more generally the symbol $\xi \cup f^*\eta$,

$$4.4 \qquad \mu_X \cap (\xi \cup f^*\eta) = (-1)^{pq}((\mu_X \cap \xi) \times (\mu_Y \cap \eta)) \cap \tau_\Gamma$$

where $\xi \in H^p(X,\mathbb{Z})$ and $\eta \in H^q(Y,\mathbb{Z})$.

Proof. Let $\gamma: X \to X \times Y$ denote the graph map and
$\delta: X \to X \times X$ the diagonal map. Notice the formula

$$\xi \cup f^*\eta = \delta^*(\xi \times f^*\eta) = \delta^*(1 \times f)^*(\xi \times \eta) = \gamma^*(\xi \times \eta)$$

Next, apply the formula 4.2 to $\gamma: X \to X \times Y$ to get

$$\mu_X \cap \gamma^*(\xi \times \eta) = (\mu_X \times \mu_Y) \cap (\xi \times \eta) \cap \tau_\Gamma$$

$$= (-1)^{pq}((\mu_X \cap \xi) \times (\mu_Y \cap \eta)) \cap \tau_\Gamma$$

where we have used IX.5.9. The result follows by combining these
two formulas.

$$\text{Q.E.D.}$$

The formula 4.4 suggest the following definition, compare Fulton (1) Ch.8

$$4.5 \qquad \alpha \cdot_f \beta = (-1)^{pq} \alpha \times \beta \cap \tau_\Gamma \quad ; \alpha \in H_p(X,\mathbb{Z}), \beta \in H_q(Y,\mathbb{Z})$$

In particular, for $f = 1_X$ the <u>intersection product</u> $\alpha \cdot \beta$ <u>is defined by</u>

$$4.6 \qquad \alpha \cdot \beta = (\beta \times \alpha) \cap \tau_X \quad ; \quad \alpha, \beta \in H.(X,\mathbb{Z})$$

The definition 4.6 may be <u>refined</u> in the following sense. Let $i: V \to X.$ and $h: W \to X$ be inclusions of closed subvarieties. The symbol

$$(\beta \times \alpha) \cap \tau_X \quad ; \quad \beta \in H.(V,\mathbb{Z}), \alpha \in H.(W,\mathbb{Z})$$

defines an element of $H.(V \cap W, \mathbb{Z})$. If we push this forward into $H.(X,\mathbb{Z})$ we obtain $i_* \beta \cdot h_* \alpha$ as it follows from the projection formula IX.3.7.

<u>Theorem 4.7</u>. Let V and W be closed subvarieties of dimension v and w. Then

$$\mu_V \cdot \mu_W = \mu_V \cap \tau_W = \mu_W \cap \tau_V$$

in $H_{2(w+v-n)}(V \cap W, \mathbb{Z})$.

<u>Proof</u>. Let $\delta: X \to X \times X$ denote the diagonal map. We start from the formula II.10.8 $\qquad \tau_V \cup \tau_W = \delta^*(\tau_V \times \tau_W)$

valid in $H_{V \cap W}^{\bullet}(X, \mathbb{Z})$ and the formula

$$\mu_X = (\mu_X \times \mu_X) \cap \tau_X$$

valid in $H_{\bullet}(X, \mathbb{Z})$. Cap these together to get

$$(\mu_X \cap \tau_V) \cap \tau_W = ((\mu_X \times \mu_X) \cap \tau_X) \cap \delta^*(\tau_V \times \tau_W)$$

and using the formula 4.1, IX.3.4 and IX.5.9

$$\mu_V \cap \tau_W = (\mu_X \times \mu_X) \cap (\tau_X \cup (\tau_V \times \tau_W))$$

$$= (\mu_X \times \mu_X) \ ((\tau_V \times \tau_W) \cup \tau_X)$$

$$= ((\mu_X \cap \tau_V) \times (\mu_X \cap \tau_W)) \cap \tau_X$$

from which the result follows.

Corollary 4.8. If the two subvarieties V and W intersect properly then

$$\mu_V \cdot \mu_W = \sum_Z i(Z, V \cdot W; X) \mu_Z$$

in $H_{2(v+w-n)}(V \cap W, \mathbb{Z})$.

Proof. In $H_{V \cap W}^{\bullet}(X, \mathbb{Z})$ we have the formula

$$\tau_V \cup \tau_W = \sum_Z i(Z, V \cdot W; X) \tau_Z$$

and the result follows by capping this with μ_X.

O.E.D.

$\underline{\text{Projection formula 4.9}}$. Let $f: X \to Y$ be a proper map between non-singular varieties. Then

$$f_*(\xi \cdot_f \eta) = f_*(\xi) \cdot \eta \qquad ; \ \xi \in H.(X, \mathbb{Z}) , \ \eta \in H.(Y, \mathbb{Z})$$

$\underline{\text{Proof}}$. With the notation of 4.5 we shall prove that

$$4.10 \qquad\qquad (f \times 1_Y)^* \tau_Y = \tau_\Gamma$$

To do so consider the commutative diagram

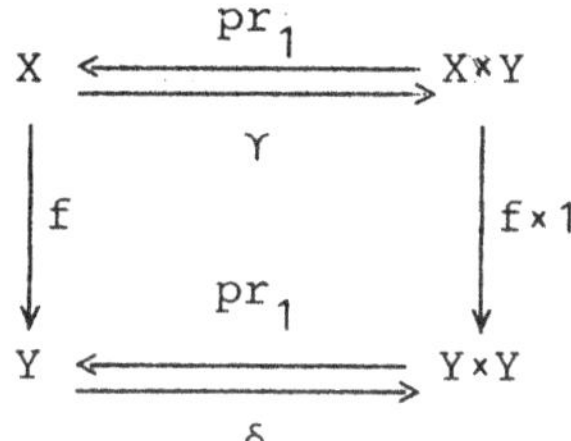

This may be viewed as a pull back diagram of $\underline{\text{microbundles}}$ and 4.10 follows from the remarks preceding VIII.3.5. combining 4.5 and 4.10 we get

$$\xi \cdot_f \eta = \alpha \times \beta \cap (f \times 1)^* \tau_Y (-1)^{pq}$$

From the projection formula IX.3.7

$$f_*(\alpha \cdot_f \beta) = (f \times 1)_* (\alpha \times \beta) \cap \tau_Y (-1)^{pq}$$

$$= (f_* \alpha \times \beta) \cap \tau_Y (-1)^{pq} = (f_* \alpha) \cdot \beta$$

where we have used the definition 4.6.

$$\text{Q.E.D.}$$

X.5. Algebraic families of cycles

Let $f: V \to T$ be a map of algebraic varities where T is non singular. Let the dimension of V and T be n and d respectively. We define the class

$$5.1 \qquad \tau_\Gamma \in H^{2d}(V \times T, \mathbb{Z})$$

to be the pull back of $\tau_T \in H^{2n}_\Delta(T \times T)$ along $f \times 1$. Let us notice that

$$5.2 \qquad \mu_V \times \mu_T \cap \tau_\Gamma = \mu_V$$

as it follows from an excision argument, compare 4.10.

Proposition 5.3. For $t \in T$ let $\tau_t \in H^{2d}_{\{t\}}(T, \mathbb{Z})$ denote the Thom class. Then we may identify

$$\mu_V \cap f^*(\tau_t) \in H_{2(n-d)}(f^{-1}(t), (\mathbb{Z}))$$

with the refined intersection class

$$\mu_T \cap p^*(\tau_t) = (\mu_V \times \mu_t) \cap \tau_T$$

where $p: V \times T \to T$ denotes the projection.

Proof. Let us first identify $\mu_V \cap f^*(\tau_t)$ with the class $(\mu_V \times \mu_t) \cap \tau_T$. To do so let $i_t: V \to V \times T$ given by $i_t(v) = (v,t)$. We must prove that

$$i_t^*(\tau_\Gamma) = f^*(\tau_t) \quad \text{in} \quad H_{f^{-1}(t)}(V,\mathbb{Z}).$$

This follows from the commutative diagram where $V_t = f^{-1}(t)$

$$
\begin{array}{ccc}
(V,V-V_t) & \xrightarrow{\ i_t\ } & (V\times T, V\times T-\Gamma) \\
\downarrow f & & \downarrow f\times 1 \\
(T,T-\{t\}) & \xrightarrow{s\ \mapsto\ (s,t)} & (T\times T, T\times T-\Delta)
\end{array}
$$

and the theory of the diagonal class IX.6.13. The second formula follows from

$$(\mu_V \times \mu_T) \cap \tau_\Gamma \cap (1\times\tau_t) = (\mu_V \times \mu_T) \cap (1\times\tau_t) \cap \tau_\Gamma$$

the formula 5.2 and $\mu_T \cap \tau_t = \mu_t$.

Q.E.D.

In case $f: V \to T$ is <u>proper</u> we have the

<u>Projection formula 5.4.</u>

$$f_*(\alpha \cdot_f \beta) = f_*(\alpha) \cdot \beta \quad ; \alpha \in H.(V,\mathbb{Z}), \ \beta \in H.(T,\mathbb{Z})$$

as it follows from the proof of 4.9.

In case the fibers of $f: V \to T$ all have dimension $n-d$ we can define $V_t \in Z_{n-d}(V)$ by

$$5.5 \qquad\qquad cl(V_t) = \mu_V \cap f^*(\tau_t)$$

in $H_{2(n-d)}(f^{-1}(t),\mathbb{Z})$. Let us remark that

5.6 $\qquad$ $cl(V_t) \in H_{2(n-d)}(V,\mathbb{Z})$ $\quad$ is independent of $t \in T$.

$\underline{\text{Proof}}$. Let $r(\tau_{\{t\}}) \in H^{2d}(T,\mathbb{Z})$ denote the image of $\tau_{\{t\}}$.
According to IX.3.5, 1. square, the image of $cl(V_t)$ in
$H_{2(n-d)}(V,\mathbb{Z})$ is $\mu_V \cap f^*(r(\tau_t))$. Conclusion by the fact that
$r(\tau_t) \in H^{2d}(T,\mathbb{Z})$ is independent of $t \in T$, as it follows from
IX.6.14 and a homotopy argument.

$$\text{Q.E.D.}$$

Consider a cycle $Y \in Z_{k+d}(X \times T)$ where X is a variety
of dimension n and T a non-singular variety of dimension d.
For a point $t \in T$ for which

5.8 $\qquad$ $\dim X \times \{t\} \cap \operatorname{Supp} Y = k$

we can define a k-cycle Y_t on X characterized by

5.8 $\qquad$ $cl(Y_t) \times \mu_t = cl(Y) \cap p^*(\tau_t)$

where $p: X \times T \to T$ is the projection and $\tau_t \in H_{\{t\}}^{2d}(T,\mathbb{Z})$ the
local Thom class.

A family $(Z_t)_{t \in T}$ of k-cycles on X is called an $\underline{\text{algebraic}}$
$\underline{\text{family}}$ if there exists Y on $Z(X \times T)$ which satisfies 5.7
for all $t \in T$ and such that $Y_t = Z_t$ for all $t \in T$.

<u>Proposition 5.9</u>. Given an algebraic family $(Z_t)_{t \in T}$ of k-cycles on the algebraic variety X. The homology class $cl(Z_t) \in H_{2k}(X, \mathbb{Z})$ is independent of $t \in T$.

<u>Proof</u>. But $\beta = cl(Z)$. For $t \in T$ define $\beta_t \in H.(X, \mathbb{Z})$ such that

$$\beta_t \times \mu_t = \beta \cap p^*(\tau_t)$$

where $\tau_t \in H_{\{t\}}(T, \mathbb{Z})$ is the local Thom class and $\mu_t \in H.(\{t\}, \mathbb{Z})$ the canonical generator. In this formulation the problem becomes local on T, thus we may assume that $T = \mathbb{R}^n$. Using the Künneth formula IX.5.8 we see that β has the form $\alpha \times \mu_T$ with $\alpha \in H.(X, \mathbb{Z})$. Thus we get

$$\beta \cap p^*(\tau_t) = (\alpha \times \mu_T) \cap (1 \times \tau_t) = \alpha \times \mu_t$$

i.e. $\beta_t = \alpha$ for all $t \in T$.

Q.E.D.

In case X is non-singular consider algebraic cycles $Y \in Z_{d+k}(X \times T)$ and $Z \in Z_{d+1}(X \times T)$ and a fixed $t \in T$ such that Y_t, Z_t and $Y_t \cdot Z_t$ are defined. Then

$$5.10 \qquad cl(Y_t \cdot Z_t) = (cl(Y) \cdot cl(Z))_t$$

in $H.(\text{Supp } Y_t \cap \text{Supp } Z_t, \mathbb{Z})$.

418

$\underline{\text{Proof}}$. Put $\alpha = cl(Y)$ and $\beta = cl(Z)$. This gives $cl(Y_t) = \alpha \cap p^*(\tau_t)$ and $cl(Z_t) = \beta \cap p^*(\tau_t)$ and the formula reads

$$(\alpha \times \beta) \cap (p^*(\tau_t) \times p^*(\tau_t)) \cap \tau_X = (\alpha \times \beta) \cap \tau_{X \times T} \cap p^*(\tau_t)$$

Thus it suffices to prove that $\tau_{X \times T} = \tau_X \times \tau_T$ and that

$$\tau_t \times \tau_t = \tau_T \cup 1 \times \tau_t$$

which is left to the reader.

Q.E.D.

We shall now prove a result which will justify more intuitive ways of defining the local intersection symbol.

$\underline{\text{Theorem 5.11}}$. Let X denote a non-singular variety of dimension n, $(Z_t)_{t \in T}$ an algebraic family of cycles on X and $(W_t)_{t \in T}$ an algebraic family of cycles of complementary dimension such that $\text{Supp } Z_t \cap \text{Supp } W_t$ is finite for all $t \in T$.

Given a $\underline{\text{compact}}$ subset K of X and a $\underline{\text{connected}}$ open sub-set D of T such that

$$\text{Supp } Z_t \cap \text{Supp } W_t \subseteq K \quad \text{for all} \quad t \in D$$

Then the sum of the local intersection numbers

$$\sum_x i(\{x\}, Z_t \cdot W_t ; X)$$

is independent of $t \in D$.

<u>Proof</u>. It follows from the hypothesis that the cycles Z and W on $X \times T$ intersect properly. Let $W \cdot Z \in Z_0(X \times T)$ be the intersection cycle. Let Y be an irreducible component of the support of $Z \cdot W$. It suffices to prove that $\deg Y_t$ is independent of $t \in D$. Let $f \colon Y \to T$ denote the restriction of the projection. Notice that $f^{-1}(D) = K \times D$ as a consequence of the second assumption and consequently that $f^{-1}(D) \to D$ is proper. Let $j \colon f^{-1}(D) \to Y$ denote the inclusion and consider the cohomology class $j^{\#}\mu_Y$. According to the principle of preservation of numbers IX.10.3

$$\deg Y_t = \deg(j^{\#}\mu_Y \cap f^*\tau_t) \qquad ; \ t \in D$$

is independent of $t \in D$.

$$\text{Q.E.D.}$$

<u>Example 5.12</u>. Let W be a plane curve, $(V_t)_{t \in \mathbb{P}^1}$ the family of lines through a fixed non singular point x of W and K a compact neighbourhood of x in the plane. Let V_{t_0} be the tangent to W at x. Variations of t in a suitable neighbourhood D of t_0 will often reveal the number $i(\{x\}, W \cdot V_{t_0})$.

X.6   Algebriac cycles and Chern classes

Let X denote an algebraic variety of dimension n, $p: E \to X$ an algebraic bundle of rank d and $s: X \to E$ an algebraic section. We shall assume that

$$6.1 \qquad \dim s^{-1}(0) = n-d$$

The pull back of the Thom class VIII.6.1

$$s^*\tau_E \in H^{2(n-d)}_{s^{-1}(0)}(X,\mathbb{Z})$$

defines a homology class

$$\mu_X \cap s^*\tau_E \in H_{2(n-d)}(s^{-1}(0),\mathbb{Z}).$$

By the assumption 6.1 this determines an algebraic cycle $Z(s) \in Z_{n-d}(X)$ by the formula

$$6.2 \qquad cl_X(Z(s)) = \mu_X \cap s^*\tau_E$$

in $H_{2(n-d)}(s^{-1}(0),\mathbb{Z})$. The image of $s^*\tau_E$ in $H^{2(n-d)}(X,\mathbb{Z})$ is $c_d(E)$: Let $r: H^{2d}_X(E,\mathbb{Z}) \to H^{2d}(E,\mathbb{Z})$ denote the restriction. We have $c_d(E) = i^*(r(\tau_E))$ where $i: X \to E$ denote the zero section, VIII.6.2. On the other hand $i^* = s^*$ since p^* is an isomorphism and i^* and s^* both are left inverses to p^*. Applying XI.3.5 to 6.2 we get

$$6.3 \qquad \boxed{cl_X(Z(s)) = \mu_X \cap c_d(E)}$$

in $H.(X,\mathbb{Z})$.

In case X is non-singular we have

6.4
$$\boxed{cl^X(Z(s)) = c_d(E)}$$

Proof. By Poincaré duality we must show that

$$\mu_X \cap cl^X(Z(s)) = \mu_X \cap c_d(E)$$

But for any algebraic cycle Z we have

$$\mu_X \cap cl^X(Z) = cl_X(Z)$$

and the result follows from 6.3.

Q.E.D.

Theorem 6.5. Let E denote a complex vector space and $Grass_i(E)$ the variety of i-subplanes in E and Q the canonical quotient bundle on $Grass_i(E)$. For a linear subspace F of E let $\sigma(F) \subseteq Grass_i(E)$ denote the set of i-planes which meet F in a non zero subspace. If $1 \le rk\ F \le rk\ Q$ then $\sigma(F)$ is a subvariety of $Grass_i(E)$ with

$$codim\ \sigma(E) = rk\ Q - rk\ F + 1$$

and for $p = codim\ F$ we have

$$c_d(Q) = cl(\sigma(F))$$

422

    <u>Proof</u>. For the first part of the statement, see Hodge-Pedoe (1) XIV.2. - The restriction of the bundle Q to the complement of $\sigma(F)$ contains F as a subbundle; consequently the restriction of $c_p(Q)$ to the complement of $\sigma(F)$ is zero. It follows that $c_p(Q)$ "comes from" $H_{\sigma(F)}^{2p}(\mathrm{Grass}_i(E), \mathbb{Z})$, i.e. is a multiple of $\mathrm{cl}(\sigma(F))$. Write

$$c_p(Q) = m\ \mathrm{cl}(\sigma(F)) \qquad ; m \in \mathbb{Z}$$

In order to prove that $m = 1$ choose an $(i-1)$-plane D with $D \cap F = 0$ and consider the canonical projection

$$p:\ \mathrm{Grass}_1(E/D) \longrightarrow \mathrm{Grass}_i(E)$$

Notice that $p^{-1}(\sigma(F+D/D)) = \sigma(F)$ and that these two subvarieties have the same codimension. Thus we can write

$$p^*\mathrm{cl}(\sigma(F)) = n\ \mathrm{cl}(\sigma(F+D/D)) \qquad ; n \in \mathbb{Z}$$

Combining the two formulas we get

$$c_p(p^*Q) = mn\ \mathrm{cl}(\sigma(F+D/D))$$

Thus we have reduced the problem to $i = 1$.

    Let L denote the canonical line bundle on $P(E) = \mathrm{Gras}_1(E)$. We have an exact sequence of bundles on $P(E)$

$$0 \longrightarrow L^{\vee} \longrightarrow E \longrightarrow Q \longrightarrow 0$$

which gives $C.(L^{\vee})c.(Q) = 1$. Writing $C.(L^{\vee}) = 1-c_1(L)$ we get immediately

$$c.(Q) = 1 + c_1(L) + c_1(L)^2 + \ldots$$

in particular $c_p(Q) = c_1(L)^p$. On the other hand $\sigma(F)$ may be identified with $P(F)$, a linear subspace of $P(E)$ a codimension p.

Q.E.D.

XI. Derived Categories

Let us consider an additive category K and a system S
of morphisms subjected to conditions FR 1, 2, 3 below. To
these data we shall associate an additive category $S^{-1}K$

FR 1. If f and g are composable morphisms belonging
to S then $g \circ f$ belongs to S. The identity of every object
of K belongs to S.

FR 2. Any diagram in K with s in S
can be completed to a commutative diagram
in K with t in S.

    Ditto, with all arrows reversed.

FR 3. For morphisms $f,g: Y \to Z$ in K and $t: Z \to W$
with $tf = tg$, there exists $s: X \to Y$ in S with $fs = gs$. —
Ditto with all arrows reversed.

We shall now introduce the additive category $S^{-1}K$

<u>Objects in $S^{-1}K$</u>: The same as those of K.

<u>Morphisms in $S^{-1}K$</u>: Given objects X and Y in K:
By a right fraction (b,s) from X to Y we understand a diagram
of the form

$$X \xleftarrow{\ \ s\ \ } \bullet \xrightarrow{\ \ b\ \ } Y$$

where s is in S. Given two right fractions (a,r) and (b,s)
from X to Y. We write (a,r) ~ (b,s)   if there exists a commutative diagram as below with u in S.

1.1

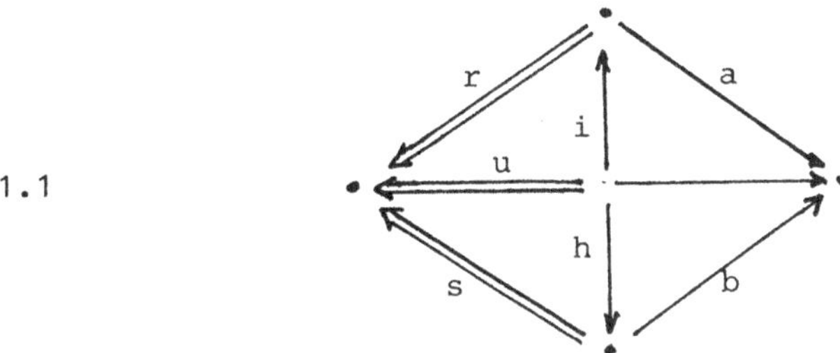

Let us prove that ~ is an equivalence relation on the fractions
from X to Y. So let there be given three fractions with
(a,r) ~ (b,s)   and (b,s) ~ (c,t). Let the former equivalence be
realized by the diagram above and the latter by the commutative
diagram below

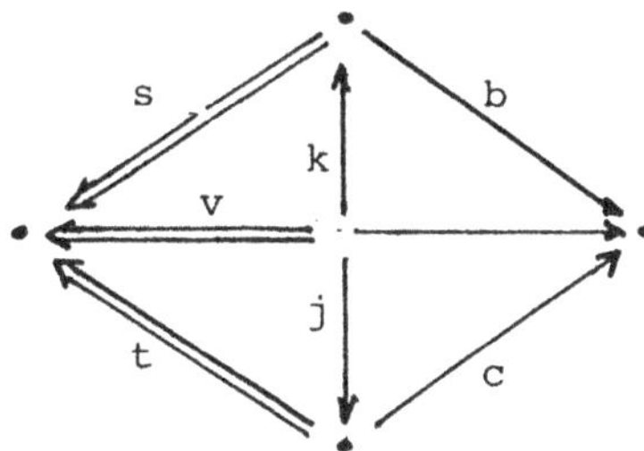

In order to construct an equivalence between (a,r) and (c,t)
consider the following diagram which is commutative except for
the middle diamond

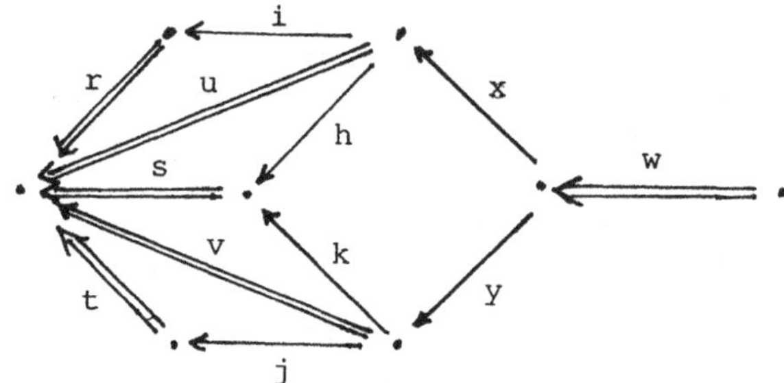

Explanation: Complete first u,v to a commutative square
u,v,x,y with x in S. This implies s(hx) = s(ky). Choose
w in S with (xh)w = (yk)w. Consequently

$$(aixw,rixw) = (bhxw,tjyw) = (bkyw,tjyw) = (cjyw,tjyw)$$

which provides an equivalence between (a,r) and (c,t).

The morphisms or arrows in $S^{-1}K$ from X to Y are the
equivalence classes of right fractions from X to Y. The equivalence class of a right fraction (a,s) is denoted a/s. The
basic relation for calculation with right fractions is

1.2 $$ac/sc = a/s \qquad ; \; s \in S, \quad sc \in S$$

<u>Composition in $S^{-1}K$</u> . Given objects X,Y,Z in K and
a right fraction (a,r) from X to Y and a right fraction
(b,s) from Y to Z. Consider the following diagram

1.3

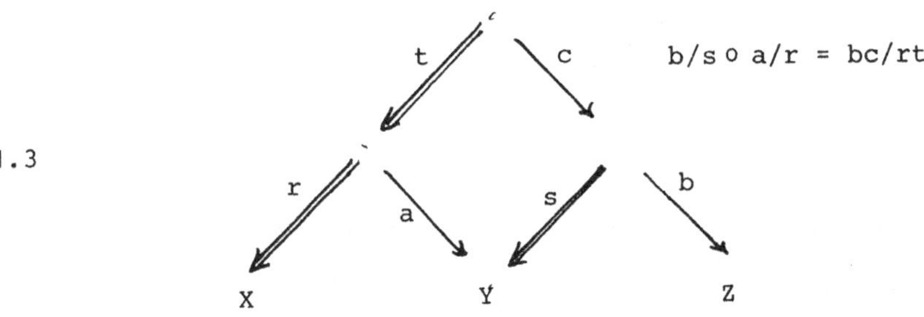

where a commutative square has been filled in according to FR 3 with t in S. The arrow bc/rt from X to Z is easily seen to be independent of the way the square has been filled in. We leave it to the reader to verify that bc/rt depends only on a/r and b/s. It is also left to the reader to verify that this composition law gives rise to a category.

Addition in $S^{-1}K$. Given objects X and Y in $S^{-1}K$. It is easy to see that two morphisms from X to Y can be represented with a common denominator a/t, b/t. We put

$$1.4 \qquad\qquad a/t + b/t = (a+b)/t$$

and leave it to the reader to verify that this defines a structure of abelian group on $\mathrm{Hom}_{S^{-1}K}(X,Y)$ which makes composition in $S^{-1}K$ bilinear.

The functor $K \to S^{-1}K$. A morphism $f: X \to Y$ in K gives rise to a morphism $f/1: X \to Y$ in $S^{-1}K$. This functor is seen to transform $X \oplus Y$ into a direct sum in $S^{-1}K$. This completes the proof of the fact that $S^{-1}K$ is an additive category. Notice that $K \longrightarrow S^{-1}K$ is an additive functor.

Proposition 1.5. Let $U: K \to L$ be an additive functor which transforms the morphisms in S into isomorphisms in L. Then there exists a unique additive functor $V: S^{-1}K \to L$ whose composite with $K \longrightarrow S^{-1}K$ is U.

Proof. Left to the reader.

$$Q.E.D.$$

428

Definition 1.6. The multiplicative system S is said to be saturated if any morphism $f: X \to Y$, for which there exists $h: W \to X$ and $k: Y \to Z$ such that fh and kf is in S, belongs to S.

Proposition 1.7. Let S be a saturated multiplicative system in K. Then, a morphism $f: X \to Y$ is an isomorphism in $S^{-1}K$ if and only if f belongs to S.

Proof. If f is an isomorphism in $S^{-1}K$ then f has a right inverse $Y \xleftarrow{\ s\ } Z \xrightarrow{\ h\ } X$. A simple consideration shows that this can be chosen such that $s = fh$ in K. The opposite relation can be obtained from the discussion below.

Q.E.D.

Left fractions. Let X and Y be objects in K. By a left fraction from X to Y we understand a pair of morphisms (s,b) with s in S

$$X \xrightarrow{\ b\ } \cdot \xLeftarrow{\ s\ } Y$$

This gives rise to a morphism

$$1.8 \qquad\qquad s\!\searrow\!b = (1/s) \circ (b/1)$$

It follows from FR 2 that any morphism in $S^{-1}K$ can be represented by a left fraction. Two left fractions (r,a) and (s,b) give $r\!\searrow\!a = s\!\searrow\!b$ if and only if there exists a commutative

diagram

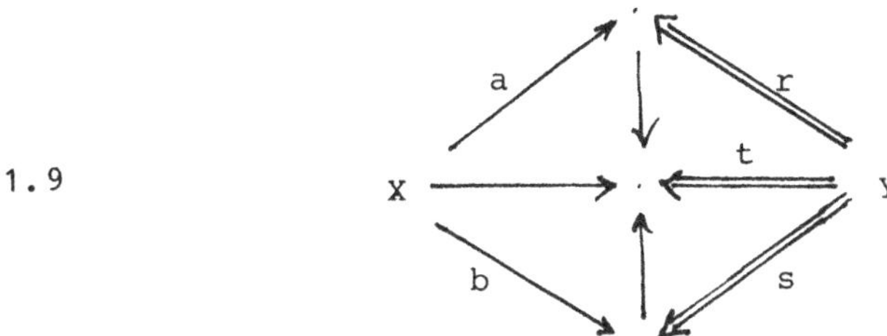

1.9

with t in S. - Let us reproduce the diagram for composing left
fractions

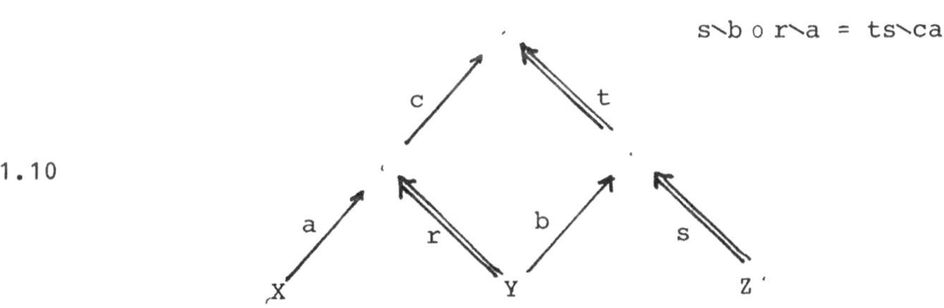

1.10

$s\backslash b \circ r\backslash a = ts\backslash ca$

XI.2 The derived category, D(A)

Let A denote an additive category and $K(A)$ the homotopy category of complexes over A. Let us recall the basic properties of the triangles in $K(A)$.

TR1 Any diagram isomorphic to a triangle is a triangle. Any morphism can be completed to a triangle. The diagram $(X,X,0,1,0,0)$ is a triangle.

TR2 A diagram of the form

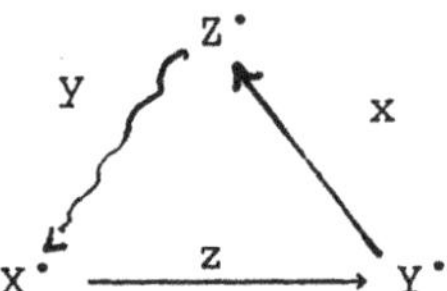

is a triangle if and only if the diagram

is a triangle.

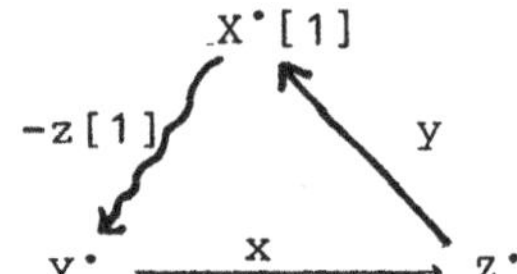

TR3 Given a diagram

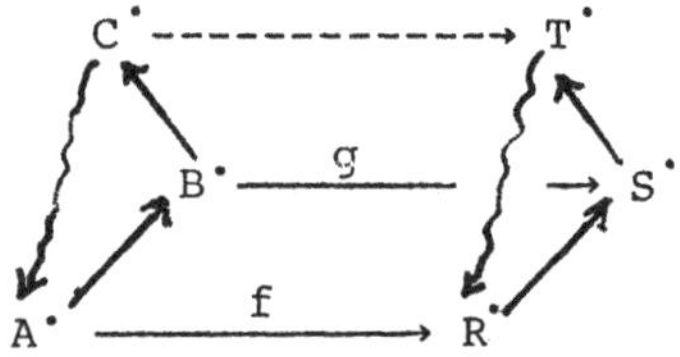

consisting of two triangles and morphisms f and g making the bottom square commutative. Then the dotted arrow may be filled in making the remaining two squares commutative.

<u>Remark 2.1</u>. The dotted arrow in TR3 is not unique in general. Here is a collection of examples

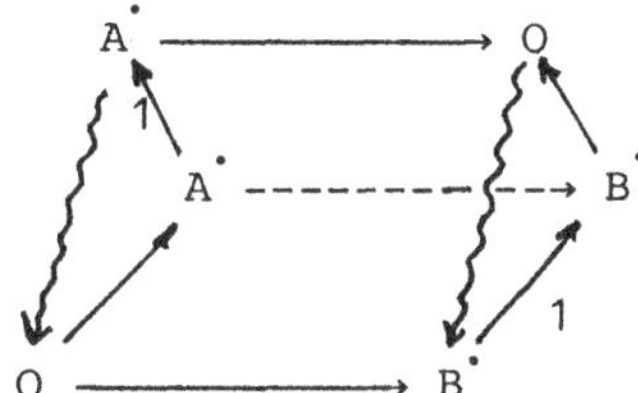

Any dotted arrow will meet the requirements.

Let us use TR 1.2.3 to prove two familiar results about triangles.

<u>Proposition 2.2</u>. From an object X and a triangle

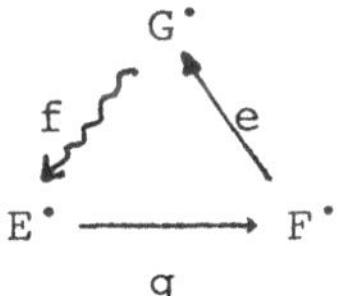

result two exact sequences

$$\longrightarrow [X^{\cdot},E^{\cdot}] \longrightarrow [X^{\cdot},F^{\cdot}] \longrightarrow [X^{\cdot},G^{\cdot}] \longrightarrow [X^{\cdot},E^{\cdot}[1]] \longrightarrow$$

$$\longrightarrow [G^{\cdot},X^{\cdot}] \longrightarrow [F^{\cdot},X^{\cdot}] \longrightarrow [E^{\cdot},X^{\cdot}] \longrightarrow [G^{\cdot}[-1],X^{\cdot}] \longrightarrow$$

<u>Proof</u>. We shall restrict ourselves to prove exactness of the following sequence

$$[X^{\cdot},E^{\cdot}] \to [X^{\cdot},F^{\cdot}] \to [X^{\cdot},G^{\cdot}]$$

Let us first notice that $g \circ e = 0$ as it follows from

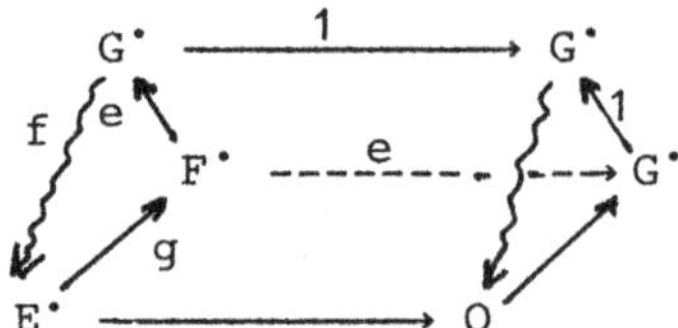

Given $x \in [X^\cdot, F^\cdot]$ with $ex = 0$ then we can find $y \in [X^\cdot, E^\cdot]$ with $x = gy$ from the following diagram

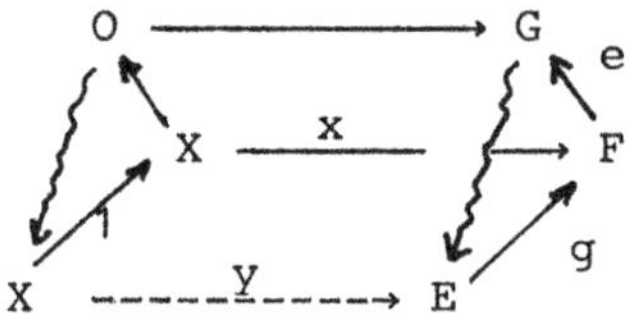

Q.E.D.

<u>Proposition 2.3</u>. Two triangles

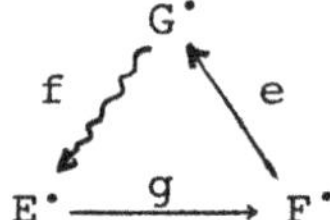 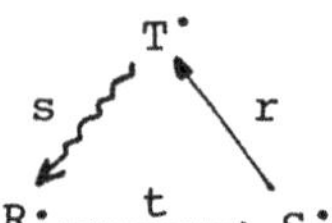

give rise to a direct sum

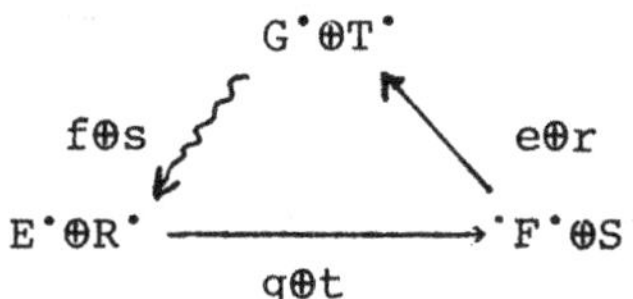

which again is a triangle.

<u>Proof</u>. Complete g⊕t to a triangle (..,W,g⊕t). Next
choose i: G˙ → W˙ to make the following diagram commutative

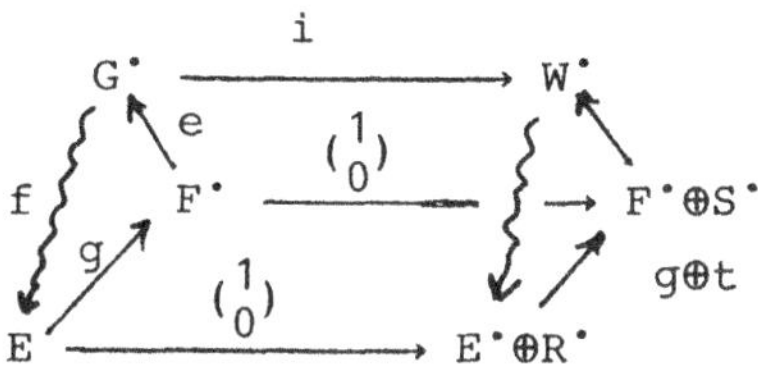

Construct j: T˙ → W˙ in a similar way. Finally prove that

(i,j) : G˙ ⊕ T˙ → W˙   is an isomorphism by using the following

commutative diagram

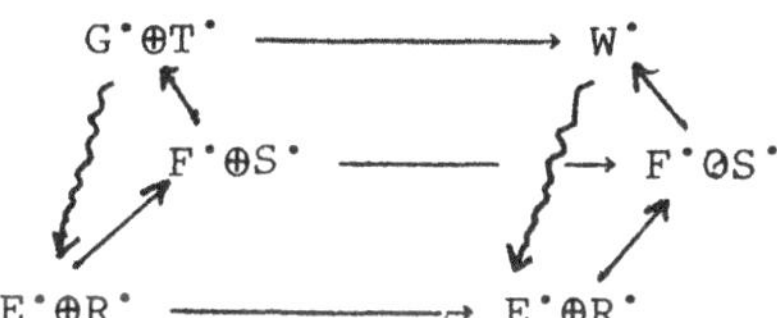

Proposition 2.2 and the 5-lemma.

Q.E.D.

Let us now assume that <u>A is an abelian category</u>. We shall
construct the derived category D(A) of A on the basis of the
following proposition.

<u>Proposition 2.4</u>. The class Q of K(A) consisting of
all quasi-isomorphisms is a multiplicative system in K(A),
i.e. satisfies FR 1,2,3 of the previous section.

    <u>Proof</u>. The axiom FR1 is clearly satisfied. The proof of FR2 is based on the observation that given a triangle

then s is in Q if and only if $H^{\cdot}(S^{\cdot}) = 0$, compare I.5.5. To prove FR2 let there be given morphisms

$$E^{\cdot} \longrightarrow Y^{\cdot} \Longleftarrow X^{\cdot} \qquad ; \quad s \quad \text{in} \quad Q$$

On this basis we can construct the following diagram

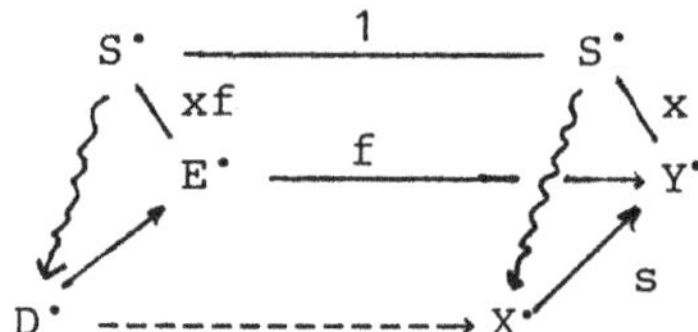

where we have used TR1 to construct the triangle $D^{\cdot}E^{\cdot}S^{\cdot}$ and TR3 to fill in the dotted arrow. The proof of the dual statement is similar.

    To prove FR3 let there be given morphisms $f,g: X^{\cdot} \to Y^{\cdot}$ and $t: Y^{\cdot} \to Z^{\cdot}$ in Q with $tf = tg$. Put $h = g-f$ to get $th = 0$. We are looking for $s: W^{\cdot} \to X^{\cdot}$ in Q with $hs = 0$.

    Let us first complete $t: Y^{\cdot} \to Z^{\cdot}$ to a triangle $Y^{\cdot}Z^{\cdot}T^{\cdot}$, next construct the diagram

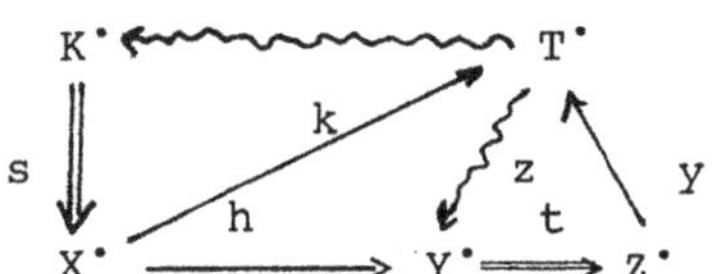

Explanation: Since th = 0 we can choose an arrow k: $X^\bullet \to T^\bullet$
with zk = h. Next, fill in the triangle $X^\bullet T^\bullet K^\bullet$ at random.
Notice that s is in Q and that hs = (zk)s = 0.

Q.E.D.

The category obtained from K(A) by inverting the class
Q of quasi-isomorphisms is <u>the derived category</u> D(A) of A

$$2.5 \qquad\qquad D(A) \ = \ Q^{-1}K(A)$$

Quite similarly we can invert the class of all quasi-isomorphisms in $K^+(A)$, the homotopy category of bounded below complexes over A to get

$$2.6 \qquad\qquad D^+(A) \ = \ Q^{-1}K^+(A)$$

<u>Proposition 2.7</u>. Let A be an abelian category and I the
full subcategory of injective objects. If every object of A
admits a monomorphism into an object of I, then

$$D^+(A) \ \xrightarrow{\ \sim\ } \ K^+(I)$$

the homotopy category of bounded below injectives.

<u>Proof</u>. Let I denote the additive category of injective
objects in A and consider the resolution functor

$$\rho: \ K^+(A) \ \to \ K^+(I)$$

The class Q is transformed into isomorphism which allows us to factor ρ through $D^+(A)$ to obtain

$$r: D^+(A) \longrightarrow K^+(I)$$

The inclusion $I \to A$ will induce a functor

$$i: K^+(I) \longrightarrow D^+(A)$$

which is a right adjoint to r, I.6.6

$$[X^{\cdot}, iI^{\cdot}] = [rX^{\cdot}, I^{\cdot}]$$

Let us notice that the adjunction morphism

$$r\,i\,I^{\cdot} \longrightarrow I^{\cdot}$$

is an isomorphism. From this we conclude that the functor
$i: K^+(I) \to D^+(A)$ is fully faithful, i.e. gives isomorphisms

$$[I_1, I_2] \xrightarrow{\sim} [i_*I_1, i_*I_2]$$

Thus it follows from I.6.1 that i is an equivalence of categories.
The same is true for r, r being a left adjoint to i.

$$\text{Q.E.D.}$$

Let us record the dual notions. The result of inverting
the quasi-isomorphisms in $K^-(A)$ is denoted

2.6° $$D^-(A) = Q^{-1}K^-(A)$$

Let P denote the full subcategory of projective objects in A.
If A has enough projectives, then

$$2.7^{\circ} \qquad\qquad D^-(A) \xrightarrow{\sim} K^-(P)$$

<u>Triangles in D(A)</u>. Let us first remark that the translation functor $X^{\bullet} \to X^{\bullet}[1]$ gives rise to an endofunctor of $D(A)$, 1.5. By a <u>triangle</u> in $D(A)$ we understand a diagram isomorphic to the transform of a triangle from $K(A)$.

<u>Proposition 2.8</u>. The triangles in $D(A)$ has the properties TR1,2,3.

<u>Proof</u>. Only TR3 needs to be treated seriously. We leave it to the reader to check that it suffices to take as point of departure a commutative diagram in $K(A)$ of the form

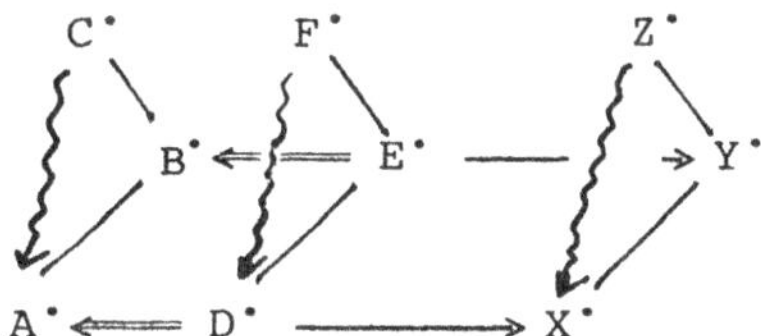

We can now fill in arrows $F^{\bullet}C^{\bullet}$ and $F^{\bullet}Z^{\bullet}$ according to TR3. The arrow $F^{\bullet}C^{\bullet}$ is necessarily a quasi-isomorphism as it follows from the long exact homology ladder resulting from the morphism $D^{\bullet}E^{\bullet}F^{\bullet} \to A^{\bullet}B^{\bullet}C^{\bullet}$ and the five lemma.

Q.E.D.

438

Homology in $\mathcal{D}(A)$. The homology functor

$$H^O: K(A) \longrightarrow A$$

transforms the multiplicative system Q into isomorphisms.
Thus by 1.5 the homology functor may be factored through $D(A)$.
The resulting functor will be denoted

$$H^O: D(A) \longrightarrow A$$

Given a triangle in $D(A)$

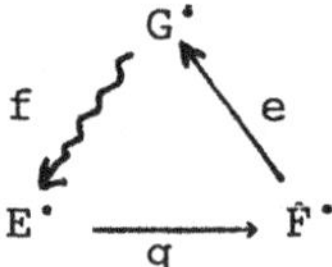

the resulting sequence in A

2.9 $\quad H^0(E^{\cdot}) \xrightarrow{H^0(g)} H^0(F^{\cdot}) \xrightarrow{H^0(e)} H^0(G^{\cdot}) \xrightarrow{H^0(f)} H^0(E^{\cdot}[1]) \xrightarrow{H^0(g[1])} H^0(F[1])$

is exact.

Ext groups. For complexes $X^{\cdot}$ and $Y^{\cdot}$ over we define
for each integer n

2.10 $\qquad \mathrm{Ext}^n(X^{\cdot},Y^{\cdot}) = \mathrm{Hom}(X^{\cdot},Y^{\cdot}[n])$

where Hom is calculated in $D(A)$. Define

2.11 $\qquad \alpha \cup \beta = \alpha[q] \circ \beta$

for $\quad \alpha \in \mathrm{Ext}^p(Y^\cdot, Z^\cdot)$ and $\quad \beta \in \mathrm{Ext}^q(X^\cdot, Y^\cdot)$.

Let us note a case where we can identify $D(A)$

<u>Proposition 2.12</u>. Let A be an abelian category in which every object has an injective resolution of length n, where n is a fixed integer. Then

$$D(A) = K(I)$$

where I denotes the category of injective objects in A.

<u>Proof</u>. According to I.7.7 every object of $K(A)$ admits a quasi-isomorphism into a complex of injective objects. Thus it suffices to prove that any quasi-isomorphism $f: I^\cdot \to J^\cdot$ in $K(I)$ is a homotopy equivalence. It suffices to treat the case where $J^\cdot = 0$ by a mapping cone argument. To prove that $I^\cdot$ is homotopic to zero it suffices to prove that $\mathrm{Ker}\ \partial^p$ is injective for all $p \in \mathbb{Z}$. By translation it suffices to treat the case $p = 0$. From the long exact sequence

$$0 \to \mathrm{Ker}\ \partial^{-n} \to I^{-n} \to \ldots I^{-n} \to \mathrm{Ker}\ \partial^0 \to 0$$

we derive for each object N in A an isomorphism

$$\mathrm{Ext}^1(N, \mathrm{Ker}\ \partial^0) \overset{\sim}{\to} \mathrm{Ext}^{n+1}(N, \mathrm{Ker}\ \partial^n)$$

Hence $\mathrm{Ext}^1(N, \mathrm{Ker}\ \partial^0) = 0$ for all N in A shows that $\mathrm{Ker}\ \partial^0$ is an injective object. For the sake of completness let us show that

2.13
$$[X^\cdot, I^\cdot] = 0 \quad \text{for} \quad I^\cdot \quad \text{in} \quad K(I) \quad \text{and}$$
$$X^\cdot \quad \text{in} \quad K(A) \quad \text{with} \quad H^\cdot(X^\cdot) = 0$$

So let there be given $f: X^\cdot \to I^\cdot$ in $K(A)$. According to FR2 applied to f and $X^\cdot \to 0$ we can find a quasi-isomorphism $k: I^\cdot \to J^\cdot$ such that $kf = 0$. By the first part of the proof we may assume that $J^\cdot$ is in $K(I)$ and by the second part k is a homotopy equivalence. Whence $f = 0$.

Q.E.D.

XI.3 Triangle associated to an exact sequence

Let us consider an abelian category A and the corresponding derived category D(A) constructed in the previous section.
To a short exact sequence of complexes

3.1
$$0 \to P^{\cdot} \xrightarrow{u} Q^{\cdot} \xrightarrow{v} R^{\cdot} \to 0$$

we are going to associate a triangle in D(A)

3.2

$$
\begin{array}{ccc}
 & R^{\cdot} & \\
{}^{w}\!\nearrow & & \nwarrow{}^{v} \\
P^{\cdot} & \xrightarrow[u]{} & Q^{\cdot}
\end{array}
$$

The arrow $w: R^{\cdot} \to P^{\cdot}[1]$ will sometimes be referred to as the
__characteristic arrow__ of the short exact sequence 3.1.

Construction: Let us notice that $(0,v): P[1] \oplus Q \to R$ in
fact defines a morphism of complexes

$$(0,v): \operatorname{Con}^{\cdot}(u) \longrightarrow R^{\cdot}$$

We are going to prove that this is a quasi-isomorphism. As in
the proof of I.4.15 consider the exact commutative diagram

$$
\begin{array}{ccccccccc}
 & & & \left(\begin{smallmatrix}0\\1\\0\end{smallmatrix}\right) & & \left(\begin{smallmatrix}-1\ 0\ 0\\ 0\ 0\ 1\end{smallmatrix}\right) & & & \\
0 & \longrightarrow & P^{\cdot} & \longrightarrow & \operatorname{Cyl}^{\cdot}(u) & \longrightarrow & \operatorname{Con}^{\cdot}(u) & \longrightarrow & 0 \\
 & & \Big\downarrow{\scriptstyle 1} & & \Big\downarrow{\scriptstyle (0,u,1)} & & \Big\downarrow{\scriptstyle (0,v)} & & \\
0 & \longrightarrow & P^{\cdot} & \xrightarrow{u} & Q^{\cdot} & \xrightarrow{v} & R^{\cdot} & \longrightarrow & 0
\end{array}
$$

It follows from the discussions preceding I.4.14 that the vertical

442

arrow in the middle is a homotopy equivalence. Now form the
long homology ladder and apply the 5-lemma to conclude that
$(0,v)$ is a quasi-isomorphism.

We can now define the characteristic arrow $w: R^{\cdot} \to P^{\cdot}[1]$
in $D(A)$ to be the fraction

3.3 $$w: R^{\cdot} \xleftarrow{\;(0,v)\;} \text{Con}^{\cdot}(u) \xrightarrow{\;(1,0)\;} P^{\cdot}[1]$$

We can conclude that 3.2 is a triangle in $D(A)$ by referring
to the standard mapping cone triangle I.4.14.

$\underline{\text{Functoriality 3.4}}$. Given an exact commutative diagram of
complexes in A

$$
\begin{array}{ccccccccc}
0 & \longrightarrow & P^{\cdot} & \xrightarrow{u} & Q^{\cdot} & \xrightarrow{v} & R^{\cdot} & \longrightarrow & 0 \\
 & & \downarrow{p} & & \downarrow{q} & & \downarrow{r} & & \\
0 & \longrightarrow & E^{\cdot} & \xrightarrow{f} & F^{\cdot} & \xrightarrow{g} & G^{\cdot} & \longrightarrow & 0
\end{array}
$$

Then (p,q,r) is a morphism from the triangle associated the
upper row to the triangle associated the lower row.

$\underline{\text{Proof}}$. This results immediately from the following diagram
of complexes in A

$$
\begin{array}{ccccc}
R^{\cdot} & \longleftarrow \text{Con}^{\cdot}(u) \longrightarrow & P^{\cdot}[1] \\
\downarrow{r} & \downarrow{\left(\begin{smallmatrix} p[1] & 0 \\ 0 & q \end{smallmatrix}\right)} & \downarrow{p[1]} \\
G^{\cdot} & \longleftarrow \text{Con}^{\cdot}(f) \longrightarrow & E^{\cdot}[1]
\end{array}
$$

which is commutative.

Q.E.D.

The formula 3.3 represents w as a right fraction. The following formula represents w as a <u>left fraction</u>

$$3.5 \qquad w: \ R^{\cdot} \xrightarrow{\ -\begin{pmatrix}0\\1\end{pmatrix}\ } \mathrm{Con}^{\cdot}(v) \xleftarrow{\ \begin{pmatrix}u[1]\\0\end{pmatrix}\ } P[1]$$

<u>Proof</u>. Consider the following diagram

$$
\begin{array}{ccccccc}
Q^{\cdot} & \xrightarrow{-\binom{0}{1}} & \mathrm{Con}^{\cdot}(u) & \xrightarrow{(1,0)} & P^{\cdot}[1] & \xrightarrow{u[1]} & Q^{\cdot}[1] \\
\big\downarrow{\scriptstyle 1} & & \big\downarrow{\scriptstyle (0,v)} & & \big\downarrow{\binom{u[1]}{0}} & & \big\downarrow{\scriptstyle 1} \\
Q^{\cdot} & \xrightarrow{\ -v\ } & R^{\cdot} & \xrightarrow{-\binom{0}{1}} & \mathrm{Con}^{\cdot}(v) & \xrightarrow{(1,0)} & Q^{\cdot}[1]
\end{array}
$$

Notice that the two rows represents triangles. In fact the diagram is a morphism of triangles in $K(A)$ since the middle square is homotopy commutative

$$
\begin{pmatrix}u[1]\\0\end{pmatrix}(1,0) - \begin{pmatrix}0\\-1\end{pmatrix}(0,v) = \begin{pmatrix}u[1] & 0\\0 & v\end{pmatrix} = \begin{pmatrix}-\partial & 0\\-v & \partial\end{pmatrix}\begin{pmatrix}0 & -1\\0 & 0\end{pmatrix} + \begin{pmatrix}0 & -1\\0 & 0\end{pmatrix}\begin{pmatrix}-\partial & 0\\-u & \partial\end{pmatrix}
$$

From the morphism of triangles results a homology ladder. Apply the 5-lemma to this to conclude that $\begin{pmatrix}u[1]\\0\end{pmatrix}$ is a quasi-isomorphism. Thus 3.5 represents indeed a left fraction. To see that this equals 3.3 we have to appeal once more to the homotopy commutative square just established.

Q.E.D.

444

<u>The 3 × 3 diagram</u>. Let us consider a commutative exact diagram in the category of complexes over A of the form

3.6

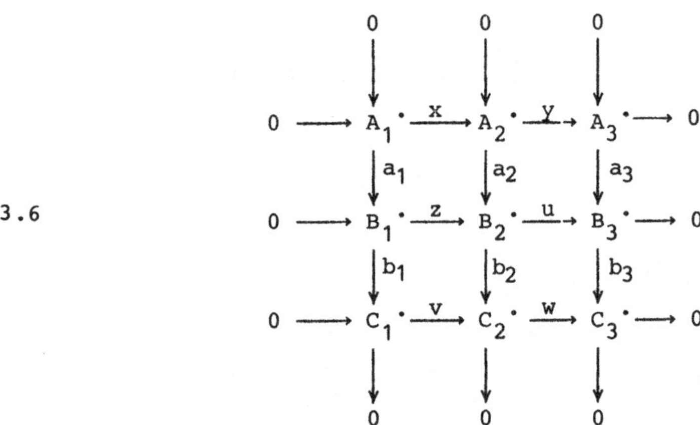

Let w_1,w_2,w_3 denote the characteristic arrows of the three columns and w_a,w_b,w_c the characteristic arrows of the rows. Then

3.7

$$w_a[1] \circ w_3 = -w_1[1] \circ w_c$$

or otherwise expressed: the diagram below is a morphism of triangles in $D(A)$.

3.8

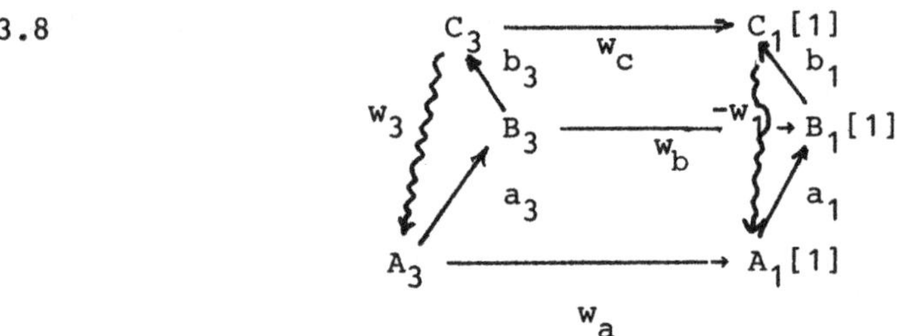

<u>Proof</u>. Consider the following commutative diagram of complexes over A

$$
\begin{array}{ccccccccc}
0 & \longrightarrow & A_1{}^{\bullet} & \xrightarrow{\ x\ } & A_2{}^{\bullet} & \xrightarrow{\ a_3 y\ } & B_3{}^{\bullet} & \xrightarrow{\ b_3\ } & C_3{}^{\bullet} & \longrightarrow & 0 \\
 & & \big\uparrow 1 & {\scriptstyle\binom{x}{a_1}} & \big\uparrow (1,0) & {\scriptstyle (a_2,-z)} & \big\uparrow u & {\scriptstyle wb_2} & \big\uparrow 1 & & \\
0 & \longrightarrow & A_1{}^{\bullet} & \longrightarrow & A_2{}^{\bullet}\!\oplus B_1{}^{\bullet} & \longrightarrow & B_2{}^{\bullet} & \xrightarrow{\ \ \ } & C_3{}^{\bullet} & \longrightarrow & 0 \\
 & & \big\downarrow 1 & & \big\downarrow (0,-1) & & \big\downarrow b_2 & & \big\downarrow 1 & & \\
0 & \longrightarrow & A_1{}^{\bullet} & \xrightarrow{\ -a_1\ } & B_1{}^{\bullet} & \xrightarrow{\ vb_1\ } & C_1{}^{\bullet} & \xrightarrow{\ w\ } & C_3{}^{\bullet} & \longrightarrow & 0
\end{array}
$$

3.9

The sequence in the middle is exact: check directly that wb_2 is a kernel for $(a_2,-z)$ and that $\binom{x}{a_1}$ is a kernel for $(a_2,-z)$.

Let us make a general remark concerning an exact sequence of complexes of the form

3.10
$$
0 \longrightarrow D_1{}^{\bullet} \longrightarrow D_2{}^{\bullet} \longrightarrow D_3{}^{\bullet} \longrightarrow D_4{}^{\bullet} \longrightarrow 0
$$

By factoring the middle arrow in an epimorphism followed by a monomorphism we obtain two short exact sequences of complexes

$$
0 \to D_1{}^{\bullet} \to D_2{}^{\bullet} \to E^{\bullet} \to 0, \quad 0 \to E^{\bullet} \to D_3{}^{\bullet} \to D_4{}^{\bullet} \to 0
$$

Let $u\colon E^{\bullet} \to D_1{}^{\bullet}[1]$ and $v\colon D_4{}^{\bullet} \to E^{\bullet}[1]$ denote respective characteristic arrows. The composite

$$
u[1] \circ v\colon D_4 \longrightarrow D_1[2]
$$

will be called <u>the characteristic class of the sequence 3.10.</u> Notice that this construction is functorial in a sense similar to 3.4. If we apply this to the diagram 3.9 we see that the top row and the bottom row has the same characteristic arrow. The characteristic arrow of the top row is $w_a[1] \circ w_3$ while the characteristic class of the bottom row is $-w_1[1] \circ w_c$.

$$Q.E.D.$$

XI.4 Yoneda extensions

Let us consider two objects M and N of an abelian category A . We shall consider <u>n-fold extensions</u> of M by N , i.e. exact sequences of the form

$$4.1 \qquad 0 \to N \to R_{n-1} \to R_{n-2} \to \cdots \to R_0 \to M \to 0$$

An n-fold extension gives rise to a complex

$$4.2 \qquad R_{\cdot}: 0 \to N \to R_{n-1} \to \cdots \to R_1 \to R_0 \to 0$$

and two morphisms of complexes over A

$$4.3 \qquad M \overset{s}{\Longleftarrow} R_{\cdot} \overset{f}{\longrightarrow} N[n]$$

As indicated, the morphism s is a quasi-isomorphism, so that f/s defines an arrow in $D(A)$. The characteristic <u>class</u> of the extension 4.1 is by definition

$$4.4 \qquad (-1)^{n(n+1)/2} \; f/s \in \mathrm{Ext}^n(M,N)$$

The reason for the sign will appear shortly.

<u>Proposition 4.5</u>. In case $n = 1$ the characteristic class of the extension $0 \to N \to R_0 \to M \to 0$ coincides with the third side in the triangle associated to the exact sequence, 3.2.

<u>Proof</u>. Consider an extension of M by N

$$0 \to N \overset{u}{\to} R_0 \overset{v}{\to} M \to 0$$

The rule from 3.3 assigns the fraction

$$M \xLeftarrow{(0,v)} \mathrm{Con}^{\boldsymbol{\cdot}}(u) \xrightarrow{(1,0)} N[1]$$

Notice that $\mathrm{Con}^{\boldsymbol{\cdot}}(u)$ and the complex $R.$ from 4.2 only differs by the sign of the differential. Consider the commutative diagram

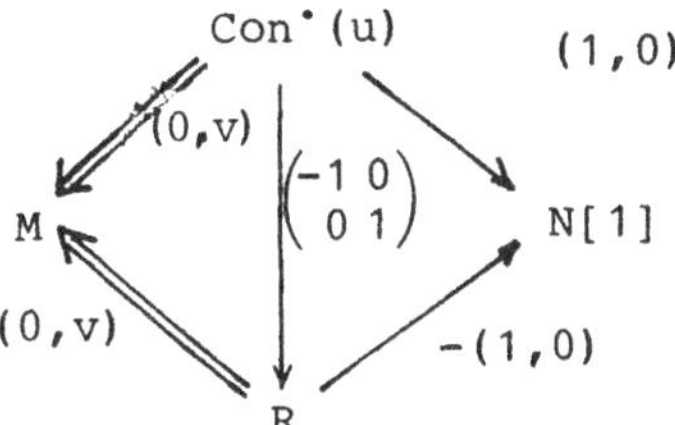

From which we conclude that the fraction above equals $-f/s$ with the notation of 4.3. This is consistent with 4.4 since $(-1)^{n(n+1)/2} = -1$ in case $n = 1$.

$$\mathrm{Q.E.D.}$$

<u>Proposition 4.6</u>. Given extensions in A

$$0 \longrightarrow P \longrightarrow S_{m-1} \xrightarrow{\partial_{m-1}} \ldots \longrightarrow S_1 \xrightarrow{\partial_1} S_0 \xrightarrow{\lambda} N \longrightarrow 0$$

$$0 \longrightarrow N \xrightarrow{\mu} R_{m-1} \xrightarrow{\partial_{n-1}} \ldots \longrightarrow R_1 \xrightarrow{\partial_1} R_0 \longrightarrow M \longrightarrow 0$$

with characteristic classes $\theta \in \mathrm{Ext}^m(N,P)$ and $\varphi \in \mathrm{Ext}^n(M,N)$ respectively. Then the spliced extension

$$0 \longrightarrow P \longrightarrow S_{m-1} \longrightarrow \ldots \longrightarrow S_0 \xrightarrow{\mu\lambda} R_{n-1} \longrightarrow \ldots \longrightarrow R_0 \longrightarrow M \longrightarrow 0$$

has characteristic class

$$\theta \cup \varphi \in \mathrm{Ext}^{n+m}(M,P)$$

Proof. Let us define three complexes

$$\begin{array}{ll}
S_{\cdot} & 0 \longrightarrow P \xrightarrow{\nu} S_{m-1} \xrightarrow{\partial_{m-1}} \cdots \longrightarrow S_1 \xrightarrow{\partial_1} S_0 \longrightarrow 0 \\[2ex]
R_{\cdot} & 0 \longrightarrow N \xrightarrow{\mu} R_{n-1} \xrightarrow{\partial_{n-1}} \cdots \longrightarrow R_1 \xrightarrow{\partial_1} R_0 \longrightarrow 0 \\[2ex]
T_{\cdot} & 0 \longrightarrow P \longrightarrow S_{m-1} \to \cdots S_0 \xrightarrow{\mu\lambda} R_{n-1} \to R_0 \longrightarrow 0
\end{array}$$

Let us first exhibit a quasi-isomorphism $u: T_{\cdot} \Rightarrow R_{\cdot}$.

$$\begin{array}{ccccccccccccccc}
0 \to & P & \to & S_{m-1} & \cdots & S_1 & \to & S_0 & \xrightarrow{\mu\lambda} & R_{n-1} & \to & & \to & R_0 & \to 0 \\
& {\scriptstyle 0}\downarrow & & {\scriptstyle 0}\downarrow & & {\scriptstyle 0}\downarrow & & {\scriptstyle \lambda}\downarrow & & {\scriptstyle 1}\downarrow & & & & {\scriptstyle 1}\downarrow & \\
0 \to & 0 & \to & 0 & \cdots & 0 & \to & N & \xrightarrow{\mu} & R_{n-1} & \to & & \to & R_0 & \to 0
\end{array}$$

and a morphism $h: T_{\cdot} \to S_{\cdot}[n]$

$$\begin{array}{ccccccccccccccccc}
0 \longrightarrow & P & \xrightarrow{\nu} & S_{m-1} & \xrightarrow{\partial_{m-1}} & \cdots & \longrightarrow & S_1 & \xrightarrow{\partial_1} & S_0 & \xrightarrow{\mu\lambda} & R_{n-1} & \longrightarrow & R_{n-2} & \longrightarrow \\
& \downarrow {\scriptstyle (-1)^{mn}} & & \downarrow {\scriptstyle (-1)^{(m-1)n}} & & & & \downarrow {\scriptstyle (-1)^{n}} & & \downarrow {\scriptstyle 1} & & \downarrow & & \downarrow & \\
0 \longrightarrow & P & \xrightarrow{(-1)^n\nu} & S_{m-1} & \xrightarrow{(-1)^n\partial_{m-1}} & \cdots & \longrightarrow & S_1 & \xrightarrow{(-1)^n\partial_1} & S_0 & \longrightarrow & 0 & \longrightarrow & 0 & \longrightarrow
\end{array}$$

Let the two extensions give rise to diagrams

$$N \xLeftarrow{t} S_{\cdot} \xrightarrow{g} P[m]$$

$$M \xLeftarrow{s} R_{\cdot} \xrightarrow{f} N[n]$$

Then we have the following commutative diagram

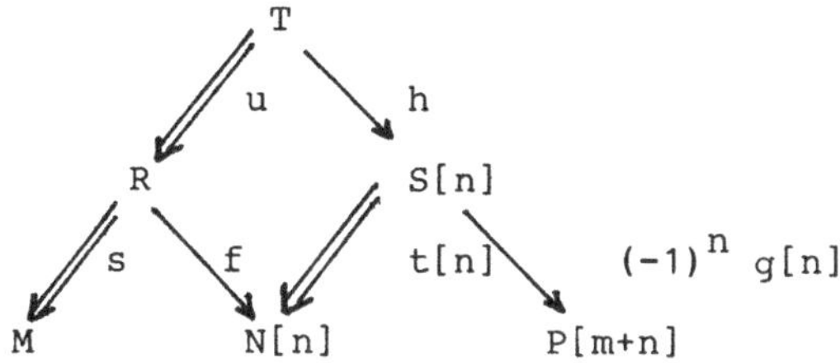

where the data 4.3 relative to the spliced sequence is represented by the composites su and $(-1)^{nm}g[n]h$. Thus the
spliced sequence has characteristic class

$$(-1)^{(n+m)(n+m+1)/2}(-1)^{nm}g[n]/t[n] \circ f/s$$

We shall finally involve the elementary identity

$$m(m+1)/2 + n(n+1)/2 = (m+n)(m+n+1)/2 - mn$$

and the formula follows.

$$\text{Q.E.D.}$$

<u>Proposition 4.7</u>. Two n-fold extensions in A

$$0 \to N \to R_{n-1} \to \ldots \to R_1 \to R_0 \to M \to 0$$

$$0 \to N \to S_{n-1} \to \ldots \to S_1 \to S_0 \to M \to 0$$

has the same characteristic class if and only if there exists
a commutative exact diagram of the form

$$0 \to N \to R_{n-1} \to \ldots \to R_1 \to R_0 \to M \to 0$$

$$0 \to N \to T_{n-1} \to \ldots \to T_1 \to T_0 \to M \to 0$$

$$0 \to N \to S_{n-1} \to \ldots \to S_1 \to S_0 \to M \to 0$$

The same statement is true with the vertical arrows reversed.

<u>Proof</u>. Given a diagram as above. Let R., S. and T. denote the corresponding complexes in the sense of 4.3. We deduce a commutative diagram

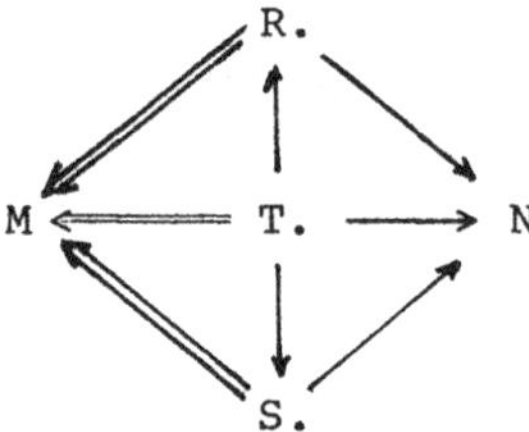

from which we conclude that the two relevant right fractions are identical. - Suppose conversely that the two extensions have the same characteristic class. Then we deduce the existence of a diagram as above where $T^{\cdot}$ is some complex. In the diagram above we can replace $T^{\cdot}$ by the truncation $\tau_{\leq 0}\, \tau_{\geq -n} T^{\cdot}$ without disturbing the diagram. This gives a diagram of the promised sort.

In order to prove the statement relative to the reversed vertical arrows we need a representation of the characteristic class by a left fraction. This is accomplieshed in 4.9 below.

Q.E.D.

Proposition 4.8. Let M and N be objects in the abelian category A, and $n \in \mathbb{N}$. Any element of $\mathrm{Ext}^n(M,N)$ can be realized as the characteristic class of an n-fold extension of M by N.

Proof. An element of $\mathrm{Ext}^n(M,N)$ can be realized as a right fraction f/s

$$M \xleftarrow{\ \ s\ \ } R_. \xrightarrow{\ \ f\ \ } N[n]$$

Replace $R_.$ by the truncation $\tau_{\leq 0}\tau_{\geq -n}R_.$ to obtain the desired n-fold extension of M by N.

$$Q.E.D.$$

Representation by a left fraction. Let there be given an n-fold extension of M by N as in 4.1. Let us introduce the complex

$$S_. : 0 \to R_{n-1} \to R_{n-2} \to \dots \to R_0 \to M \to 0$$

and the two morphisms

$$M \xrightarrow{\ \ g\ \ } S_. \xleftarrow{\ \ t\ \ } N[n]$$

With the notation of 4.3 let us prove the formula

$$4.9 \qquad\qquad f/s = (-1)^n \, t \diagdown g$$

<u>Proof</u>. We shall prove that the following diagram

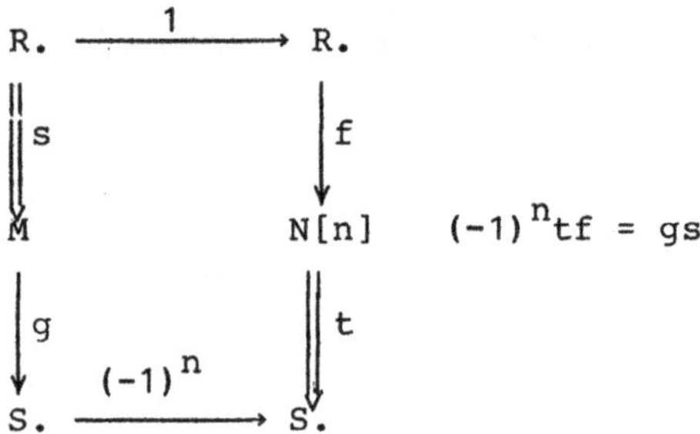

is homotopy commutative. Put $\theta_0 = gs$ and $\theta_n = tf$. For $i = 1,\ldots,n-1$ let $\theta_i \colon R. \to S.$ be given by

The single dotted arrow above provides a homotopy from θ_i to θ_{i+1}. A simple induction will complete the proof.

Q.E.D.

XI.5. Octahedra

Let A denote an additive category by an octahedra in
$K = K(A)$ we understand a diagram of the form

5.1

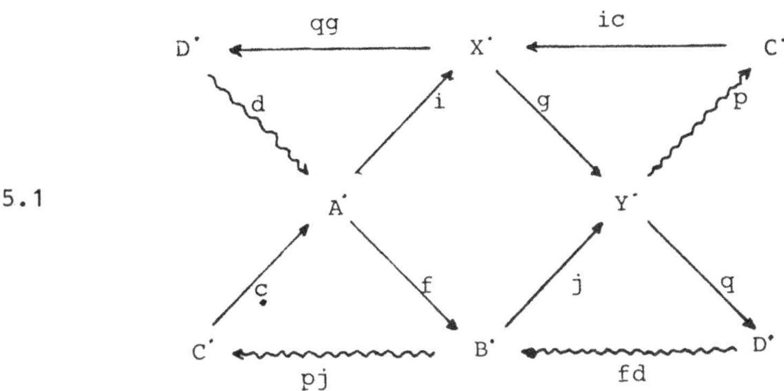

consisting of four triangles and such that

5.2 $\qquad\qquad gi = jf \qquad cp = dq$

The first of these relations expresses commutativity of the
middle diamond in 5.1. The second relation is nicely realized by
reflecting the diagram 5.1 in the line AY and attaching it to
the old diagram

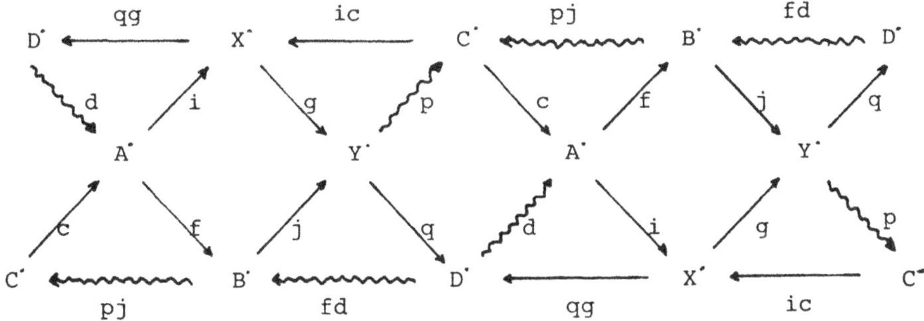

Alternatively the diagram can be represented as a <u>braid</u>

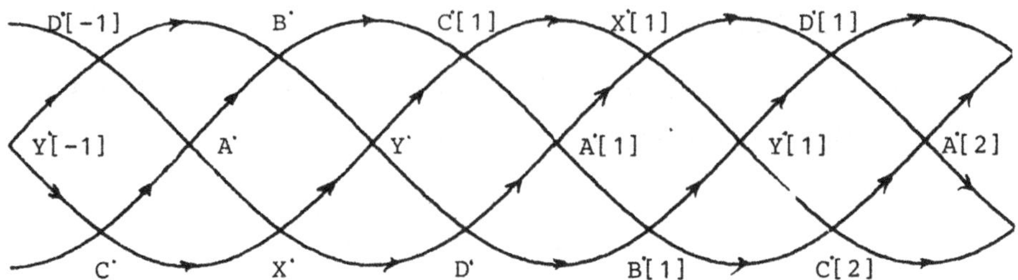

This representation reveals that an octahedron contains four morphisms of triangles.

To explain the name of the diagram 5.1 we shall represent it as an octahedron

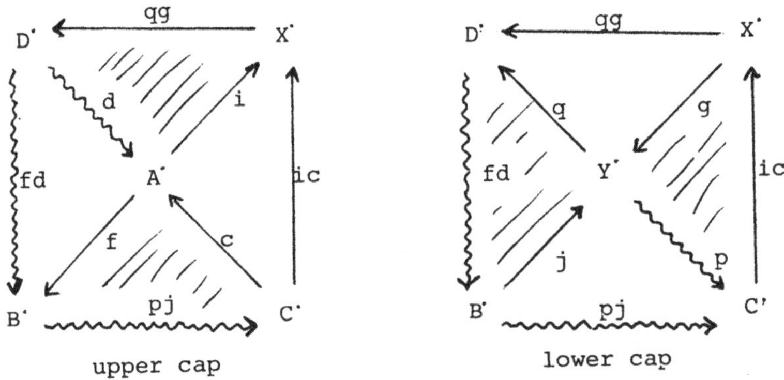

where the shaded areas represent triangles and the unshaded areas represent commutative diagrams. To the general octahedron we can associate two sequences

$$5.3 \qquad A^{\cdot} \xrightarrow{\binom{i}{-f}} X^{\cdot} \oplus B^{\cdot} \xrightarrow{(g,j)} Y^{\cdot} \xrightarrow{\quad cp \quad} A^{\cdot}[1]$$

$$5.4 \qquad A^{\cdot} \xrightarrow{gi} Y^{\cdot} \xrightarrow{\binom{-p}{q}} C^{\cdot}[1] \oplus D^{\cdot} \xrightarrow{(c,d)} A^{\cdot}[1]$$

which we call the <u>Mayer-Vietoris sequences</u> of the octahedron.

Theorem 5.5. Any diagram consisting
of two triangles with a common vertex can be
completed to an octahedron whose Mayer-Vietoris sequences are triangles.

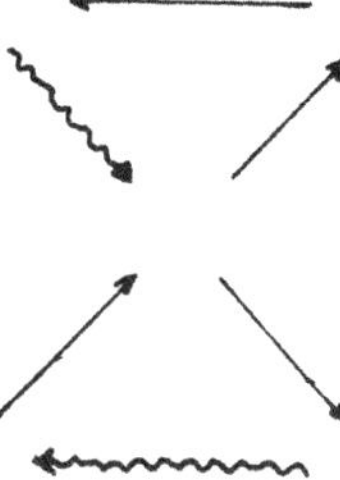

Proof. The triangle I.4.22 can be completet to the diagram

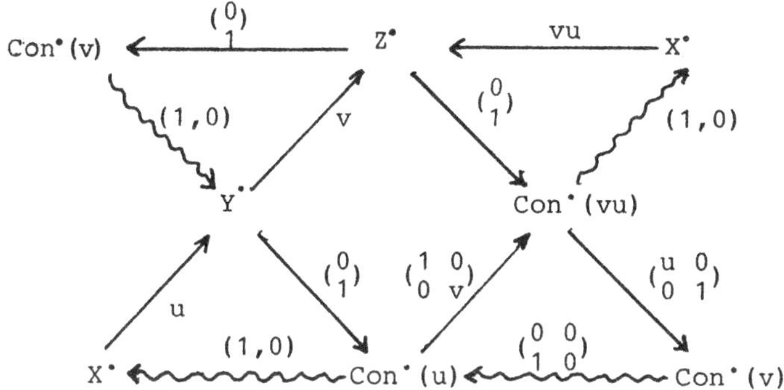

which is easily verified to be an octahedron.

To establish the Mayer-Vietoris sequence 5.3 we notice
that the chainwise split exact sequence

$$Y^{\bullet} \xrightarrow{\begin{pmatrix} 0 \\ -1 \\ v \end{pmatrix}} \operatorname{Con}^{\bullet}(u) \oplus Z^{\bullet} \xrightarrow{\begin{pmatrix} 1 0 0 \\ 0 v 1 \end{pmatrix}} \operatorname{Con}^{\bullet}(vu)$$

has homotopy invariant $(u,0)$ as it follows from

$$\begin{pmatrix} -\partial & 0 & 0 \\ -u & \partial & 0 \\ 0 & 0 & \partial \end{pmatrix}\begin{pmatrix} 1 & 0 \\ 0 & 0 \\ 0 & 1 \end{pmatrix} - \begin{pmatrix} 1 & 0 \\ 0 & 0 \\ 0 & 1 \end{pmatrix}\begin{pmatrix} -\partial & 0 \\ -vu & \partial \end{pmatrix} = \begin{pmatrix} 0 & 0 \\ -u & 0 \\ vu & 0 \end{pmatrix} = \begin{pmatrix} 0 \\ -1 \\ v \end{pmatrix}(u,0)$$

We leave the sequence 5.4 to the reader.

$$Q.E.D.$$

<u>Proposition 5.6</u>. Any commutative square $A_1A_2B_1B_2$ can be completed to a 9 diagram

$$
\begin{array}{ccccccc}
A_1 & \xrightarrow{\;x_1\;} & B_1 & \xrightarrow{\;y_1\;} & C_1 & \xrightarrow{\;z_1\;} & A_1[1] \\
\downarrow{a_1} & & \downarrow{b_1} & & \downarrow{c_1} & & \downarrow{a_1[1]} \\
A_2 & \xrightarrow{\;x_2\;} & B_2 & \xrightarrow{\;y_2\;} & C_2 & \xrightarrow{\;z_2\;} & A_2[1] \\
\downarrow{a_2} & & \downarrow{b_2} & & \downarrow{c_2} & & \downarrow{a_2[1]} \\
A_3 & \xrightarrow{\;x_3\;} & B_3 & \xrightarrow{\;y_3\;} & C_3 & \xrightarrow{\;z_3\;} & A_3 \\
\downarrow{a_3} & & \downarrow{b_3} & & \downarrow{c_3} & - & \downarrow{b_3[1]} \\
A_1[1] & \xrightarrow{\;x_1[1]\;} & B_1[1] & \xrightarrow{\;y_1[1]\;} & C_1[1] & \xrightarrow{\;z_1[1]\;} & A_1[2]
\end{array}
$$

in this diagram the three first rows and the
three first columns are triangles and all squares are commutative,
except the one marked - anticommutative.

<u>Proof</u>. Choose first triangles based on x_1, b_1, a_1, x_2 and
complete the two octahedrons below, finally complete the third octahedron. We let the diagrams speak for themselves.

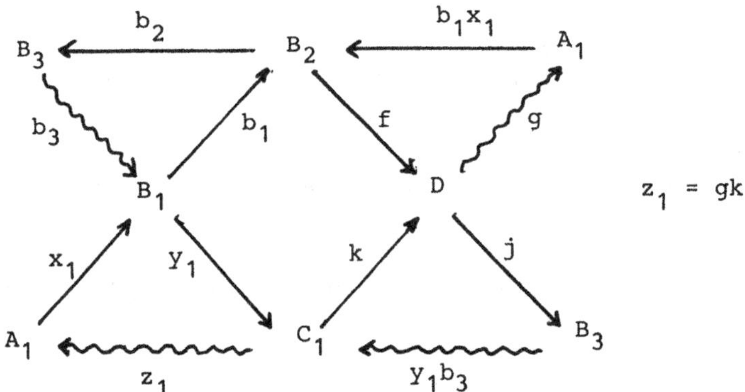

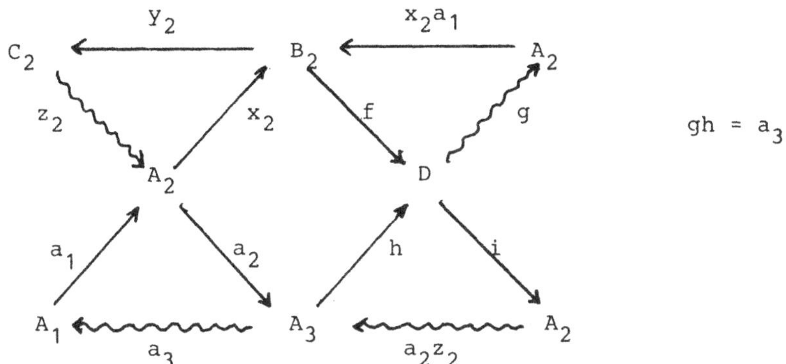

$$gh = a_3$$

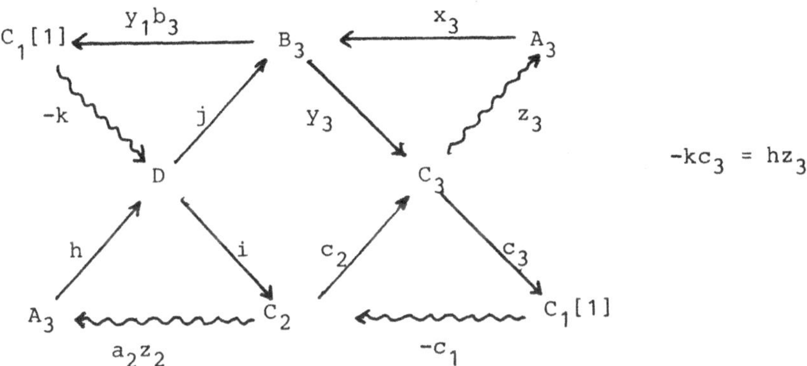

$$-kc_3 = hz_3$$

Let us verify the required anti commutativity

$$z_1 c_3 = gkc_3 = g(-hz_3) = -a_3 z_3$$

The commutativity relations are left to the reader.

Q.E.D.

XI.6 Localization

Let us consider the homotopy category $K(A)$ of complexes over an additive category and M a subcategory of $K = K(A)$, which satisfies the following two conditions

6.1   The category M is stable under translation.
           If two of the vertices of a triangle belongs to
           M then the third vertex belongs to M.

6.2   If $X \oplus Y$ belongs to M, then X and Y
           belongs to M.

Let S denote the class of morphisms s in K which are the basis for a triangle in K whose third vertex are in M. We are going to prove that S is a multiplicative system in K i.e. that it satisfies FR1,2,3 from XI.1 . Let us add

FR 4 S is stable under decalage.

FR 5 Given a diagram consisting of two triangles and a commutative square

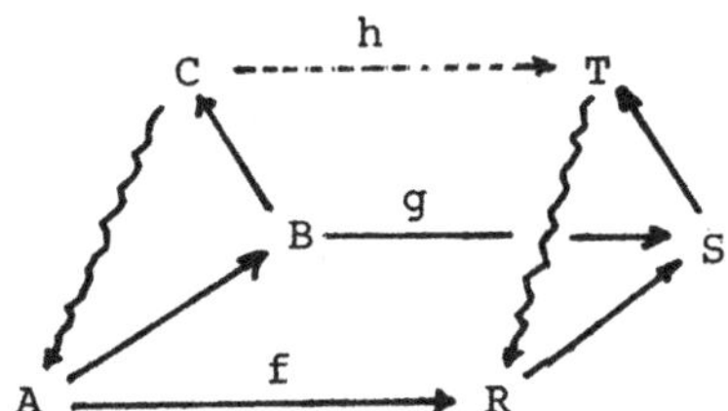

with f and g in S. Then there exists h: C $\rightarrow$ T in S making the remaining two squares commutative.

<u>Theorem 6.3</u>. The class S of morphisms in K is a multiplicative system in the sense that it satisfies FR1,2,3 from XI.1 and FR 4,5 above. Moreover, the system S is <u>saturated</u>, 1.6.

<u>Proof</u>. Let $s: X \to Y$ and $t: Y \to Z$ be morphisms from S . Consider the octahedron

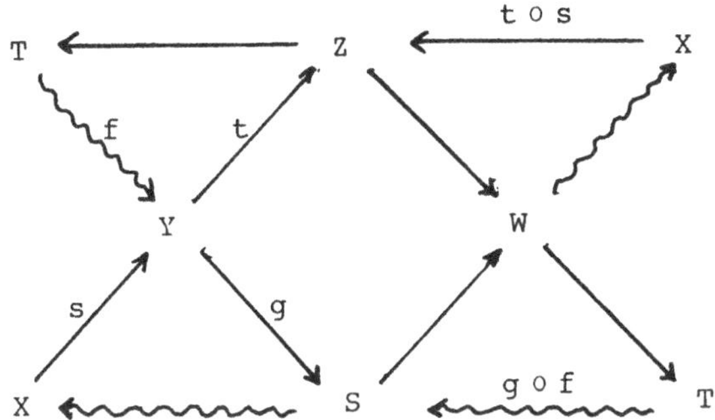

Notice that S and T belong to M to conclude that W belongs to M i.e. that t o s belongs to S. This proves FR1. The proof of FR2 and FR3 can be read off the proof of 2.4. The condition FR4 is a consequence of 6.1 while FR5 follows from 5.6.

We shall now prove that S is saturated, so let there be given morphisms $r: V \to X$, $s: X \to Y$ $t: Y \to Z$ such that sr and ts are in S. With the notation from the octahedron above remark that g(sr) = (gs)r = 0. From this follows that g can be factored through the third vertex of a triangle based on sr. In particular we conclude that g o f can be factored through an object of M. Apply the lemma below to the morphism g o f and conclude that S is in M, i.e. that s belongs to S.

Q.E.D.

<u>Lemma 6.4</u>. Given a triangle with W in M

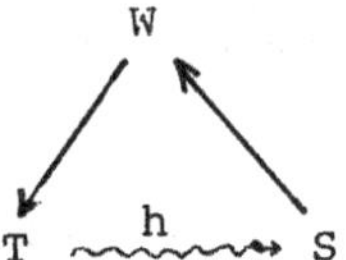

If the morphism h can be factored through an object of M,
then T and S belong to M.

   <u>Proof</u>. Let $T \xrightarrow{f} Y \xrightarrow{g} S$ be a factorization of h with Y
in M. Consider the octahedron above and notice that Y and W
belong to M. It follows from the <u>Mayer-Vietoris sequences</u> that
Z ⊕ S and T ⊕ X belong to M. We conclude from 6.2 that
S and T belong to M.

                                                        Q.E.D.

   On the basis of this theorem we can construct the category
$S^{-1}K$. Let us say that a diagram in $S^{-1}K$ is a <u>triangle</u> if it is
isomorphic to the transform of a triangle from K. The category
$S^{-1}K$ is a <u>triangulated category</u> in that the triangles satisfy
TR1,2,3 from XI.2.

   It is easy to see that the construction $M \to S$ gives a
complete description of the saturated multiplicative ystems in K
in terms of subcategories of K satisfying 6.1 and 6.2.

Bibliography

A. A. Beilinson, J. Bernstein, P. Deligne, Faisceaux pervers,
in Analyse et topologie sur les espaces singuliers I - II,
Astérisque 100 - 101 (1982).

A. Borel et al., Intersection Cohomology,
Progress in Math. Vol. 50 . Birkhäuser, Basel 1984.

A. Borel, Seminar on transformation groups,
Annals of Math. Studies 46, Princeton Univ. Press,
Princeton 1960.

A. Borel and J. C. Moore, Homology theory for locally
compact spaces,
Michigan Math. J. 7 (1960), p.137 - 159.

A. Borel and A. Haefliger, La classe d'homologie fondamen-
tale d'un espace analytique,
Bull. Soc. Math. France 89 (1961), p.461 - 513.

R. Bott and L. Tu, Differential forms in algebraic topology,
Springer-Verlag, Heidelberg 1982.

N. Bourbaki, Topologie Générale 1 - 2, Hermann, Paris 1961.

N. Bourbaki, Algèbre 10, Algèbre Homologique, Masson, Paris 1980.

G. E. Bredon, Sheaf theory, McGraw - Hill, New York 1965.

R. B. Burckel, Introduction to Classical Complex Analysis I,
Academic Press, New York 1979.

H. Cartan and S. Eilenberg, Homological Algebra,
Princeton Univ. Press, Princeton 1956.

P. Deligne, Equations différentielles à points singuliers
   réguliers, Lecture Notes in Math. 163.
   Springer-Verlag, Heidelberg 1970.

P. DEligne, Séminaire de Geométrie Algébrique du Bois - Marie
   SGA $4\frac{1}{2}$. Lecture Notes in Math. 569.
   Springer-Verlag 1977.

W. Fulton, Intersection theory,
   Ergebnisse der Mathematik und ihrer Grenzgebiete 3, 2,
   Springer-Verlag, Berlin 1984.

R. Godement, Topologie Algébrique et théorie des faisceaux,
   Hermann, Paris 1958.

M. Goresky and R. MacPherson, Intersection homology II,
   Inv. Math. 71, 1 (1983), p.77 - 129.

P. Griffiths and J. Harris, Principles of algebraic geometry,
   John Wiley & Sons, New York 1978.

A. Grothendieck, Sur quelques points d'algèbre homologique,
   Tohoku Math. J. IX (1957), p.119 - 221.

A. Grothendieck, Cohomologie Locale,
   Lecture Notes in Math. 41. Springer-Verlag, Berlin 1967.

A. Grothendieck, La théorie des classes de Chern,
   Bull. Soc. Math. France 86 (1958), p.137 - 154.

R. C. Gunning and H. Rossi, Analytic Functions of several
   Complex variables,
   Prentice Hall, 1965.

R. Hartshorne, Residues and Duality,
   Springer Lecture Notes 20, Springer-Verlag, Berlin 1966.

W. V. D. Hodge and D. Pedoe, Methods of algebraic geometry II,
   Cambridge University Press, Cambridge 1952.

P. Holm, The microbundle representation theorem,
   Acta Mathematica 117 (1967), p.191 - 213.

B. Iversen, Local Chern classes,
   Ann. Sci. Ec. Norm. Sup. 9 (1976), p.155 - 169.

B. Iversen, Octahedra and braids,
   Bull. Soc. Math. France, to appear.

J. Kister, Microbundles are fibre bundles,
   Annals of Math. 80 (1964), p.190 - 199.

J. W. Milnor, Microbundles I,
   Topology 3 (Suppl. 1) 1964, p.53 - 80.

J. W. Milnor and J. D. Stasheff, Characteristic classes,
   Annals of Math. Studies 76. Princeton Univ. Press 1974.

R. Narasimhan, Complex analysis in one variable,
   Birkhäuser, Basel 1985.

G. Nöbeling, Verallgemeinerung eines Satzes von Herrn E. Specker,
   Inventiones Math. 6 (1968), p.41 - 44.

G. de Rham, Variétés différentiables, formes, courants,
   formes harmoniques,
   Hermann, Paris, 1955.

E. H. Spanier, Algebraic Topology,
   McGraw-Hill, New York 1966, Springer-Verlag, New York 1981.

R. C. Swan, The theory of sheaves,
   University of Chicago Press, 1964.

J. - L. Verdier, Catégories dérivées,
   SGA $4\frac{1}{2}$, Lecture Notes in Math. 569.
   Springer-Verlag, Berlin 1976.

J. - L. Verdier, Théorème de dualité de Poincaré,
   Comtes Rendus 256 (1963), p.2084 - 2086.

J.-L. Verdier, Dualité dans la cohomologie des espaces
    localement compacts,
    Séminaire Bourbaki, 18e année 65/66, no. 300.

J.-L. Verdier, M. Zisman et al.,
    Séminaire Heidelberg - Strasbourg 1966/67.
    Publication I.R.M.A., Strasbourg 1969.

J.-L. Verdier, A duality theorem in the étale cohomology
    of schemes,
    In Proc. of a conference on local fields, p.184 - 198,
    Springer-Verlag, Berlin 1967.

J.-L. Verdier, Class d'homologie associée á un cycle,
    Astérisque 36 - 37 (1976), p.101 - 151.

H. Grauert, R. Remmert

Coherent Analytic Sheaves

1984. XVIII, 249 pages. (Grundlehren der mathematischen Wissenschaften, Band 265). ISBN 3-540-13178-7

Contents: Complex Spaces. – Local **Weierstrass** Theory. – Finite Holomorphic Maps. – Analytic Sets. Coherence of Ideal Sheaves. – Dimension Theory. – Analyticity of the Singular Locus. Normalization of the Structure Sheaf. – **Riemann** Extension Theorem and Analytic Coverings. – Normalization of Complex Spaces. – Irreducibility and Connectivity. Extension of Analytic Sets. – Direct Image Theorem. – Annex. Theory of Sheaves. Notion of Coherence. – Bibliography. – Index of Names. – Index.

The theory of coherent analytic sheaves revolutionized complex analysis. Function theory based on such sheaves reached its first high point in 1953 at the Brussels Colloquium, when H. Cartan and J-P. Serre presented what has since then been termed Theorems A and B for Stein Manifolds. There are four fundamental coherence theorems in complex analysis: Coherence of the structure sheaf $\mathcal{O}_X$ of any complex space X; coherence of the ideal sheaf i(A) of any analytic set A; coherence of the normalization sheaf $\mathcal{O}_X$ of any structure sheaf; and coherence of the direct image sheaves under proper holomorphic maps. Complex analysis is developed in this book so to speak ab ovo; general sheaf theory, as far as it is needed, is given in an appendix. Along with the basic theorems on coherent sheaves the following theorems necessary either as tools or as applications are included in the text:
local description lemma, on which the proof of the coherence of the structure sheaf is based; Rückert's Nullstellensatz; a profound discussion of finite holomorphic maps and dimension theory of complex spaces; Riemann's extension theorem; singular locus and construction of normalization spaces; theorem on removable singularities of analytic sets; theorems of Chow, Levi and Hurewitz-Weierstrass.
The book concludes with a proof of the coherence of all direct image sheaves under proper holomorphic maps and its applications to the proper mapping theorem, families of complex spaces and the **Stein** factorization. The approach taken is goemetric; detailed historical notes provide further insight into the theory of coherent analytic sheaves. The authors who played a prominent part in the development of modern complex analysis give proofs of today's point of view. This aspect and the self-contained exposition make the theory of coherent sheaves, which actually means the theory of complex spaces, easily accessible to a wide readership.

Springer-Verlag
Berlin Heidelberg
New York Tokyo

H. Delfs, M. Knebusch

Locally Semialgebraic Spaces

1985. XVI, 329 pages. (Lecture Notes in Mathematics, Volume 1173). ISBN 3-540-16060-4

Contents:

Chapter I – The basic definitions:
Locally semialgebraic spaces and maps. – Inductive limits, some examples of locally semialgebraic spaces. – Locally semialgebraic subsets. – Regular and para-compact spaces. – Semialgebraic maps and proper maps. – Partially proper maps. – Locally complete spaces.

Chapter II – Completions and triangulations:
Gluing paracompact spaces. – Existence of comple-tions. – Abstract simplicial complexes. – Triangulation of regular paracompact spaces. – Triangulation of weakly simplicial maps, maximal complexes. – Trian-gulation of amenable partially finite maps. – Stars and shells. – Pure hulls of dense pairs. – Ends of spaces, the LC-stratification. – Some proper quotients. – Modifica-tion of pure ends. – The Stein factorization of a semial-gebraic map. – Semialgebraic spreads. – Huber's theorem on open mappings.

Chapter III – Homotopies:
Some strong deformation retracts. – Simplicial approxi-mations. – The first main theorem on homotopy sets; mapping spaces. – Relative homotopy sets. – The second main theorem; contiguity classes. – Homotopy groups. – Homology; the Huréwicz theorems. – Homo-topy groups of ends. – Appendix A–B. – References. – List of symbols. – Glossary.

Springer-Verlag
Berlin Heidelberg
New York Tokyo

Made in the USA
Monee, IL
07 July 2026

56552005R00267